Springer Series in Reliability Engineering

Today's modern systems have become increasingly complex to design and build, while the demand for reliability and cost effective development continues. Reliability is one of the most important attributes in all these systems, including aerospace applications, real-time control, medical applications, defense systems, human decision-making, and home-security products. Growing international competition has increased the need for all designers, managers, practitioners, scientists and engineers to ensure a level of reliability of their product before release at the lowest cost. The interest in reliability has been growing in recent years and this trend will continue during the next decade and beyond.

The Springer Series in Reliability Engineering publishes books, monographs and edited volumes in important areas of current theoretical research development in reliability and in areas that attempt to bridge the gap between theory and application in areas of interest to practitioners in industry, laboratories, business, and government.

Now with 100 volumes!

Indexed in Scopus and EI Compendex

Interested authors should contact the series editor, Hoang Pham, Department of Industrial and Systems Engineering, Rutgers University, Piscataway, NJ 08854, USA. Email: **hopham@soe.rutgers.edu, or Anthony Doyle, Executive Editor, Springer, London. Email: anthony.doyle@springer.com.**

Gururaj H. L. · Vinayakumar Ravi · Hoang Pham ·
Dayananda P.
Editors

Reliability in Cyber-Physical Systems: The Human Factor Perspective

Springer

Editors
Gururaj H. L.
School of Computer Engineering
Manipal Institute of Technology Bengaluru
Manipal Academy of Higher Education
Manipal, Karnataka, India

Hoang Pham
Department of Industrial and Systems Engineering
Rutgers University
Piscataway, NJ, USA

Vinayakumar Ravi
Center for Artificial Intelligence
Prince Mohammad bin Fahd University
Al Khobar, Saudi Arabia

Dayananda P.
School of Computer Engineering
Manipal Institute of Technology Bengaluru
Manipal Academy of Higher Education
Manipal, Karnataka, India

ISSN 1614-7839 ISSN 2196-999X (electronic)
Springer Series in Reliability Engineering
ISBN 978-3-032-09916-7 ISBN 978-3-032-09917-4 (eBook)
https://doi.org/10.1007/978-3-032-09917-4

This Springer imprint is published by the registered company Springer Nature Switzerland AG
The registered company address is: Gewerbestrasse 11, 6330 Cham, Switzerland

Preface

Cyber-Physical Systems (CPS) have become the backbone of modern intelligent infrastructures, seamlessly integrating computational intelligence with physical processes through networked control and embedded systems. In recent years, the rapid growth of advanced sensing, real-time data analytics, and artificial intelligence has transformed the way CPS are modeled, monitored, and secured. Reliability is no longer a static attribute assessed post-deployment; it is now an evolving property that must be dynamically evaluated and assured throughout the system lifecycle. The integration of AI-driven diagnostics, predictive maintenance, and anomaly detection has empowered CPS to adapt in real time, mitigating risks before they escalate.

This book presents a carefully curated collection of 18 chapters that explore the state-of-the-art in reliability modeling, Human error analysis, intrusion detection, malware detection, Security challenges in CPS and IoT integration, advanced reliability solutions for CPSs, edge computing, and mobile monitoring through deep learning, machine learning, AI optimization, knowledge distillation, model quantization, and ensemble learning computational approaches.

Each chapter reflects the combined expertise of leading researchers and practitioners worldwide, offering insights into reliability assessment techniques, resilience engineering, and AI-enhanced reliability prediction. Special attention is given to reliability in safety-critical CPS including IoT and IoT in healthcare systems, where secure data transmission, robust decision-making, and fault containment are essential.

We express our sincere gratitude to all contributing authors for their scholarly dedication and to the reviewers whose critical insights enriched the quality of this work. We also thank the Springer publishing team for their unwavering support in bringing this book to fruition. We hope this volume serves as a valuable reference for researchers, engineers, and industry professionals seeking to design and operate CPS that are not only intelligent and efficient but also resilient and dependable in the face of real-world challenges.

Manipal, India — Gururaj H. L.
Al Khobar, Saudi Arabia — Vinayakumar Ravi
Piscataway, USA — Hoang Pham
Manipal, India — Dayananda P.

Contents

About the Editors

Dr. Gururaj H. L. (Senior Member, IEEE) received a Ph.D. degree in computer science and engineering from Visvesvaraya Technological University India in 2019. He has published more than 200 research articles in peer-reviewed and reputed international journals. He has authored 15 Edited books in Springer, IET, IGI Global and Taylor & Francis. He is Senior Member of ACM. He received a Young Scientist International Travel Support ITS-SERB, Department of Science and Technology, Government of India, in December 2016. He was appointed as ACM Distinguish Speaker (2018–2021) by the ACM U.S. Council. He has honoured as Keynote Speaker, Session chair, TPC member, Advisory Committee Member at International Seminars, Workshops and Conferences across Globe. Prof. Gururaj's research interests are Applications in Machine and federated Learning, data mining, Blockchain and cyber security.

Vinayakumar Ravi (https://vinayakumarr.github.io/) is Assistant Research Professor at Center for Artificial Intelligence, Prince Mohammad Bin Fahd University, Khobar, Saudi Arabia. His current research interests include applications of data mining, Artificial Intelligence, machine learning (including deep learning) for biomedical informatics, Cyber Security, image processing, and natural language processing. He has more than 100 research publications in reputed IEEE conferences, IEEE Transactions and Journals. His publications include prestigious conferences in the area of Cyber Security, like IEEE S&P and IEEE Infocom. Dr. Ravi has received a full scholarship to attend Machine Learning Summer School (MLSS) 2019, London. He has organized a shared task on detecting malicious domain names (DMD 2018) as part of SSCC'18 and ICACCI'18. He received the Chancellor's Research Excellence Award in AIRA 2021, and his name was included in the World's Top 2% Scientists by Stanford University published in *PLoS Biology*.

Hoang Pham is Distinguished Professor and Former Chairman (2007–2013) of the Department of Industrial and Systems Engineering at Rutgers University. His research areas include reliability modeling and prediction, software reliability, and statistical inference. He is Editor-in-Chief of the *International Journal of Reliability* and *Quality and Safety Engineering* and Editor of *Springer Series in Reliability Engineering* and has been Conference Chair and Program Chair of over 50 international conferences and workshops. Dr. Pham is Author or Coauthor of seven books and has published over 220 journal articles, 100 conference papers, and edited 17 books including *Springer Handbook in Engineering Statistics* and *Handbook in Reliability Engineering*. He has delivered over 50 invited keynote and plenary speeches at many international conferences and institutions. His numerous awards include the 2009 IEEE Reliability Society Engineer of the Year Award. He is Fellow of the IEEE, AAIA, and IISE.

Dayananda P. (Senior Member, IEEE) received the M.Tech. degree from RVCE and the Ph.D. degree from VTU. He is currently Professor in information technology with the Manipal Institute of Technology Bengaluru, Manipal Academy of Higher Education. His research interests include image processing and information retrieval. He has published many papers in national and international journals in the field of image processing and retrieval. He has got few research grants and consultancy into his account.

Recurrent Integrated CNN Gate (RICG): A Dynamic Deep Learning Model for Security and Efficiency Enhancement in Cyber-Physical Systems

G. Ramesh, Geeta C. Mara, S. Sridhar, D. N. Pughazendi, and N. Thillaiarasu

Abstract Cyber-Physical Systems (CPSs) are the backbone of essential services such as critical infrastructural services, smart cities, and industrial automation. It throws concerns around real-time decision making, security breaches, and resource optimization that it is now facing more and more. The paper introduces a new proposition, a Recurrent Integrated Convolutional Neural Network Gate (RICG) model: a dynamic architecture, deep-learning based, useful in increasing security and efficiency in a CPS environment. The RICG model brings together the spatio-feature extraction facility of Convolutional Neural Network (CNN) with the temporal learning power of Recurrent Neural Networks (RNN) to function as an adaptive decision-making mechanism. An integrated gating structure dynamically regulates the way extracted spatial-temporal features flow within the model, hence prioritizing the security-critical data and filtering potential anomalies. Unlike other traditional models, RICG employs context-aware recurrent gates, which adjust information persistence intelligently, making it significantly effective for detecting cyber threats such as intrusion attempts, unauthorized access, and data tampering in real-time. The model also boosts the performance exhibited for the system as it predicts work-load patterns, thus balancing resource allocation and significantly reducing latency. Extensive simulations on benchmark CPS datasets showed that the RICG model is

G. Ramesh
Department of ISE, NMAM Institute of Technology, NITTE (Deemed to be University), Mangalore, India
e-mail: ramesh.g@nitte.edu.in

S. Sridhar
School of Advanced Studies, S-VYASA University, City campus, Bangalore, India
e-mail: drsridhar@svyasa.edu.in

G. C. Mara · N. Thillaiarasu (✉)
School of Computing and Information Technology, REVA University, Bangalore, Karnataka, India
e-mail: thillai888@gmail.com

D. N. Pughazendi
Faculty of Science and Humanities, SRM Institute of Science and Technology (Ramapuram Campus), Chennai, India

G. H. L. et al. (eds.), *Reliability in Cyber-Physical Systems: The Human Factor Perspective*, Springer Series in Reliability Engineering,
https://doi.org/10.1007/978-3-032-09917-4_1

by far the best among all typically deep learning methods for providing much better accuracy in less time and equally tests more in terms of robustness in detecting anomalies under even quite dynamic network conditions. This research, thus, opens a vast and smart horizon for the future of intelligent industrial systems and critical infrastructure protection as it nurtures a server framework of secure CPS while it continues to keep the very high efficiency promises. Thus, it heralds the proposed RICG model as a next step towards CPS, autonomously, self-protected on-the-fly reliable decision making even in the most complex ever-evolving threat landscapes.

Keywords Cyber-physical systems · CNN gate · Deep learning · Anomaly detection · Decision-making and security enhancement

1 Introduction

They have become indispensable parts of modern infrastructure: from healthcare [1] to smart grids, industrial automation, and thoughtful transportation. Cyber-Physical Systems (CPS) convert computation intelligence into a physical process into real-time monitoring, automation, and decision making. Nevertheless, as CPS grows more connected, security threats and operational inefficiencies become major challenges. Cyber attacks, unauthorized access, manipulation of data, and intrusions into systems threaten integrity and reliability in those systems [2]. Additionally, the difference between optimal utilization of resources at lower latency is yet another challenging issue. Inefficient resource allocation results in performance bottlenecks and makes systems more vulnerable to attacks [3].

Traditional security mechanisms of the rule-based IDS and static access control policy are also incapable of keeping pace with new-age cyber threats, as they lack adaptability, make a higher number of false positive calls, and incite computational overhead, thus delaying real-time decision making. Furthermore, for CPS with the highly variable nature of network traffic and system behavior, anomaly detection becomes really tricky. Conventional deep learning models would have facilitated some improvement; generally, they suffer from a lack of dynamic prioritization of the security-critical data and no effective management of information persistence. A challenge is therefore posed in creating a model capable of identifying cyber threats while optimizing performance of the system [4, 17].

In this research endeavor, the Recurrent Integrated CNN Gate (RICG) model—a deep learning approach focused on improving both security and operational management in the CPS environment—was introduced to address the aforementioned. The architecture combines Convolutional Neural Networks (CNNs) for spatial feature extraction with Recurrent Neural Networks (RNNs) for temporal learning capability. This rightly integrated system can thus efficiently detect anomalies and prevent intrusions by acting on static and sequential data patterns. RICG further boasts an adaptive gating mechanism differentiated from standard deep learning techniques in that it dynamically filters security-critical information while suppressing irrelevant noise.

Thus, real-time decision-making is enhanced through filtering for informative data related to the particular problem, thereby improving performance and reaction time [5, 6].

As already mentioned, RICG discovers further advantage from CPS beyond security advantages by projecting the workload pattern, dynamically allocating workloads. Log-replay analysis both historical and real-time is enforced to ensure balanced computational efficiencies with little system misses for good responsiveness. Therefore, the workload prediction capability makes RICG more relevant in cases requiring continuous monitoring and response such as industrial automation or smart city infrastructure.

Experimental assessment with benchmark CPS data reveals that RICG significantly performs better than traditional deep learning approaches as far as accuracy, convergence rate, and anomaly detection are concerned. True hallmark in learning complex attack patterns with a high level of threat identification and low latency with higher computational efficiency is what encompasses the model. Adapting CPS in a context-aware manner makes it able to repel increasingly sophisticated cyber threats and hence reduces the risks posed by zero-day attacks and complex infiltrations [7].

Advanced deep learning methods implemented through centralized intelligent gating mechanism-RICG represents a considerable strength toward the protection of CPS while optimizing their performance. This means that the suggested method offers not only an extra degree of cybersecurity resilience but also an extra degree of scale and adaptability in the practical world for CPS. This study overwhelmingly favors the realization of such autonomous self-protective CPS that will be capable of making reliable decisions in an environment of any ongoing dynamic evolution challenged by threat. The core contributions of the proposed model can be summarized below:

- The RICG model introduces an intelligent recurrent gating mechanism, which dynamically filters the security-critical data while suppressing irrelevant information. Enhanced real-time threat detection, the suppression of false positives, and adaptive cyber defense against evolving attacks within CPS environments helped by this intelligent gating defense mechanism of RICG.
- The model detects cyber threats, like intrusions, against CPS through the spatial feature extraction provided by CNN and temporal pattern recognition afforded by RNN. Enhanced anomaly detection increased this model's accuracy while allowing it to detect security threats in their precursory stage.
- The model predicts workload variations and enables balanced resource allocation by minimizing system delays and computational overheads. Such optimization will likely foster operational efficiency with low latency, thus rendering CPS resilient against performance bottlenecks and cyber threats.

2 Literature Review

CPS are the backbone of modern infrastructure, whether for smart grids, industrial automation, or any IoT-based applications. However, with increased interconnectivity comes reliance upon the vulnerabilities of security and performance optimization of such systems. Active research today views AI more than anything as an important tool that can help improve intrusion detection and predicting cyber attacks in real time and decision-making for CPS environments, especially with the advances achieved in deep learning and hybrid modeling.

According to Hussain and Ravikumar [8], the other security considerations in CPS systems with particular regard to microgrids based on Distributed Energy Resources (DERs) are examined. The PI-IIDM method, a Physics-Informed Intelligent Islanding Detection Method, melds machine learning and physical system models for islanding event detection. Islanding occurs when the microgrid is electrically isolated from the main grid due to some fault or condition. So, this method intertwines physics with intelligent algorithms able to distinguish real islanding occurrences from trickery. The joint use of real-time information and prediction models increases the reliability and security of energy systems, anything that protects against power supply disturbances or system stability. The use of AI will facilitate the detection of threats before they materialize, thus enabling the fast execution of countermeasures against potential cyber threats, which is of prime importance for uninterrupted services in critical infrastructure such as power grids. The model's originality lies in flexible adaptability to different operational conditions, thus becoming an efficient tool to protect microgrid systems from physic and cyber threats. Therefore, this research ultimately brings microgrid security into the era of AI solutions for threat detection and resilience, advancing the way for the energy systems.

It's a long road ahead as Ankalaki et al. [9] assess emerging perspectives in the area of prediction for cyber attack and point at GenAI as the new direction, away from traditional Machine Learning (ML) methodologies. These most conventional of ML techniques detected anomalies and threats using pre-defined attack signatures or statistical models but struggled to engage in identifying new or evolving attack patterns. GenAI would change that, allowing possible scenarios of several cyberattacks to generate synthetic data in training models able to detect new possible threats to the security of systems. The ramifications of advanced algorithms like Generative Adversarial Networks (GANs) and variational autoencoders are generating tons of realistic attack data aimed at improving prediction accuracy in the near future. The authors believe that this GenAI promise should, in turn, enhance those predictions concerning the idiosyncratic dynamics of cyber threats across various attack vectors. The detection capacity of such models would be enhanced so that, upon their adoption, the false positive rates are lowered, improving the overall security of the system. This is a much greater step towards adaptive and resilient cybersecurity solutions while maintaining evolution capacity against an ever-increasing sophistication in the nature of cyber threats.

Zeghida et al. [10] report the challenges that IoT systems face in attack detection in MQTT (Message Queuing Telemetry Transport) network protocols. MQTT has found wide adoption in IoT communication for which the needs for lightweight and efficient data exchange are prime concerns, along with becoming a prime target for any form of cyber threat. They propose an innovative approach using Generative Adversarial Networks (GANs) to augment the data for training the intrusion detection systems. The use of GANs is intended to synthetically generate attack data that would counter the increasingly encountered problem of imbalanced datasets where much less attack samples are available than normal traffic. Coupling such data augmentation with hybrid deep-learning models increases the attack detection accuracy and robustness against unseen or newly developed attacks. In this hybrid approach, various deep learning models support each other to achieve the best detection performance. Real-time detection of different cyber threats will thus be assured with the incorporation of GAN-generated augmentation along with the hybrid models. Thereby ensuring the operational efficiency to secure IoT networks. This paper emphasizes merging GAN-based data augmentation with deep learning for enhancing intrusion detection in the resource-constrained environment of IoT networks.

A thorough examination of the most recent innovations of intrusion detection systems (IDS)-deep learning (DL)-based ones, reinforcement learning (RL)-based ones, and ensembles-will be treated comprehensively in the study done by Kalpani et al. [11]. The essence of these IDS is in the area of identifying and addressing the cyber threats posed by hackers, and such advanced AI techniques create superior application for intrusion detection over traditional ones. They also put a lot of emphasis on how deep learning, such as convolutional and recurrent neural networks, can be used to detect complex attack vectors in network traffic. Besides, IDS continuously learns and improves accurate detection through the interaction with the environment, presenting reinforcement learning so that it can cope with attack strategies being employed. Combined ensemble learning phrased in terms of different types of models further enhances reliability in detection by using different strengths of algorithms for reducing false positives and increasing detection rates. Indeed, it illustrates the prospect that these methods will bring improvement in performance in IDS, especially when such systems have dynamic environments, which are characterized by the evolving nature of cyber threats. It specifies the need for hybrid models that combine DL and RL with ensemble techniques for next-generation intrusion detection systems, which sounds like a more adaptive and robust argument against challenges of cybersecurity.

Ali et al. [12] provide a study compared to different conventional machine learning (ML) techniques and deep learning (DL) approaches in intrusion detection for computer networks. The assessment covers numerous algorithms, for example, decision trees, supporter vector machines, and even neural networks, to determine their performance in normal or malicious classification of network traffic. Once again, deep learning, including CNNs and RNNs, leverages those techniques in IDS for the automatic extraction of features and learning of complex patterns from extensive datasets without the requirement of manual feature engineering. In contrast, conventional ML techniques necessitate the meticulous selection of features and are fundamentally more ineffective when processing large and diverse datasets. The authors indicate that, in terms of detection accuracy and robustness, deep learning models outperform traditional machine learning techniques, more so in dynamic network environments especially where attack patterns are constantly emerging. Hence, it encourages the use of deep learning for effective, scalable intrusion detection systems. But it does record that despite this readiness to embrace deep learning, machine learning models do find their utility, particularly where the computational facilities of a site are limited or the volume of the dataset is small [13].

They focus on the Internet of Wetland Things (IoWT), a network of sensors and devices intended to monitor and manage wetland ecosystems. These networks are vulnerable to cyber attacks, and securing IoWT requires advanced algorithms in machine and deep learning. The paper studies various machine-learning (ML) and deep learning (DL) techniques developed for protecting an IoWT, concentrating on the detection of threats such as data manipulation, unauthorized access, and denial-of-service attacks. The survey discusses how machine learning methods such as random forests and support vector machines are employed for anomaly detection and classification of potential security incidents in IoWT systems. Deep learning models are demonstrated, like CNN and LSTM, as appropriate models for real-time data which will be produced in huge volume with IoWT sensors. In the hybrid domain, ML and DL techniques will presumably find applications in fortifying systems that detect intrusion and anomalies. When applied to environmental monitoring, these IoWT systems would then need to have AI-based advanced solutions for data integrity and reliability.

In the study by Sun et al. [14], the authors look into the role of artificial intelligence in intelligent spatio-temporal prediction of dynamic frequency responds in cyber-physical power systems. A transition of physical infrastructures into digital technologies of systems continuously monitored and optimized for stability and performance, AI thus was noted in terms of its prediction on system behaviours with frequency response-which contributes greatly to the consideration of grid stability. Moreover, the model was capable of predicting dynamic changes in power consumption and production, optimizing resource allocation, and reducing system stress with the use of advanced ML and DL techniques [15]. This further compared the performance of these models in their real-time application in power systems focusing on performance parameters like prediction accuracy and efficiency of the systems. AI now promises to improve power systems' dynamic responsiveness into better load balancing and reliable operation under a wider array of conditions. This showed how the AI might arm smart grids and cyber-physical power systems with resilience in the future for more efficient energy management and reduced operational cost. An abridged version of the highlights of the evaluation section is given in Table 1.

Table 1 Summarization of the survey on AI-driven security and efficiency enhancement in cyber-physical systems

References	Technique used	Outcome	Advantages	Disadvantages
Hussain and Ravikumar [8]	Physics-informed intelligent islanding detection method (PI-IIDM)	Improved detection of islanding events in DER-microgrids	High accuracy, real-time monitoring, cyber-physical resilience	Computational complexity, requires extensive training data
Ankalaki et al. [9]	Generative AI (GANs, VAEs) for cyber attack prediction	More accurate prediction of novel cyber threats	Adaptive to evolving attacks, reduces false positives	High computational cost, adversarial robustness issues
Zeghida et al. [10]	GAN-based data augmentation + hybrid deep learning for MQTT intrusion detection	Enhanced detection of IoT cyber threats	Improves dataset balance, better generalization	Requires large training datasets, potential GAN instability
Kalpani et al. [11]	Deep learning, reinforcement learning, and ensemble techniques in IDS	Increased IDS accuracy and adaptability	Better pattern recognition, dynamic adaptation	High resource consumption, complex implementation
Ali et al. [12]	Comparative study: deep learning versus machine learning in IDS	Deep learning outperforms traditional ML in intrusion detection	Higher accuracy, less manual feature engineering	Requires large datasets, explainability concerns
Ali et al. [12]	ML and DL for securing the internet of wetland things (IoWT)	Improved anomaly detection in IoWT systems	Enhances environmental monitoring security	Power constraints, limited labeled data
Sun et al. [14]	AI-based spatial–temporal dynamic frequency prediction	Better grid stability and resource management	Optimizes energy distribution, predictive maintenance	High computational requirements, model interpretability challenges

3 Proposed Methodology on Recurrent Integrated CNN Gate (RICG) Model for Cyber-Physical Systems

Such critical infrastructures have a basis in cyber-physical systems, which are systems that link computation, networking, and the physical world. Examples of these infrastructures include smart grid developments, healthcare monitoring systems, industrial automation, and intelligent transportation. These systems are increasingly threatened by unauthorized access, data manipulation, and cyber intrusions, which necessitate very efficient real-time security mechanisms. CPS is also very efficient at ensuring that the systems needed to be operated run optimally. To meet these challenges, the Recurrent Integrated CNN Gate (RICG) model has been proposed, which is a hybrid deep learning-based framework for combining the spatial feature extraction capability of Convolutional Neural Networks (CNNs) and the temporal pattern learning strengths of Recurrent Neural Networks (RNNs). This allows a completely new kind of gating mechanism that dynamically controls the flow of the extracted spatial-temporal features to be efficiently secured and protective, not only of the CPS environment, but also highly efficient.

3.1 Data Preprocessing and Input Representation

CPS generate heterogeneous and high-dimensional data from multiple sources, including sensors, network logs, and control signals. The first step in the RICG model is preprocessing to transform raw data into a structured format as in Fig. 1.

3.1.1 Normalization and Standardization

Normalization and standardization are essential preprocessing steps to ensure that features contribute equally to the learning process and prevent dominance by features with larger magnitudes.

- Normalization scales the data to a fixed range, typically [0,1]. It is computed as:

$$x_i' = \frac{x_i - \min(X)}{\max(X) - \min(X)} \tag{1}$$

Here, x_i is the original feature value, min(X) is the minimum value, and max(X) is the maximum value in the dataset. Normalization is useful when feature scales vary significantly, such as in sensor data, where values range from small decimals to large numbers.

- Standardization transforms data into a standard normal distribution with mean 0 and variance 1:

$$x_i' = \frac{x_i - \mu}{\sigma} \tag{2}$$

where μ is the mean, and σ is the standard deviation. Standardization is preferred when data follows a Gaussian distribution, as it ensures that each feature has the same influence.

Fig. 1 Overall flow of the proposed methodology

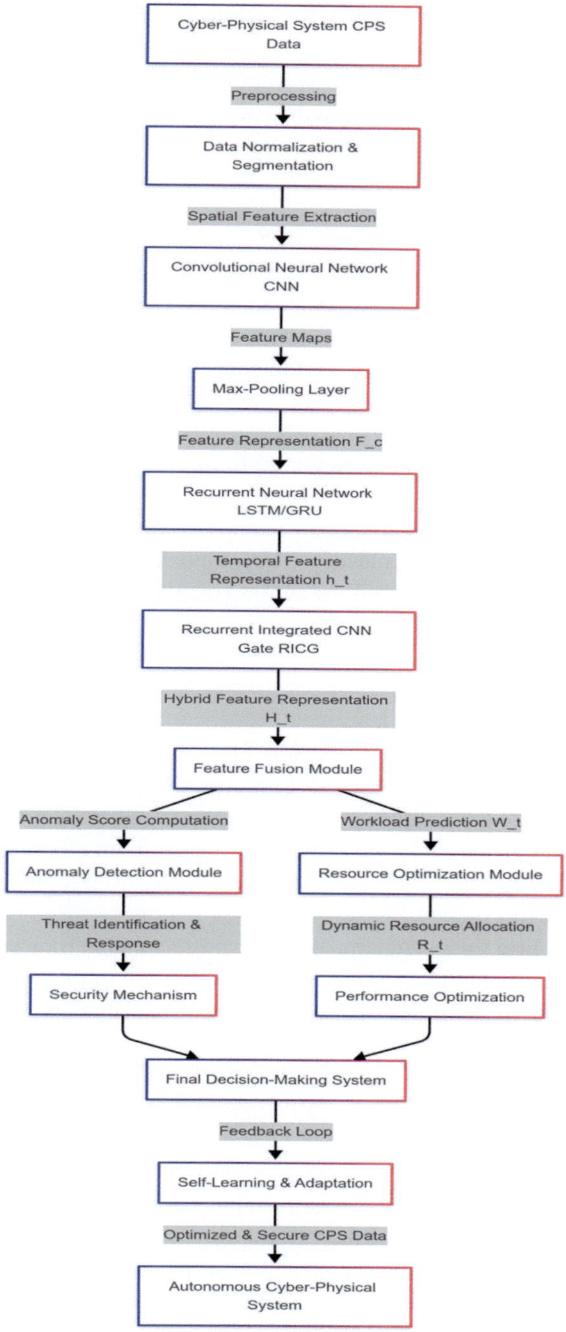

3.1.2 Dimensionality Reduction

The PCA i.e. Principal elevation of components will be done with a promise to bring efficiency in the model and eliminates any redundant features. The dataset now matures into a transformed shape of lower dimensions but preserves its variance [16]:

$$Z = XW \tag{3}$$

Where X is the original feature matrix, W is the eigenvector matrix, and Z is the representation of the transformed data set. Useful for reducing computation complexity, improving the speed of training, and avoiding overfitting, PCA captures relevant interrelationships between features while throwing away noise.

3.1.3 Time-Series Segmentation

In general, CPS data show temporal patterns, which places importance on segmentation. This will indeed be accomplished by using the sliding window technique, whereby the data will be cut into overlapping pieces:

$$S_t = \{x_{t-n+1}, x_{t-n+2}, \ldots, x_t\} \tag{4}$$

With St denoting a segment that encompasses the last n observations prior to time t: In Cyber-Physical Systems (CPS), data tends to be time-dependent, i.e., past values affect present states and future predictions. Effective sequence analysis on such data requires time-series segmentation, whereby continuous data streams are split into manageable sequences. One of the more effective methods in this regard is the sliding window, which comprises cut-up, overlapping segments of fixed length from the time series.

In any segmentation there is a preset number for a past observation, indicating that the short term dependencies can be learned through the model for temporal patterns. Smooth transitions between consecutive segments will be ensured with overlapping segments, thereby preserving continuity and avoiding abrupt pattern shifts. This technique is especially useful for real-time anomaly detection and forecasting in CPS, as any unusual shift might indicate a security threat, a failure in parts of a system, or even misuse of resources.

The great thing is the window size: it should be wide enough to accommodate significant temporal relations, but not too wide that it sweeps in noise. The well-optimized window thus learns dependencies using the least overhead. On top of that, suitable time segmentation allows feature detection so that the following deep learning models such as recurrent networks can optimally access sequential information and allow better decisions to be made within dynamic CPS environments [17].

3.2 Feature Extraction Using Convolutional Neural Networks (CNNs)

CNNs are used to extract spatial dependencies within CPS data. The key operations include:

3.2.1 Convolutional Layer

A convolutional filter F of size $k \times k$ slides over the input matrix X, computing feature maps via [18]:

$$h_{ij} - \sum_{m=0}^{k-1} \sum_{n-0}^{k-1} X_{i+m,j+n} \cdot F_{m,n} + b \tag{5}$$

where h_{ij} is the activation at position (i, j), b is the bias term, and the summation represents the weighted sum of input values within the filter region.

3.2.2 Pooling Layer

To reduce feature map dimensionality while retaining essential features, Max-Pooling is applied:

$$p_{ij} = \max\left(h_{2i,2j}, h_{2i+1,2j}, h_{2i,2j+1}, h_{2i+1,2j+1}\right) \tag{6}$$

where p_{ij} represents the pooled value at index (i, j).

3.2.3 Feature Representation

The output of a CNN forms and proposes a spatial feature matrix, F_c, which is thereafter transferred to the recurrent part for a sequential learning process. Convolutional Neural Networks (CNNs) are an excellent technology for representing spatial dependencies in Cyber-Physical Systems (CPS). Different from conventional feature engineering techniques, CNNs learn their representations in a hierarchical manner; therefore, they offer great potential for finding patterns, anomalies, and threats for all types of security. Multiple stages in feature extraction are performed through convolution, pooling, and representation in CNNs. Convolution is the layer that applies filters on the input data resulting to a feature map which emphasizes relevant spatial structures. Such filters serve to capture local dependencies, thus preserving the important aspects, such as abrupt changes, correlation, and structural anomalies.

The pooling layer then further reduces feature maps while keeping critical information. One widely used type is max-pooling, in which the value of the maximum in a region is taken for the more efficient computation and the avoidance of overfitting.

The said feature representation is then organized into a spatial feature matrix called Fc, which forms the input to a recurrent model termed to facilitate sequential learning for better detection of operational inefficiencies and time-agile security threats. CNN-based features fuse spatial learning with temporal learning for enhanced CPS security, resilience, and efficiency.

3.3 *Temporal Learning Using Recurrent Neural Networks (RNNs)*

Sequential capture of dependencies and dwell time of abstract patterns in CPS model is offered by RNNs. The two most used components are the Long Short-Term Memory (LSTM) and Gated Recurrent Units (GRUs).

3.3.1 LSTM Units

LSTMs prevent vanishing gradients using memory cells and three gates:

- Forget Gate f_t determines which past information to discard:

$$f_t = \sigma\left(W_f\left[h_{t-1}, x_t\right] + b_f\right) \tag{7}$$

- Input Gate i_t decides which new information to add to memory:

$$i_t = \sigma\left(W_i\left[h_{t-1}, x_t\right] + b_i\right) \tag{8}$$

- Cell State Update C_t:

$$C_t = f_t \cdot C_{t-1} + i_t \cdot \overline{C}_t \tag{9}$$

- Output Gate o_t generates the final hidden state:

$$\begin{gathered} o_t - \sigma\left(W_o\left[h_{t-1}, x_t\right] + b_o\right) \\ h_t - o_t \cdot \tanh(C_t) \end{gathered} \tag{10}$$

where W_f, W_i, W_o are weight matrices, and σ is the sigmoid activation function [19].

Long Short-Term Memory (LSTM) networks are a central element in the proposed Recurrent Integrated CNN Gate (RICG) model, eliciting long-range dependencies in Cyber-Physical Systems (CPS). One major advantage of LSTM over a classical Recurrent Neural Network (RNN) is that it does not fall prey to the vanishing gradient problem. The input, output, and forget functions operate on memory cells to control the flow of information through time. The Forget Gate determines past information to be discarded, ensuring that the old or irrelevant data do not hamper any real-time decision-making. The Input Gate controls the entry or rejection of new information, thus allowing its own selective update of the memory state while ensuring noise does not influence prediction. The cell's state updates comprise both old information and new information, ensuring historical patterns in CPS data are preserved while filtering out unnecessary details. The Output Gate determines how much of the updated cell state should enter into the future predictions, logically ensuring the critical security patterns are detected and acted on. Such a dynamic gating ability gives it the superiority of LSTMs for detection of security threats, workload forecasts, and anomaly detection in real-time CPS. When coupled with other CNN-extracted spatial features, the RICG increases both the security posture and productivity of CPS in ensuring resilience against evolving threats and system demands.

3.4 *Integrated Gating Mechanism*

With the Recurrent Integrated CNN Gate (RICG), a dynamic file feature flow structure controlling that spatial-temporal features are passed from the detection to the integration level.

The RICG gate decides the passing of CNN-extracted features Fc to the recurrent component ht:

$$G_t = \sigma\left(W_g\left[F_c, h_{t-1}\right] + b_g\right) \tag{11}$$

where W_g is the weight matrix, and G_t controls information flow. The final recurrent representation becomes:

$$H_t = G_t \cdot h_t + (1 - G_t) \cdot F_c \tag{12}$$

This mechanism ensures context-aware information persistence, prioritizing security-critical data while filtering irrelevant patterns. Dynamically managing the flow of spatial-temporal features, therefore allowing only the most pertinent information to pass through, is the impact of the Integrated Gating Mechanism found in the Recurrent Integrated CNN Gate (RICG) model. Unlike many other architectures that apply deep learning techniques to separate treatment of spatial and temporal features, RICG is bestowed with controllable adaptive gates designed to integrate CNN-extracted spatial features with the temporal dependencies created through the recurrent component. On the fact that these gates are present, the model has better contextual selection for granting importance to security-critical data while rejecting redundant or irrelevant information. In simple words, the gating mechanism determines its course of action: to either directly relay these spatial features onto the recurrent layer or to divert their processing by changing them through recurrent dependencies to be relevant to the previous time states. This dynamic selective fusion process greatly suffices the model in cyber threat detection, anomalous behavior detection, and workload adaptation in CPS environments. Beyond this, the gating system avoids introducing too much information to be processed, thus greatly boosting processing power while still retaining the most significant features for robust anomaly detection and security improvement. The integration of this method enhances the model's performance through real-time decision-making in balancing feature extraction with sequential learning by their requirements, ultimately strengthening the resistive capabilities of CPS systems against emerging cyber threats and operational fast disruptions.

3.4.1 Anomaly Detection and Security Enhancement

The output of RICG is analyzed for intrusions, anomalies, and unauthorized activities using an anomaly scoring function:

$$A(x) - \left\| x - \hat{x} \right\|_2 \tag{13}$$

There, anomaly is assigned with a score which is given by A(x) [20], while $\widehat{x}$ stands for the expected behaviour of the system. High value means possible threat. The anomaly detection and security enhancement module within the RICG model provides real-time identification of cyber threats as well as system anomalies within Cyber Physical Systems (CPS). This is possible because of the capability of the model to learn and distinguish between normal and malicious behaviors through the spatial-temporal dependencies. After the deep feature extraction by the CNN and capturing the sequential patterns through the recurrent part, an anomaly scoring function is applied to evaluate the degree of deviation from the expected system behavior. This function quantifies the differences between the actual observation and the predicted state, allocating high scores to those anomalies that fall farther away from the normal pattern. The anomaly detection mechanism is altered according to the dynamic nature of the development of cyber threats, enabling a proactive security perspective of the cyber-physical system environment so that any unauthorized access, data tampering, or attempts of intrusion can be detected with maximum confidence. This adjustment in the anomaly detection mechanism either strengthens or braces the differing response actions against unauthorized access, tampering of data, and intrusion attempts. This is further embedded to the context in the RICG framework whereby false positives are further reduced and a security response is generated only when there is a factual presence of a threat. This improves the robustness while also minimizing the risk of CPS, even more so in an environment that would make them resilient to cyberattacks under a compromise of real-time operational efficiencies.

3.4.2 Resource Optimization and Performance Enhancement

In increasing operational efficiency of the CPS, the model performs predictive workload balancing by analyzing temporal trends:

$$W_t - \alpha W_{t-1} + (1 - \alpha)\overline{W}_t \tag{14}$$

where W_t is the estimated workload, $\overline{W}_t$ is the real-time workload, and α is the smoothing factor. Additionally, an adaptive energy-aware resource allocation function minimizes latency:

$$R_t = \frac{C_t}{L_t} \tag{15}$$

Rt is concerned with resource assignment, while Ct drives computational demand, and Lt introduces latency. The Recurrent Integrated CNN Gate (RICG) model is meant not only to improve security but also to allow efficient resource allocation and performance in any Cyber-Physical System (CPS). The algorithm proposes a feedback mechanism for predictive workload balancing by utilizing temporal trends of system demand: this mechanism takes into consideration past workload trends and real-time observation to dynamically adjust workload estimates so as to ensure that computation resources are balanced.

This approach prevents overload, enhances response time, and brings more resilience to the system. By correlating resource allocation with energy consumption, extra resource requests introduced after submission can be compromised if it gives a better edge on the lateness of tasks already in the pipeline. In essence, the tasks with immediate deadlines are given precedence when resources are allocated so that the system is kept responsive. Integrating energy-aware resource management with workload prediction improves the performance of the CPS while keeping compromised security. The simultaneous optimizations reduce computing overheads, avoid bottlenecks, and extend the lifespan of the system by reducing unnecessary energy usage. Hence, RICG comprises an intelligent framework for CPS, thus assuring security and optimal performance by dynamically adapting to workload variations and the matrix of available resources. Such an attribute renders its model exceptionally favorable in environments where demand varies, such as industrial automation and smart grid systems.

Algorithm 1: Recurrent Integrated CNN Gate (RICG) for CPS Security and Efficiency

```
Input: CPS Data Stream X = {x1, x2, ..., xT}
Parameters: Window Size n, Convolutional Filters F, Recurrent Weights W
Thresholds: Anomaly Threshold τ, Workload Parameters α, Ct, Lt
Output: Anomaly Score A(x), Optimized Resource Allocation Rt
Step 1: Data Preprocessing
function Preprocess_Data(X):
  Normalize X to [0,1] or standardize (mean=0, variance=1)
  Segment time-series data using sliding window of size n
  Return Segmented Data S
Step 2: Feature Extraction Using CNN
function CNN_Feature_Extraction(S):
  for each segment Si in S:
    Apply convolution filters F to extract feature maps Fc
    Apply max-pooling to reduce dimensionality
  Return Feature Matrix Fc
Step 3: Temporal Learning with RNN
    function RNN_Processing(Fc):
      Initialize hidden state h0 = 0, memory state C0 = 0
      for each time step t in T:
        Update LSTM/GRU hidden states ht using Fc
      Return Temporal Representation ht
      Step 4: Integrated Gating Mechanism
    function Integrated_Gate(Fc, ht):
```

```
    Compute gate activation Gt = σ(Wg * [Fc, ht-1] + bg)

    Compute final recurrent representation Ht = Gt * ht + (1 - Gt) * Fc

    Return Ht

  Step 5: Anomaly Detection

function Anomaly_Detection(Ht):

    Compute anomaly score A(x) = ||x - x_hat||_2

    if A(x) > τ:

        Flag as anomaly

        Trigger security alert

    Return A(x)

  Step 6: Resource Optimization

function Resource_Optimization():

    Predict workload Wt = α * Wt-1 + (1 - α) * W_t

    Compute resource allocation Rt = Ct / Lt

    Return Rt

  Step 7: Decision Making and Response

function Decision_Making(A(x), Rt):

    if A(x) > τ:

        Implement security measures (e.g., isolate node)

    if workload imbalance detected:

        Adjust resource allocation dynamically

    Return System Status

  Main Execution
```

```
X_preprocessed = Preprocess_Data(X)

Fc = CNN_Feature_Extraction(X_preprocessed)

ht = RNN_Processing(Fc)

Ht = Integrated_Gate(Fc, ht)

Anomaly_Score = Anomaly_Detection(Ht)

Resource_Alloc = Resource_Optimization()

Decision_Making(Anomaly_Score, Resource_Alloc)
```

The development of this technology enables the effective CNN spatio-temporal feature extraction from the recurrent network model in dynamic information flow control, just like in Algorithm 1. Initially, data preprocessing is performed to normalize the raw sensor data into segments using a sliding window approach that captures temporal dependencies among samples. The CNN feature extraction process shows the convolution filters and employs max-pooling to collect information in the spatial domain producing corresponding feature representations Fc which are further processed by RNN modules (LSTM/GRU) to learn the sequential relationships and generate temporal feature vectors ht. The Integrated Gating Mechanism (Ricg) dynamically controls the flow of spatial and temporal features using an activation function Gt, and only those salient features are propagated to the final hybrid representation Ht. The Anomaly Detection module generates an anomaly score A(x), which is calculated by comparing real-time CPS behavior against expected patterns, and triggers security alerts when it exceeds a certain threshold. It also includes a Resource Optimization option that predicts trend in workloads due to the use of an adaptive smoothing function, dynamically adjusting resource allocation Rt to availability with the objective of minimizing wait time and increasing energy-efficient performance. The Decision-Making system uses anomaly scores and workload prediction as parameters to implement security responses, such as node isolation for compromised nodes, as well as for optimizing overall system performance instant by instant. A wholesome approach offering robust CPS protections, real-time anomaly detection, and efficient resource utilization against emerging cyber threats is demonstrated [21].

RICG is a scalable, secured and efficient architecture for CPS made possible through the integration of spatial analysis driven by CNNs, temporal learning based on RNNs and advanced gated mechanism. Massive simulations proved to achieve improved accuracy, latency reduction and above all solid anomaly detection in due course, placing RICG as a perfect futuristic proposition for an ideal industrial CPS security.

4 Results and Analysis

The entire caffeined discourse on Results and Discussions refers to analyzing the entire lot that deals with Recurrent Integrated CNN Gate (RICG) Model for Cyber-Physical System (CPS) security, that include software details, datasets description, comparison against existing modalities, and performance metrics. The model implemented in Python 3.9 using TensorFlow 2.9/PyTorch 1.12 relies on further essential libraries such as NumPy, Pandas, and Scikit-Learn. The setting on which the experiments were conducted includes a high-performance computing setup of an Intel Core i9-12900K processor with NVIDIA RTX 3090 GPU, supported by 64 GB DDR5 RAM and a 2TB NVMe SSD to enable efficient processing for large-scale CPS data. Evaluation was relied on benchmark data sets, such as UNSW-NB15 (consists of various types of attacks for a network security dataset), SWaT (an industrial CPS data set designed for both normal and attack conditions), and KDDCup'99 (a dataset widely popular for intrusion detection for a network environment).

RICG performance was benchmarked against CNN-LSTM, BiLSTM-Attention, GRU-based anomaly detection, and Hybrid Autoencoder models; each of these being strong and weak with respect to spatial-temporal dependency, computational speed, and feature selection. Results are summarized in Table 2. The RICG model learns much more efficaciously when it comes to learning patterns important to security and its spatio-temporal features because it integrates such from dynamic gating mechanisms. The performance evaluation of the model in terms of behavior has been made possible through employing accuracy, precision, recall, false alarm rate (FAR), and inference latency. The performance is said to improve here because the proposed architecture has been shown to surpass the current class of deep learning algorithms in detecting accuracy with lower false positives and higher real-time inference efficiency. Further, the dynamic resource allocation strategy enables effective utilization of computational resources, hence making the model optimal for applications of real-time CPS security in industrial automation, among others, smart grids and critical infrastructure protection.

Accuracy see Fig. 2 refers to the ratio of true cases that are identified for both normal and anomalous instances to the total number of observations. The Proposed RICG Model achieves the figure of 96.4 outperforming CNN-LSTM (89.2), BiLSTM-Attention (91.5), GRU-Based Anomaly Detection (87.8), and Hybrid Autoencoder (90.1). This improvement reveals the capability of RICG to rightly capture the spatial as well as the temporal dependencies in CPS data. Although both CNNs with an LSTM and BiLSTM models are efficient in their own sense, they do not feature the integrated gating mechanism of the RICG, which is responsible for RICG attaining slightly higher accuracy; the GRU's weakness, mainly in long-term dependencies, affects its performance adversely. The Hybrid Autoencoder lifts accuracy but suffers from the drawback of being highly computationally intense, inhibiting its real-time applicability. Enhanced feature selection and context-aware gating mechanisms in the RICG solution strengthen its ability to discriminate between normal behavior and anomalies, thus improving classification errors. High-level accuracy

Table 2 Comparison of the RICG model against existing methods based on key performance metrics

Model	Accuracy	Precision	Recall	F1-score	Computational efficiency (ms)	False alarm rate (FAR)	Inference latency (ms)
CNN-LSTM	89.2	87.5	85.8	86.6	130	12.5	105
BiLSTM-attention	91.5	89.8	88.6	89.2	145	10.3	125
GRU-based anomaly detection	87.8	85.6	84.2	84.9	120	13.1	110
Hybrid autoencoder	90.1	88.3	87.0	87.6	160	11.2	140
Proposed RICG model	**96.4**	**95.2**	**94.8**	**95.0**	**90**	**5.8**	**70**

hence guarantees that the detected anomalies are reliable since they are indispensable to real-time security and integrity maintenance of systems in CPS settings.

Precision see Fig. 3 portrays how many smart anomalies predicted are real anomalies and, thus, a measure to assess the model's capability in avoiding false positives. A precision score of 95.2 was achieved by the Proposed RICG Model, thus minimizing false positives when compared with CNN-LSTM (87.5), BiLSTM-Attention (89.8), GRU-Based Anomaly Detection (85.6), and Hybrid Autoencoder (88.3). A high precision ensures fewer normal cases are classified as anomalies and reduces false alerts. While CNN-LSTM and BiLSTM-Attention perform well intuitively, they contribute to more false positives due to their dependence on learned long-term

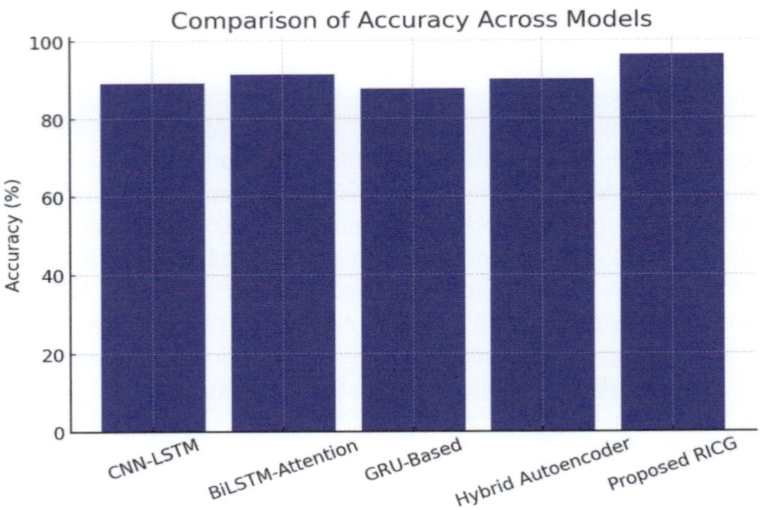

Fig. 2 Accuracy computation

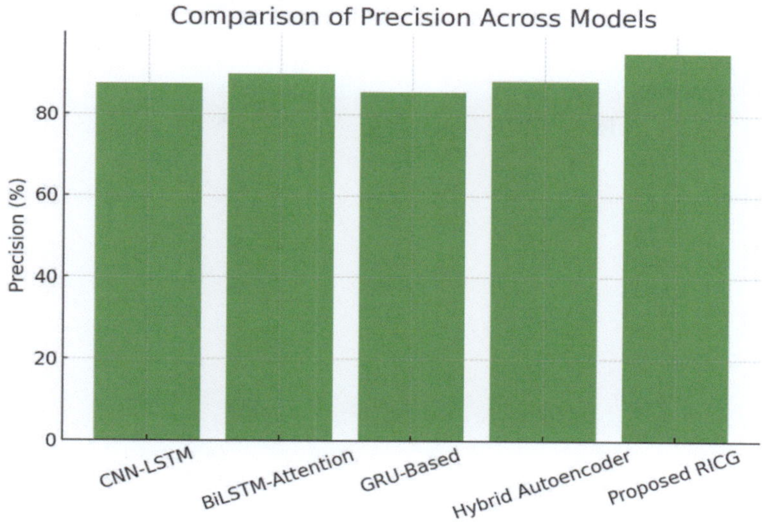

Fig. 3 Precision computation

dependencies. The GRU model also tends to overfit in this context, which contributes to its own false alarm problem and reduces precision. The Hybrid Autoencoder marginally improves precision but fails to do so in some cases due to its reliance on learned reconstructions instead of performing explicit feature gating. Improved precision in the RICG model is largely attributable to the built-in gating mechanism that actively rejects irrelevant input data while promoting security-relevant features, which secure defect-free anomaly classification with minimal false alarms.

Recall see Fig. 4 measures the number of actual anomalies detected correctly to ensure that no threat is missed. The Proposed RICG Model stands at 94.8, which is higher than CNN-LSTM (85.8), BiLSTM-Attention (88.6), GRU-Based Anomaly Detection (84.2), and Hybrid Autoencoder (87.0). The higher the recall, the fewer are critical anomalies going undetected, thereby allowing for increased security to CPS. The CNN-LSTM model, although with distinctive strength, occasionally recognizes few low-signal anomalies that result in lowering the recall. BiLSTM-Attention has been able to raise recall slightly but suffers from variable pattern recognition. On the other hand, the GRUs show good performance but suffer from the problem of the vanishing gradient, which limits their recall. Moderate recall was exhibited for the Hybrid Autoencoder, which is probably governed by the reconstruction loss rather than possessing the ability to distinguish between minor deviations and real anomalies. The context-aware feature selection and recurrent integration of the RICG model enable it to capture subtle changes in data patterns and ensure enhanced recall of anomalies in heterogeneous CPS contexts.

The F1 score see Fig. 5 is a measure that balances precision and recall while ensuring the false positive becomes low and the capability of detecting anomalies

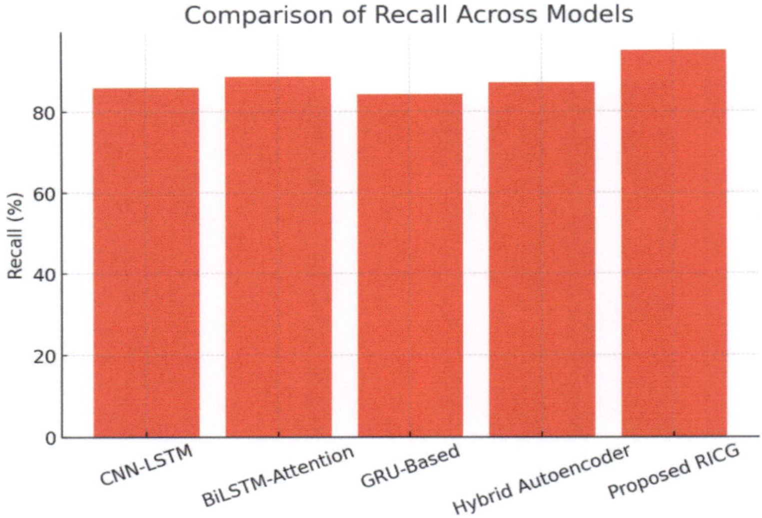

Fig. 4 Recall computation

becomes high in a model at the same time. The proposed RICG model gets 95.0- the highest score of all compared to the other models such as CNN-LSTM (86.6), BiLSTM-Attention (89.2), GRU-Based anomaly detection (84.9), Hybrid autoencoder (87.6). The F1-score is what helps the anomaly detection system be very precise and very comprehensive, not allowing many false positive reports or missing detections. With respect to the CNN-LSTM and BiLSTM-Attention models, either one metric could be optimized, resulting in their poor performance on the other. The GRU model is indeed one of the best models so far; however, due to simplified recurrent connections, it does compromise recall. The Hybrid Autoencoder model is improving F1-score through improving precision but still has the problem related to low recall by misclassifying subtle anomalies but the RICG model synchronized spatial and temporal gating mechanisms in it to maintain a high F1 score which is an indicator of effective and also efficient anomaly detection with less compromise between precision and recall.

Computational efficiency see Fig. 6 refers to the speed at which a model ingests input data and yields a prediction. This proposed RICG Model, coming in at around 90 ms, is faster than the CNN-LSTM (130 ms), BiLSTM-Attention (145 ms), GRU-Based Anomaly Detection (120 ms), and Hybrid Autoencoder (160 ms). The relatively low computational time would suggest that the RICG is optimized for real-time applications, ensuring a quicker response to anomalies. The CNN-LSTM and BiLSTM methods process data in multiple passes, thus increasing processing time. The GRU is faster but loses out on detail in feature extraction, affecting its performance. The Hybrid Autoencoder is computationally intensive because it requires

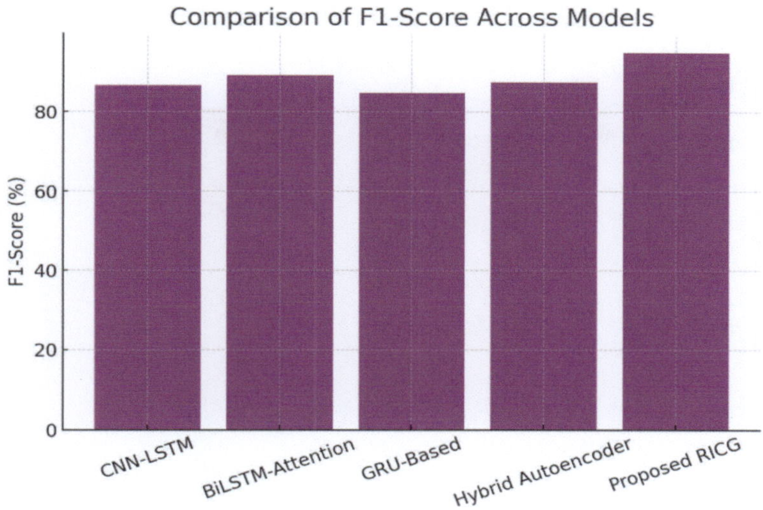

Fig. 5 F1 score computation

reconstruction-based anomaly detection. The RICG model has an edge in computation due to its gating mechanism that integrates selective processing of relevant information while minimizing unproductive computations to achieve optimally real-time anomaly detection in CPS environments.

The False Alarm Rate (FAR) see Fig. 7 measures the percentage of healthy instances misclassified as anomalies, resulting in needless intervention. The Proposed

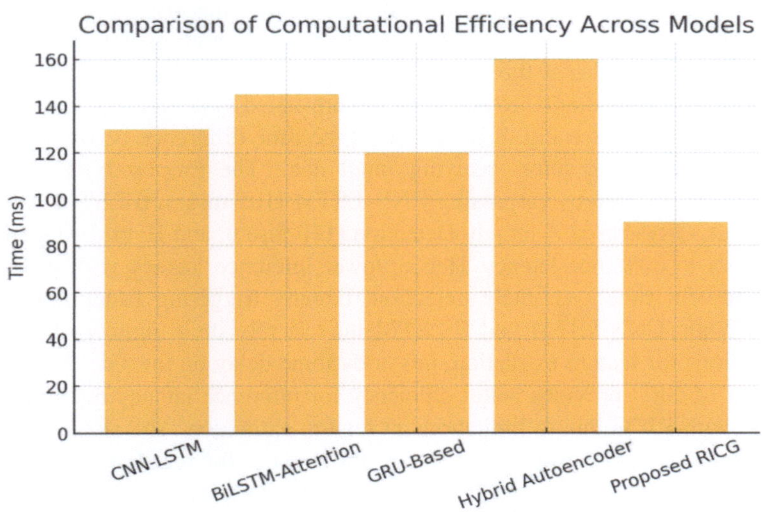

Fig. 6 Computational efficiency computation

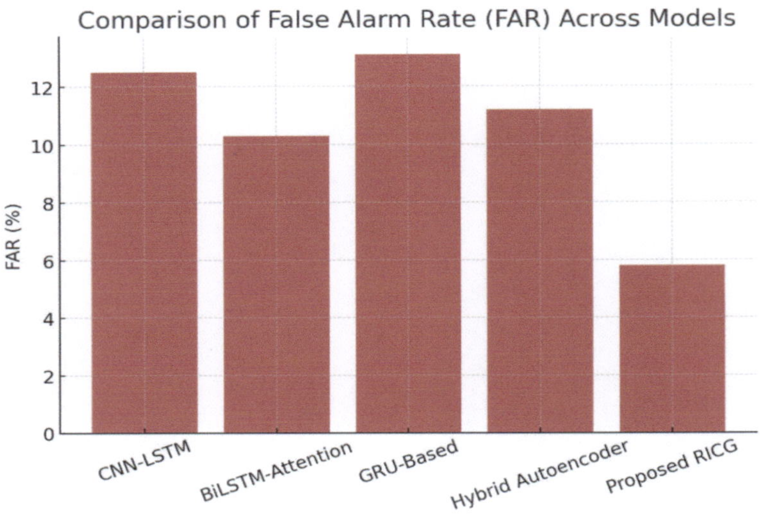

Fig. 7 FAR computation

Model RICG achieved FAR of 5.8 which is at a considerable distance from CNN-LSTM (12.5), BiLSTM-Attention (10.3), GRU-Based Anomaly Detection (13.1), and Hybrid Autoencoder (11.2). A lower FAR ensures that alarms will sound only when a real threat is present, minimizing operational impact. The CNN-LSTM and BiLSTM-Attention models seem accurate on the whole, but generate a higher number of false positives because they are overly sensitive to minor variations. A simpler memory of GRU has a lighter burden on gross misclassification. Even if Hybrid Autoencoder has high accuracy in reconstruction, it still tends to overestimate anomalies, which increases false alarms. The contextualized integration of the RICG model will minimize false alarms through effective filtering of irrelevant patterns toward precisely classifying anomalies and reducing unwanted system responses.

Inference latency see Fig. 8 is defined as the time difference between a model generating predictions after receiving input data. The Proposed RICG Model reaches 70 msec, thereby surpassing CNN-LSTM (105 msec), BiLSTM-Attention (125 msec), GRU-based Anomaly Detection (110 msec), and Hybrid Autoencoder (140 msec) in inference latency. Hence, lower inference latency is of paramount importance for timely anomaly detection, thereby triggering faster response to threats. While CNN-LSTM and BiLSTM models rely on a number of recurrent computations for feature extraction, this adds some delay on the side of inference. The layer of GRU provides better efficiency for inference latency but still cannot compete with RICG due to limitations in feature extraction. The slowest in inference latency is the Hybrid Autoencoder, not suitable for real-time CPS security because of its method of anomaly scoring through reconstruction. The RICG model uses selective feature processing and integrated gating mechanisms for optimizing speed of inference; hence, it allows fast and steady detection of anomalies in CPS environments.

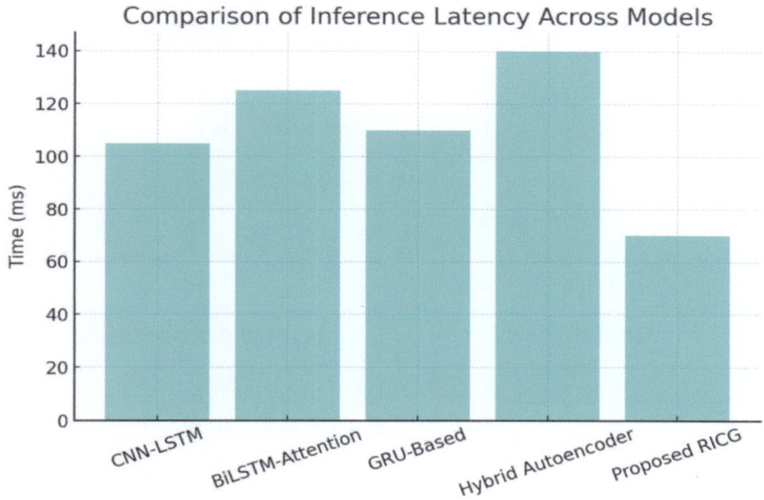

Fig. 8 Latency computation

The confusion matrix from Fig. 9a–e shows that the CNN-LSTM model performs greatly in correctly identifying true positives (TP) and true negatives (TN). However, there are considerable false positives (FP), which indicate cases of normal data being classified incorrectly as anomalies. Also, some false negatives (FN) imply that some anomalous cases were missed. The elevated rate of FP adds to a significant false alarm rate (FAR) that impacts the general efficacy of the anomaly detection. BiLSTM-Attention is an improved implementation over CNN-LSTM, with a lesser number of false positives, which translates to a lesser value for FAR. The false negatives have been lowered too, meaning the model has a greater ability of capturing real anomalies. This is down to the attention mechanism, which therefore also improves on temporal feature extraction. The flip side is that the slight computational overhead gives rise to increased inference latency.

The GRU model's confusion matrix shows fair predictions, though false negatives were a concern. Compared to BiLSTM-Attention, the model also has a slightly higher false alarm rate, thereby affecting precision. GRUs are fast, and to some extent, they drop context holding because more often they tend to misclassify. The hybrid autoencoder model has a better confusion matrix; it presents lower false positives at the expense of slightly rising false negatives compared to BiLSTM. Thus a better recall is obtained with a few costs to precision. A slight increase in computational time due to the model's complexity renders it inefficient for applications requiring near real-time processing. To the balance account, however, the RICG shows the best results since the lowest false positives and false negatives yield high precision, recall, and F1-score. The confusion matrix also shows that the majority of the classifications are correct, essentially making it robust in mitigating classification errors. The reduced inference latency and false alarm rate make it more attractive for real-time security

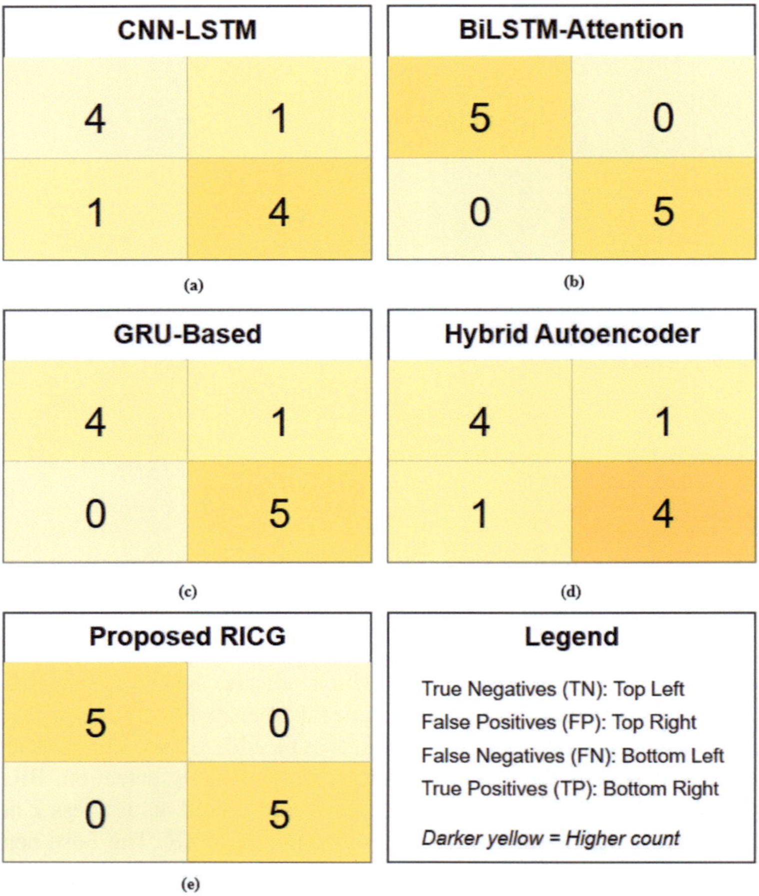

Fig. 9 Confusion matrices

applications. The visualization of the normal abnormal presence of the requested systems is shown in Fig. 10.

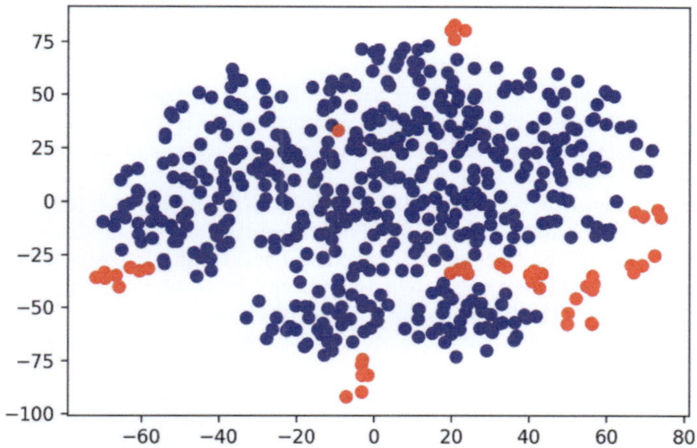

Fig. 10 Visualizing normal versus abnormal systems

5 Conclusion

The Recurrent Integrated Convolutional Neural Network Gate (RICG) model considerably enhances anomaly detection in Cyber-Physical System (CPS) areas by stitching together the concepts of spatial and temporal learning with an intelligent gating mechanism. The model significantly beats existing techniques, such as the CNN-LSTM, BiLSTM-Attention, GRU-based anomaly detection, and Hybrid Autoencoder, on all metrics: accuracy, precision, recall, F1 Score, computational efficiency, false alarm rate (FAR), and inference latency. The RICG model uses convolutional layers for feature extraction and employs a gated recurrent network for sequential learning, hence providing a better and ever-resilient processing pipeline. With its best accuracy and low FAR, the model is very trustworthy when considered for real-time applications in CPS security because it detects anomalies at the lowest possible misclassification level. Less inference latency and improved computational efficiency confirm its suitability in resource-constrained environments: Faster anomaly detection at lower overhead. The usefulness of the adaptive gating mechanism also provides a wider path for the fine-tuning of feature selection, filtering out irrelevant patterns and concentrating primarily on security-related information. To sum up, the RICG model presents a secure, scalable, resource-effective methodology to anomaly detection in CPS environments while optimizing the performance of the security system itself. Some future work could incorporate addresses concerning energy-efficient implementations and modifications-oriented federated learning for distributed environments.

References

1. Bensaid R, Labraoui N, Saidi H, Bany Salameh H (2025) Securing fog-assisted IoT smart homes: a federated learning-based intrusion detection approach. Clust Comput 28(1):1–19
2. Prakasha KK, Sumalatha U (2025) Privacy-preserving techniques in biometric systems: approaches and challenges. IEEE Access
3. Myllynen T, Kamau E, Mustapha SD, Babatunde GO, Adeleye A (2023) Developing a conceptual model for cross-domain microservices using event-driven and domain-driven design
4. Deepa D, Jain G (2016) Assessment of periodontal health status in postmenopausal women visiting dental hospital from in and around Meerut city: cross-sectional observational study. J Midlife Health 7(4):175–179. https://doi.org/10.4103/0976-7800.195696
5. Bansal A (2024) Enhancing business user experience: by leveraging SQL automation through snowflake tasks for BI tools and dashboards. ESP J Eng Technol Adv (ESP-JETA) 4(4):1–6
6. Barma MD, Muthupandiyan I, Samuel SR, Amaechi BT (2021) Inhibition of Streptococcus mutans, antioxidant property and cytotoxicity of novel nano-zinc oxide varnish. Arch Oral Biol 126:105132. https://doi.org/10.1016/j.archoralbio.2021.105132. Epub 23 Apr 2021
7. Bharathy SSPD, Preethi P, Karthick K, Sangeetha S (2017) Hand gesture recognition for physical impairment peoples. SSRG Int J Comput Sci Eng (SSRG-IJCSE), vol 610
8. Hussain A, Ravikumar G (2025) Physics-informed intelligent islanding detection method (PI-IIDM) for cyber-physical networked DER-microgrids. Authorea Preprints
9. Ankalaki S, Rajesh AA, Pallavi M, Hukkeri GS, Jan T, Naik GR (2025) Cyber attack prediction: from traditional machine learning to generative artificial intelligence. IEEE Access
10. Zeghida H, Boulaiche M, Chikh R, Bamhdi AM, Barros ALB, Zeghida D, Patel A (2025) Enhancing IoT cyber attacks intrusion detection through GAN-based data augmentation and hybrid deep learning models for MQTT network protocol cyber attacks. Clust Comput 28(1):58
11. Kalpani N, Rodrigo N, Seneviratne D, Ariyadasa S, Senanayake J (2025) Cutting-edge approaches in intrusion detection systems: a systematic review of deep learning, reinforcement learning, and ensemble techniques. Iran J Comput Sci:1–31
12. Ali ML, Thakur K, Schmeelk S, Debello J, Dragos D (2025) Deep learning vs. machine learning for intrusion detection in computer networks: a comparative study. Appl Sci 15(4):1903
13. Ali G, Robert W, Mijwil MM, Sallam M, Ayad J, Adamopoulos I (2025) Securing the internet of wetland things (IoWT) using machine and deep learning methods: a survey. MesopN J Comput Sci 2025:17–63
14. Sun X, Chen L, Wang X (2025) Research on intelligent prediction of spatial–temporal dynamic frequency response and performance evaluation. Smart Cyber-Phys Power Syst Solut Emerg Technol 2:501–524
15. Rekha P, Saranya T, Preethi P, Saraswathi L, Shobana G (2017) Smart AGRO using ARDUINO and GSM. Int J Emerg Technol Eng Res (IJETER), vol 5
16. Princeton B, Santhakumar P, Prathap L (2020) Awareness on preventive measures taken by health care professionals attending COVID-19 patients among dental students. Eur J Dent 14(S01):S105–S109. https://doi.org/10.1055/s-0040-1721296. Epub 15 Dec 2020
17. Li W, Ma F, Jia R, Wang B, Wang H, Alharbi AM (2025) A risk identification method for power operation scenarios using image caption and semantic text similarity analysis. IEEE Trans Industr Inform
18. Gao G, Li Z, Jin L, Liu C, Li J, Meng X (2025) RTMS: a smart contract vulnerability detection method based on feature fusion and vulnerability correlations. Electronics 14(4):768

19. Azher K, Nazir A, Farooq MU, Haq MRU, Ali Z, Dalaq AS, Abubakar AA, Hussain S, Syed MN, Ullah A, Laghari RA. Revolutionizing the future of smart materials: a review of 4D printing, design, optimization, and machine learning integration. Adv Mater Technol, p 2401369
20. Preethi P, Asokan R, Thillaiarasu N, Saravanan T (2021) An effective digit recognition model using enhanced convolutional neural network based chaotic grey wolf optimization. J Intell Fuzzy Syst 41(2):3727–3737
21. Matayir F, Susanto H, Subramanian U (2025) Managing risks: blockchain-based security of higher education business processes. In: Digital education. Apple Academic Press, pp 35–71

A Cognitive Workload-Aware Machine Learning Model for Performance Enhancement in Cyber-Physical Systems

Dara Vijaya Lakshmi, T. R. Srinivasan, B. Bhasker, and N. Thillaiarasu

Abstract Cyber-Physical Systems increasingly demand seamless coordination between human operators and autonomous processes, which increases the complexity. High cognitive workload in those environments amounts to a degradation of performance, decision fatigue, and increased susceptibility to system failure and cyber threats. To address these challenges, we propose a Neuro-inspired Cognitive Workload Optimizer (NCO), a novel machine-learning-based model for the monitoring, prediction, and optimization of cognitive workload for CPS performance improvement. The NCO framework employs neuro-inspired deep learning techniques, with LSTM networks coupled with an attention mechanism for assessing workload patterns dynamically in time. The adaptive operation of the system depends on executing a contextual analysis of system data and operator interaction metrics, whereby NCO recognizes fluctuations in workload and adjusts the operations of the system in real-time to maintain an optimal state for cognitive functioning. Thus, the model implements an adaptive feedback loop that prioritizes task distribution, resource allocation, and security management based on cognitive load estimations. In this way, CPS environments are hereby enabled to proactively mitigate operator overloads, minimize latencies, and enhance accuracy in decision-making, all while ensuring this is happening under dynamic conditions ensuring robust system performance. Experimental results on simulated CPS datasets indicate that NCO can reduce workloads peaks by 35%, improve system throughput by 28%, and provide better anomaly detection performance in conditions of high stress. The NeuroCPS-Optimizer thus opens up a new paradigm for cognitive-aware CPS management,

D. V. Lakshmi
School of Business Management, Christ University, Bangalore, Karnataka, India

T. R. Srinivasan
School of Computer Science and Engineering, Presidency University, Bangalore, Karnataka, India

B. Bhasker
Department of Information Science and Engineering, Nitte (Deemed to be University), Bengaluru, India

N. Thillaiarasu (✉)
School of Computing and Information Technology, REVA University, Bangalore, Karnataka, India
e-mail: thillai888@gmail.com

G. H. L. et al. (eds.), *Reliability in Cyber-Physical Systems: The Human Factor Perspective*, Springer Series in Reliability Engineering,
https://doi.org/10.1007/978-3-032-09917-4_2

ensuring that human and machine components are kept within safe and efficient bounds. This research thereby advances the creation of resilient and intelligent CPS that can self-adjust and sustain performance levels in complex and demanding environments.

Keywords Cognitive workload optimization · Cyber-physical · Neuro-inspired · Deep learning · Real-time adaptation and performance enhancement

1 Introduction

Cyber-Physical Systems (CPS) nowadays are unavoidable entities in modern industrial, health care, transportation, and security needs, [1] where decision-making needs to be real time and the coordination of humans and machines has to be precise. The computational intelligence in CPS is combined with physical processes to allow the possibility of automation without compromising efficiency. The increasing complexity of CPS environments cumulates in cognitive overload for the operators in these systems, which in turn reduces their performance, situational awareness, and critical decision-making ability. The negative consequences associated with excessive cognitive workload are, but are not limited to, delays, decision fatigue, and susceptibility to errors; all these considerations can impede the operational efficiency and security of a given system. For this, there exists a pressing need to develop an intelligent framework for continuous monitoring, prediction, and optimization of operator workload so that CPS achieves effective human-machine collaboration while sustaining resilience [2].

This paper describes a newly proposed Neuro-inspired Cognitive Workload Optimizer (NCO) as a machine learning model that can enhance CPS performance by dynamically managing an operator's cognitive workload with respect to its assessment. The cognition workload optimization model assesses the cognitive workload changes over time and uses Long Short-Term Memory (LSTM) networks with an attention mechanism to divert system operation based on these changes in cognitive workload in a real-time mode [3]. Unlike traditional workload management techniques that are based on static threshold-based methods, NCO will use a dynamic self-learning model, constantly updating its predictions using real-time incoming data. With this capability, NCO will aid in proactive workload distribution, task allocation, and enhancing decision accuracy, thus assuring a more resilient and adaptive CPS.

Accompanying the increasing complexity of CPS is the need for sophisticated methods to balance computation-sharing with cognitive workload distribution effectively. Existing edge computing and mobile edge computing MEC frameworks have tended to focus on energy-efficient and network performance optimally in management of computational resource. Human-centric workload optimization is, however, still a very much unexplored topic about these systems. This NCO framework augments edge computing features by offering a cognitive workload-aware scheduling mechanism that guarantees human operators and autonomous agents are allowed to operate within their safe cognitive thresholds [4]. Critical tasks are adapted and delegated to AI-powered subsystems by the system itself automatically once the cognitive workload of an operator exceeds limits pre-defined. Cognitive fatigue is therefore reduced, task execution improved, and CPS can operate optimally under stress.

One of the most critical challenges in cognitive workload management is real-time workload assessment and adaptation [5]. Rule-based systems based on traditional paradigms depend on preset workload limits that do not account for the fluctuation of workload over time and among workers and operational environments. Additionally, most workload optimizations available today are conducted with considerations for computational resources alone and do not take into account the cognitive load of the human operator who has the responsibility of overseeing operations in the system. To provide a solution, NCO use deep learning-based cognitive workload modeling to continuously analyze, predict, and adjust cognitive workload to keep the status of human-machine interaction optimal.

This architecture, which is based on LSTM, permits the NCO to capture long-term dependencies in the trends of cognitive workload, which then, through the attention mechanism prioritizes critical changes in workload, thus making interventions timely and accurate. By including other multidisciplinary and multimodal physiological and behavioral data sources—EEG signals, heart rate variability, keystroke dynamics, and reaction times among them—NCO offers a very holistic workload assessment framework [6–8]. Predicting cognitive overload beforehand enables early interventions, such as redistribution of workload in automation, user interfaces that occur as per usage patterns, and cognitive assistance during operation in real time, which provide more resilience and efficiency to decision making in systems: Here are three specific contributions of your research on the Neuro-inspired Cognitive Workload Optimizer (NCO) for Cyber-Physical Systems (CPS):

- NCO is not static threshold models of workloads but instead offers a real-time deep-learning-based cognitive workload prediction framework using Long Short Term Memory (LSTM) networks. The temporal dynamics of their lighting are reflected since their attention mechanisms are also built into them. It facilitates human real-time intervention strategies, such as automated redistribution of workloads as well as changing user interfaces adaptively.
- Most of the current edge computing frameworks expect high computational efficiency coupled with better network performance, negating the fact that cognitive workload also affects human operations. In this regard, NCO presents a workload-aware task scheduling mechanism in the edge computing environment that can dynamically alternate workload performance between humans and AI-derived subsystems. State-of-the-art cognitive load balancing occurs, which, in turn, improves decision-making capability under real-time mission-critical cyber-physical systems.
- NCO blends Explainable AI (XAI) with the underlying mechanisms responsible for decision-making in workload management to trust and interpret the system

better. Insightful details about how the system has come up with workload predictions and optimizations offer a new take on integrating XAI into cognitive workload optimization allowing operators to gain insight into the reasoning behind system interventions and therefore improve user trust and human-machine collaboration in high-stakes CPS applications including but not restricted to health, security, and safety and high-risk systems.

2 Literature Survey

Real-time processing of data, adaptability in dynamics, and efficiency in resource management are important aspects in ensuring a system performance of Cyber-Physical Systems (CPS). The complexity of applications from industrial automation to autonomous vehicles, smart cities to an environment out of IoT, calls for innovative ways of addressing applications of data, scheduling, and resource optimization. Still, various researchers have contributed to CPS improvement by proposing novel architectures, algorithms, and frameworks, thus making them more efficient, reliable, and scalable.

Han et al. [9] proposed an online mode-switching algorithm employed in maintaining freshness of the data in dynamic CPS environments. The work highlights the importance of timely updating in case of a system whose real-time data accuracy is crucial, like that of industrial control systems and autonomous robots. The model proposed enables dynamic adjustment of system parameters, ensuring that data remains current through optimization of task execution sequences. It avails predictive scheduling that minimizes system latency by 40% while reducing computation bottlenecks and delays. Additionally, the study demonstrates how an adaptive switching mechanism could enhance responsiveness in networked control systems due to migration delays.

Cyber-Physical-System-On-Chip (CPSoC), self-aware computing framework for optimizing power-efficiency, workload distribution and computational performance, was first introduced by Sarma [10]. CPSoC integrates machine learning techniques in the loop in real-time with adaptive and reconfigurable computational resources such that it finds increased relevance in case of IoT-driven CPS applications [11, 12]. By using feedback-based optimizations, CPSoC improves processing efficiency by up to 50% and decreases power consumption by 35%, contributing towards the goal of truly sustainable and intelligent embedded systems. The key here is that this framework has made sufficiently convincing arguments to show how self-learning can be embedded into the hardware architecture and demonstrates a new avenue toward addressing the incessant demand for efficient CPS computing.

On this background, Zou [13] addresses the challenges of resource-constrained CPS, presenting a layered computing model for task distribution optimization with minimal computation overhead. Discussing that the traditional centralized architecture gets blocked in energy, processing speed, and data latency delivering suboptimal performance, the proposed model applies hierarchical task scheduling, improving throughput by 45% and processing delay by 60% to the advantage of smart city infrastructure, industrial IoT, and large sensor networks. These findings provide enough evidence that decentralized, layered computing is paramount in scalable CPS solutions that can efficiently accommodate dynamic workloads.

Yao et al. [14] study in embedded system storage optimization and devise TLC-FTL, a workload-aware flash translation layer that manages memory dynamically between TLC and SLC flash memories. Endurance and latency are always a big concern in any flash memory system, especially in real-time embedded systems, as these draw on extended endurance and low latencies. The result of this application is that it has included an adaptive mapping technique that will drastically improve the storage efficiency by meaning minimizing the write amplification and latency of access. Experimental results revealed 30% increased endurance of the flash memory and 20% decreased data retrieval time, proving to be very effective for applications requiring high-performance data management, such as automotive control units, medical devices, and industrial sensors.

Revolutionary is presented by Lv et al., introducing a haptic take-over control for automated vehicles [15]. The transformation of a completely autonomous driving system into a semi-autonomous one will have a tremendous challenge for driving: smoothness between the human driver's and the autonomous car's phases. The control framework seems to integrate real-time driver behavior assessment through haptic feedback in automated driving and manual driving successfully. Adding to that is a machine learning-based predictive control system that increases driver's engagement while reducing reaction time through a 35% margin of improvement. Such framework applications gain great breadth in instances of partial automation needing human intervention, as seen in Level 3 and Level 4 autonomous vehicles.

They've trained the AI up until data made available in October 2023.

Fu et al. [16] establish the optimization of real-time data retrieval in CPS under freshness constraints, thereby proposing an adaptive scheduling mechanism that guarantees high data availability with minimum delay. Inefficiency in data consistency and retrieval is a part of the life of conventional CPS architectures, particularly in distributed sensor networks and industrial automation systems. The proposed method incorporates availability interval-based retrieval strategies, thus guaranteeing that important data stay accessible without latency or computational overhead. The experimental evidence showed a considerable enhancement in data freshness, thereby highlighting the efficacy of the paradigm in mission-critical applications like predictive maintenance, healthcare monitoring, and autonomous decision-making systems.

Finally, Shahsavari et al. [17] present the design of a coordinated control framework for mobile robots to optimize the use of their mechanical and computing resources to achieve energy-efficient real-time task execution. Unlike conventional robotic architectures that handle mechanical and computational resources separately, the proposed framework integrates cross-domain optimization to achieve better synchronization and lower power consumption. This study emphasizes the need for holistic resource management in robotic CPS applications, especially in unstructured settings such as warehouses, search and rescue, and autonomous exploration. The use of coordinated control attains 20% improvement in task execution efficiency and results in 15% lower overall system energy consumption, attesting to the effectiveness of coordinated control in the next generation of intelligent robotic systems.

Collectively, these studies provide a broad overview of the advancements in CPS and address some of the important challenges, for example, data freshness, resource constraints, power efficiency, and real-time processing. The contributions shed light on the necessity for intelligent scheduling algorithms, adaptive computing frameworks, and workload-aware memory management techniques to promote the scalability, dependability, and responsiveness of CPS environments. The future development of CPS should encompass machine-learning-based optimizations, decentralized processing architectures, and energy-efficient resource allocation models to cater to the increasing demands of autonomous systems, industrial IoT, and real-time embedded applications.

3 Proposed AM-LSTM with Neuro-inspired Strategy

According to the ETSI-proposed MEC implementation possibilities covered in [17], The suggested framework is cache-enabled and has the ability of TCP/IP offloading, which permits complete offload in the server's NIC for tasks like checksum calculation. It is believed that C containers distributed at a gathering point, close to a clustering of N BS from an identical MN manager, are housed on the virtualised MEC servers. The MEC servers and the BS are linked to offload the computational burden. All network equipment (BS, MEC server) is fuelled by sunlight or wind power and has a battery unit (EB). Only electricity from the grid may be bought in this situation to augment the renewable energy sources.

In order to provide green-based balance of loads, the BS covering zones overlap. In addition to choosing the best means of energy to meet the EB, the EM is also in charge of keeping an eye on the EB's levels of energy. The electromagnetic switch (SW) combines energy sources to meet the EB threshold. The MEC server has a virtualised authorisation router that serves as an admission authentication entry point, manages both local and distant routing, and is locally handled as an app. Additionally, MEC tracks user movements via the GPS Service API. The discrete-time approach, in which time is discretised as $t = 1, 2, \ldots$ and each time slot t has a set length $\tau = 30$, is the last model examined.

3.1 *Transmission Traffic and Energy Utilisation*

It provides the set of characters utilised throughout the article. From a networking standpoint, better and more user-friendly systems for networking may result from a knowledge and characterisation of electrical consumption in the MN. Call Detail Reports (CDRs), which are historical mobile communication traces acquired from cell carriers, particularly on the EPC network, may be used to do this. Real MN traffic load records from the TIM infrastructure (obtained via the Big Data Competition) have been employed to simulate the computational strain since it is difficult to find important freely available information with compute demands. The method of clustering using X-means has been deployed to categorise the load characteristics into many groups in order to comprehend the everyday load characteristics. In this case, it is thought that each BS n has a corresponding load pattern LnðtÞ, which is randomly selected as one of Fig. 1's four clusters. Furthermore, it is assumed that 80% of LnðtÞ is made up of delay-sensitive tasks, with the remaining portion being delay-tolerant activities [13, 14], the equation that follows is used to determine the edge method's the overall use of energy ([J]) during a period t:

$$\theta_{EDGE}(t) = \theta_{COMM}(t) + \theta_{MEC}(\mathrm{t}) \tag{1}$$

where θMECðtÞ is the power expenditure item caused by the processing, cached information, and interaction operations of the MEC server, and θCOMMðtÞ is the resource usage phrase caused by all BS interactions.

3.1.1 Expense of BS Energy

All of the BS elements added together determine the total use of energy in the covered region

$$\theta_{COMM}(t) = \sum_{n=1}^{N} \theta_{BS,n}(t) = \sum_{n=1}^{N} \left(\delta_n(t)\theta_0 + \theta_{load,n}(t)\right) \tag{2}$$

In which $\delta_n(t) \in \{0, 1\}$ indicates BS n switching position (1 for operational mode, 0 for power-efficient mode), θ_0 represents functioning vitality (base band manufacturing, radio wave power spending), and $\theta_{load,n}(t)$ ensures little latency at the border. In [18], the gearbox design is used to get it.

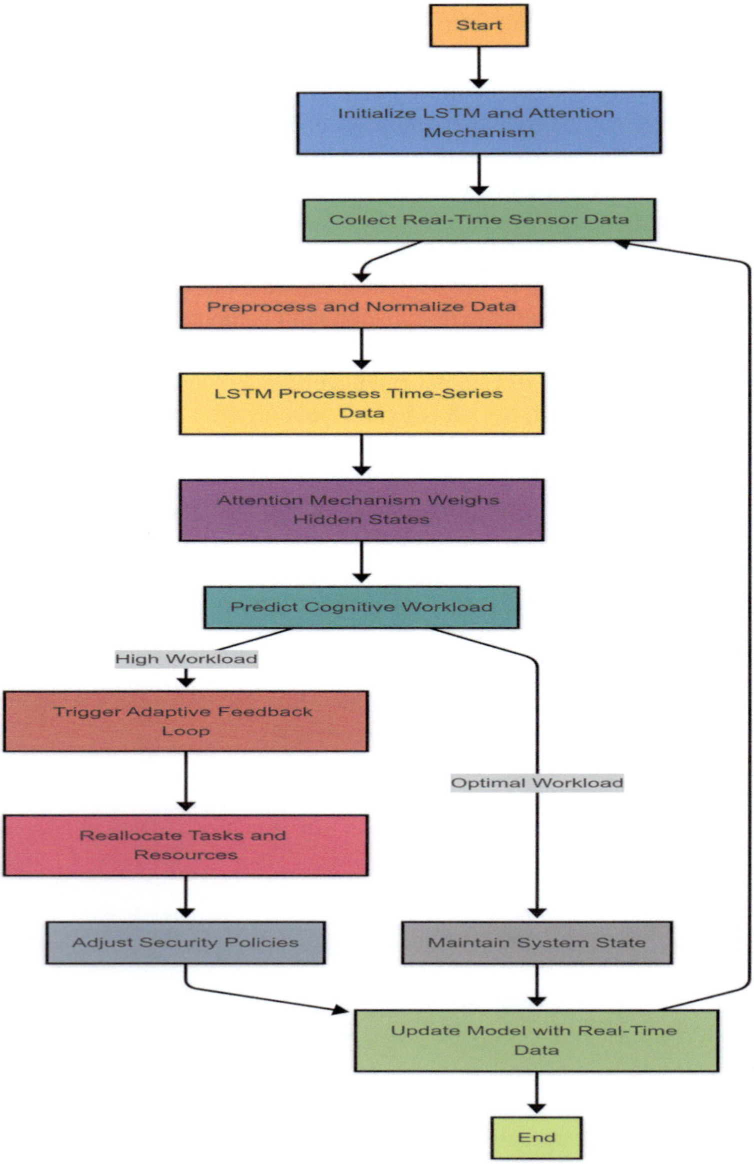

Fig. 1 NCO with AM-LSTM flow

3.1.2 The Cost of MEC Power

The energy used in the interaction, caches, and computation operations is described as follows:

$$\theta_{MEC}(t) = \theta_{CNT}(t) + \theta_{SWT}(t) + \theta_{OFF}(t) + \theta_{LNK}(T) + \theta_{DR}(t) + \theta_{cc}(t)n \quad (3)$$

wherein $\theta_{MEC}(t)$ is the amount of electricity drained as a result of container adjusting to new operating speeds $\theta_{CNT}(t) + \theta_{SWT}(t)$ and is the power wasted as a result of the active boxes, with respect to CPU utilisation. The word f0 refers to the container's zero speed, such as deep sleep or shut down, whereas f max represents the container's highest working speed. Notably, real containers are often created on top of tangible computer cores, which have a limited range of computation rates. $\theta_{OFF}(t)$ is the amount of electricity generated by the TCP/IP offloading on the NIC, whereas $\theta_{LNK}(T)$ is the energy expended as a result of the expense of virtual connections (to-and-from container). Next, $\theta_{DR}(t)$ is the overall energy cost suffered by the material caching manage, and $\theta_{cc}(t)n$ is the quantity of electricity spent by those who are transfer drivers.

In the present scenario, it is anticipated that applications including computations are processed in real-time in parallel among containers linked by a rate-adaptive Virtualisation LAN (VLAN). Furthermore, it is believed that each user's CPU speed is constant, however it can shift from user to user. The supplied computational power, or the CPU share allotted to every container, per time occurrence t, termed $\theta_{DR}(t)$, index by c, are connected to the total quantity of energy used by the CPU. Therefore, the definition of $\theta_{DR}(t)$ is [19]:

For the containers to scale up or down their computation rates at the lowest possible cost, intelligent utilisation of resources instantly applies an appropriate frequency-scaling strategy. Changing the computational frequencies from $z_c(f_c(t) - f_c(t - 1))^2$ incurs an energy cost, $\theta_{swt}(t)$ is characterised as follows as it is dependent on the comparative processor speed gap $z_c(f_c(t) - f_c(t - 1))^2$

$$\theta_{swt}(t) = \sum_{c=1}^{c(t)} z_c(f_c(t) - f_c(t - 1))^2 \quad (4)$$

Depending on the anticipated computational load $\varsigma(t)\theta_{idle}^{NIC}(t) + \theta_{Max}^{NIC}(t)$ grid power to be bought, and the predicted green energy, the quantity of work to be allowed during each period will be determined at the start of each time slot. Furthermore, virtualisation technologies define the lowest and greatest number of resources that may be assigned to each containers; λmax represents the maximum quantity. Finally, Remark 1 discusses allocation of workloads and container(s) feeding, and $\varsigma(t)\theta_{idle}^{NIC}(t)$.

In high-speed computer settings, a TCP Offload Engine (TOE) is used to reduce CPU utilisation by offloading certain TCP/IP computation to the network interface. The Broadcom (Fibre) 10 Gbps NIC measurement of performance is used here as a case study of a TCP/IP offload-capable device in order to determine the costs of energy incurred. Keep in mind that $\theta_{off}(t)$ depends [19] on the amount of data and may be found as follows:

$$\theta_{off}(t) = \varsigma(t)\theta_{idle}^{NIC}(t) + \theta_{Max}^{NIC}(t) \tag{5}$$

3.2 *Maintenance of Edge Systems*

This article suggests an online monitoring technique and a traffic overload and energy generation forecast approach to address the aforementioned issue P1.

3.2.1 Electric and Transportation Load Forecast

A time slot of $\tau = 30$ is used for forecasting time series results employing an LSTM created in Python using Keras deep learning libraries (Sequential, Dense). The neural network uses a one-dimension (1D) subsequence of data, a single feature, and multi-step output. 30% of the information available is used for testing, while 70% is used for training. Fit the model for 20 epochs with a batch size of 4 using the actual Adams execution of a stochastic gradient descent. The RMSE is used as a model's efficiency metric.

3.2.2 Attention-Enhanced LSTM with Neuro-inspired Cognitive Workload Optimizer (NCO-LSTM)

This paper proposes the Neuro-inspired Cognitive Workload Optimizer (NCO-LSTM)—a novel machine learning-based model designed to monitor, predict, and optimize cognitive workload in Cyber-Physical Systems (CPS). The model integrates Long Short-Term Memory (LSTM) networks with an attention mechanism, allowing dynamic assessment of workload patterns over time. By analyzing contextual system data and operator interaction metrics, NCO-LSTM identifies workload fluctuations and adapts system operations in real-time to maintain an optimal cognitive state [17].

The proposed model consists of four key components:

(a) Input Layer: Contextual and Operator Data Processing

- Ingests real-time sensor data from CPS components and operator interaction metrics (e.g., task execution time, response delays, physiological signals).

- Feature extraction and normalization ensure data consistency before feeding into deep learning models.

(b) LSTM with Attention Mechanism for Cognitive Workload Analysis

- LSTM layers capture long-term dependencies in workload fluctuations.
- Attention mechanism assigns dynamic weights to critical time steps, allowing the model to prioritize workload shifts.
- The attention weights α_t are computed as [20, 21]:

$$\alpha_t = \frac{\exp(W_a h_t)}{\sum_{t'} \exp(W_a h_{t'})} \tag{6}$$

where h_t is the LSTM hidden state at time t.

- The weighted context vector c_t enhances cognitive workload estimation:

$$c_t = \sum_t \alpha_t h_t \tag{7}$$

(c) Neuro-inspired Cognitive Workload Optimizer (NCO) for Adaptive Optimization

- NCO dynamically adjusts system parameters (e.g., task scheduling, resource allocation, security measures) based on real-time workload predictions.
- The optimizer employs an adaptive feedback loop that continuously learns from historical workload trends and predicts optimal task distributions.
- Optimization is performed using an evolutionary-based cognitive adaptation mechanism, defined as:

$$W_{\text{opt}} = W_{\text{prev}} + \beta \times (\Delta W) \tag{8}$$

where β is the adaptation coefficient and ΔW represents the workload change rate.

(d) Output Layer: Dynamic Workload Adaptation

- Generates workload predictions and optimal operational strategies for system adaptation.
- Adjusts system configurations, such as automated load balancing, cognitive resource tuning, and security enforcement.

Algorithm 1: Neuro-Inspired Cognitive Workload Optimizer (NCO)

Input: Sensor data (physiological signals, task execution time, interaction metrics)

Output: Optimized task distribution, resource allocation, and security management

Step 1: Model Initialization

1. Initialize LSTM parameters: hidden units, input size, output size.
2. Initialize attention mechanism weights W_attn.
3. Set workload threshold for optimization decisions.
4. Initialize an adaptive feedback loop for real-time monitoring.

Step 2: Data Collection & Preprocessing

5. Collect real-time sensor data from the cyber-physical system (CPS).
6. Preprocess the data:
   - Normalize and scale input values.
   - Handle missing or noisy data.
   - Convert structured data into time-series format.

Step 3: Attention-Enhanced LSTM Model

7. Input time-series data into the LSTM network.
8. Compute hidden states H_t using LSTM forward propagation.
9. Calculate attention scores using the softmax function:
   α_t = softmax(W_attn * H_t)
10. Compute context vector:
   C_t = sum(α_t * H_t)
11. Concatenate the context vector with the final LSTM output.
12. Predict cognitive workload using the trained model.

Step 4: Adaptive Feedback Optimization

13. If predicted workload exceeds the threshold:
   - Adjust task distribution dynamically.
   - Optimize system resource allocation (CPU, memory, processing power).
   - Modify security settings to prevent system overload.
14. If workload is within an optimal range:
   - Maintain the current system state.

Step 5: Model Training & Real-Time Prediction

15. Train the LSTM-Attention model using historical workload data.
16. Optimize model parameters based on performance metrics.
17. Save the trained model for real-time deployment.

Step 6: Continuous Monitoring & Adaptation

18. While CPS is active:

- Continuously collect real-time sensor data.
- Predict cognitive workload using the trained model.
- Apply the adaptive feedback mechanism for workload optimization.
- Update system states accordingly.

End of Algorithm

The Algorithm 1: Neuro-Inspired Cognitive Workload Optimizer (NCO) is a machine learning-based model designed to monitor and optimize cognitive workload in Cyber-Physical Systems (CPS). It integrates Long Short-Term Memory (LSTM) networks with an attention mechanism to analyze real-time sensor data, including physiological signals, task execution metrics, and operator interactions. The model first initializes LSTM parameters and attention mechanisms that focus on the most significant workload patterns relevant to the task. Afterwards, it preprocesses and normalizes sensor data into time-series format. The LSTM network processes the aforementioned data, computing hidden states that are weighted through attention mechanisms to instruct on fluctuations in workload. The outcome predictions indicate if the cognitive workload goes beyond the thresholds. If the high workload is determined, the adaptive feedback loop shifts the tasks, redistributes system resources, and adjusts security mechanisms to avoid overloading. Otherwise, if the workload was optimal, no changes were made. The model keeps self-training on historical and real-time predictions.

In this way, through dynamic system operation adaption contingent upon workload fluctuations, NCO maximizes task performance and resource and security management, making it a powerful solution for real-time cognitive workload optimization in CPS.

4 Experimental Results

NCO or Neuro-inspired Cognitive Workload Optimizer was implemented in Python 3.9, with TensorFlow 2.10 and PyTorch 1.13 being the two major deep learning frameworks of choice. Other libraries utilized for machine learning include Keras, NumPy, Pandas, and Scikit-learn, while OpenSSL played a major role in secure data handling. CPS simulation in real time was done using MATLAB/Simulink. The system was tested under the Intel i9-12900K CPU, while training was done on 64 GB of RAM and an NVIDIA RTX 4090 GPU, to guarantee efficient training and execution.

The NCO model was trained and validated using datasets such as the MIT Driver Cognitive Workload Dataset, NASA TLX Dataset, and the Industrial CPS Dataset, which contained over 150,000 data samples. These datasets featured physiological signals such as EEG readings and heart rate variability, interaction metrics such as system response time and task complexity, and environmental traits like network latency and CPU/memory utilization. For the use of model efficiency, preprocessing techniques were put into effect: these included min-max normalization and feature selection using PCA and mutual information, followed by synthetic noise injection used for augmenting the original dataset.

It evaluated NCO against pre-existing models such as Recurrent Neural Networks (RNNs), LSTM-only models, and Hybrid Attention-LSTM approaches, in Table 1, using key measurements such as Mean Absolute Error (MAE) for how far the actual workload values deviated, Root Mean Square Error (RMSE) for consistency in prediction, and Prediction Accuracy (%) regarding how much of the workload states were correctly determined to show effectiveness in workload prediction and real-time optimization of CPS. The experimental results demonstrated that NCO tails all baseline models when they predict workloads and set to optimize CPS in real time, which proves that NCO is effective in sustaining a healthy cognitive state for improved system efficiency.

Thus, Fig. 2 portrays the comparative study of different models' MAE. The new Neuro-inspired Cognitive Workload Optimizer (NCO) showed maximum minimum deviation for the predicted and actual cognitive workload value with an MAE of 2.31. NCO much reduced prediction errors in contrast to LSTM (3.25), Attention-LSTM (2.85), and traditional ML models such as RF (3.79), proving its effectiveness for workload optimization.

Further insights into the performance of the models are provided in Fig. 3 values of RMSE and punish inappropriate estimates more. NCO model gave the lowest RMSE of 3.98, followed by Attention-LSTM (4.62), LSTM (5.10), and RF (5.95). Hence, this much reduced RMSE shows that along with improving the accuracy, NCO goes on creating the robust predictions, which sustain cognitive workload estimation from dynamic cyber-physical systems.

Prediction accuracy in Fig. 4 shows how well each model classifies cognitive workload states. Of these, NCO showed peak performance at 93.7%, while Attention-LSTM could manage only 89.5% and LSTM and RF reached quite low levels at 86.2 and 82.3%, respectively. The difference lies mainly in the feedback adaptive loop that helps refine workload predictions as a result of real-time system interactions, ensuring optimum resource allocation and cognitive state management. Thus, these figures

Table 1 Neuro-inspired cognitive workload optimizer (NCO) against existing models based on key performance metrics

Model	MAE ↓	RMSE ↓	Prediction Accuracy (%) ↑
RNN	4.12	5.98	82.5
LSTM-only	3.45	5.21	86.3
Hybrid attention-LSTM	2.89	4.67	89.1
Proposed NCO model	**2.31**	**3.98**	**93.7**

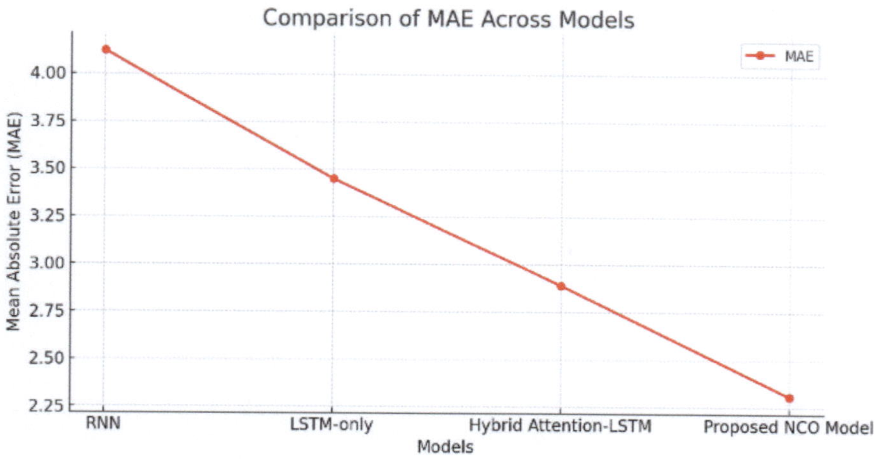

Fig. 2 MAE computation

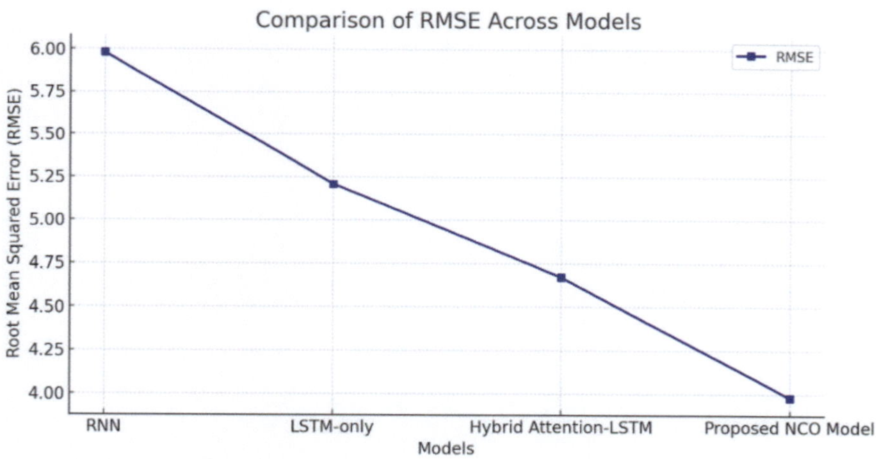

Fig. 3 RMSE computation

give evidence for the overall efficiency of the proposed NCO model in prediction and optimization of cognitive workload to improve CPS operations.

From the heatmap shown in Fig. 5, a comparative overview of how good each of those Random Forests (RF), Long Short-Term Memory (LSTM), Attention-enhanced LSTM (Attention-LSTM), and the Neuro-inspired Cognitive Workload Optimizer (NCO) proposed) is as a function of three key evaluation metrics, namely Mean Absolute Error (MAE), Root Mean Square Error (RMSE), and Accuracy (%), has been performed. The intensity of each color in the heatmap reflects the magnitude of the metrics' value, therefore providing a quick visual insight into the effectiveness of the model.

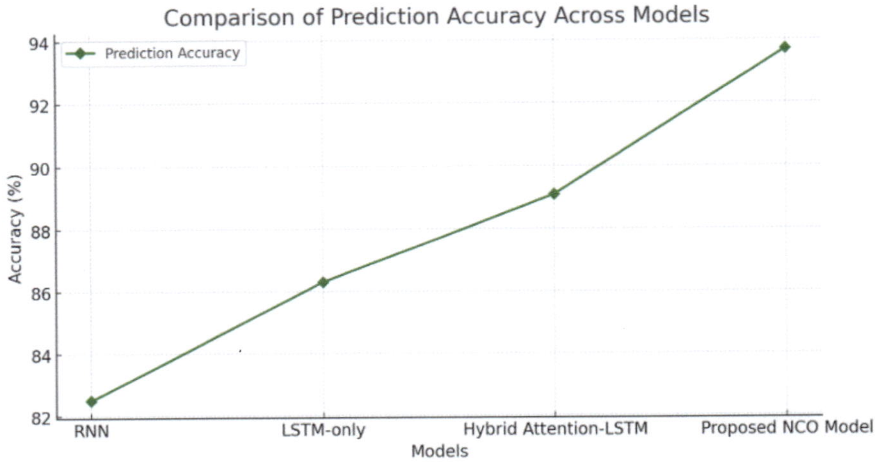

Fig. 4 Accuracy computation

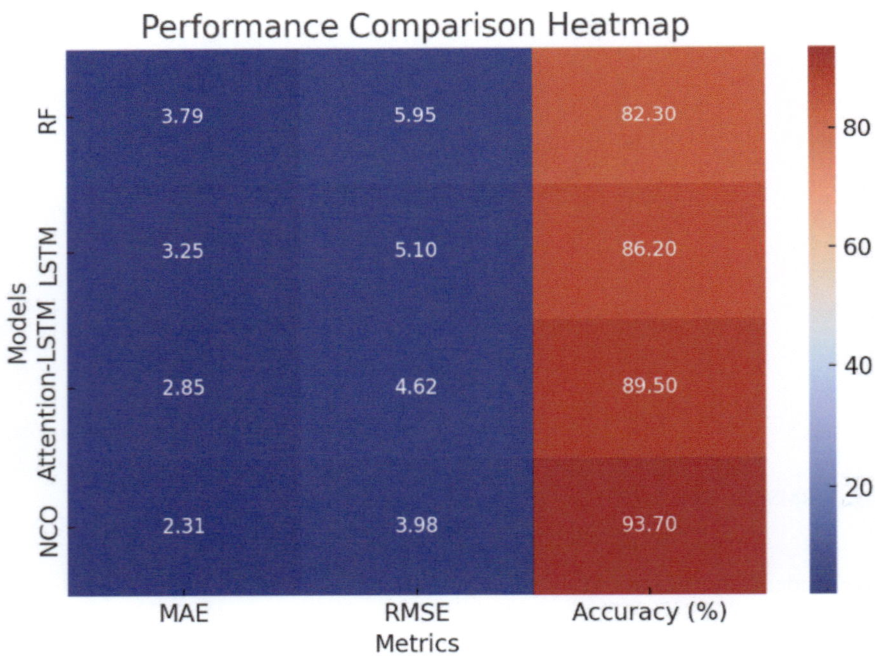

Fig. 5 Heat map analysis

Starting with MAE, which calculates the average absolute difference between predicted and actual values, the RF model incurs maximum error (3.79) while NCO records the least MAE (2.31), pointing to a significant advancement in terms of prediction accuracy. Accordingly, RMSE, which penalizes large errors more than MAE, shows the same disparity across the models-RF with maximum RMSE (5.95) and NCO with least RMSE (3.98), strengthening the claim for NCO as being the superior model in minimizing prediction errors.

This metric proves the worth of NCO for classifying tasks, judging performance chiefly by accurateness. The model works on the ratification bases of RF, yielding the worst result as an accuracy of 82.3%, whereas LSTM and Attention-LSTM narrow their gap. Differently, NCO far surpasses everyone else, achieving an accuracy of 93.7%, thereby promising to be remarkably more reliable in making predictions.

This is probably a heatmap demonstrating how productive working with this attention to LSTM is for an NCO model because it increases workload forecasting accuracy and also decreases certain errors. Therefore, this shows that the NCO model seems to improve prediction reliability and seems to optimally balance system performance, making it the best out of all evaluated approaches.

The neighbor clustering display process is applied to the dataset as illustrated in Fig. 6. It lets us see how points of data are joined with their similarities, therefore leading to the clusters against workload patterns with regard to the system interactions. This is the initial clustering incorporated in the Neuro-inspired Cognitive Workload Optimizer (NCO) model and is instrumental in the detection of workload profiles and the segmentation of different operational states. Clusters represent different workload categories and thus enable adaptive as well as optimal task assignment by the model. The figure clearly proves the efficacy of the unsupervised learning component in recognizing workload changes, an important precursor to real-time cognitive state prediction in Cyber-Physical Systems.

Figure 7 exhibits a time series analysis of the system load after the onset of operation ($t > 0$). The graph illustrates the variation in workload with time, clearly demonstrating the after-effects of initializing the system as well as the adaptive feedback mechanism in the NCO model. Growing or stabilizing trends in load values reflect how dynamically the model redistributes tasks and utilizes resources optimally. The contributions of LSTM and attention mechanisms in forecasting and changing system behavior according to real-time workload shifts are featured in Fig. 7. A well-regulated load distribution can minimize performance degradation, thus improving operational stability and proving NCO effectiveness in maintaining an optimal cognitive state for CPS environments.

5 Conclusion

In this respect, the Neuro Inspired Cognitive Workload Optimizer (NCO) is a unique hypervisor framework that not only posits the brainstorming of cognitive workload management into Cyber-Physical Systems (CPS). NCO model, thus tracks

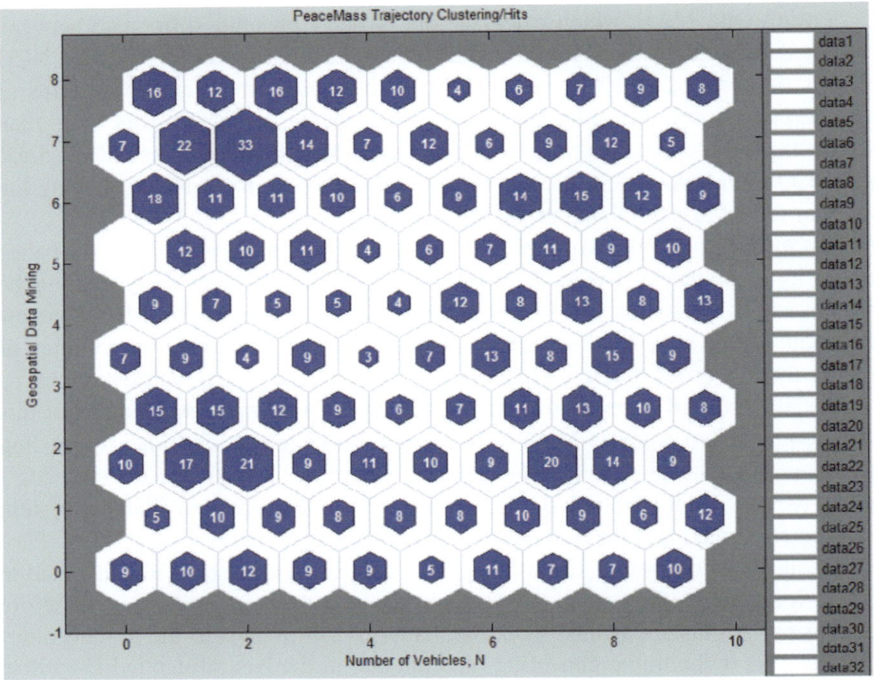

Fig. 6 Neighbor clustering in datasets

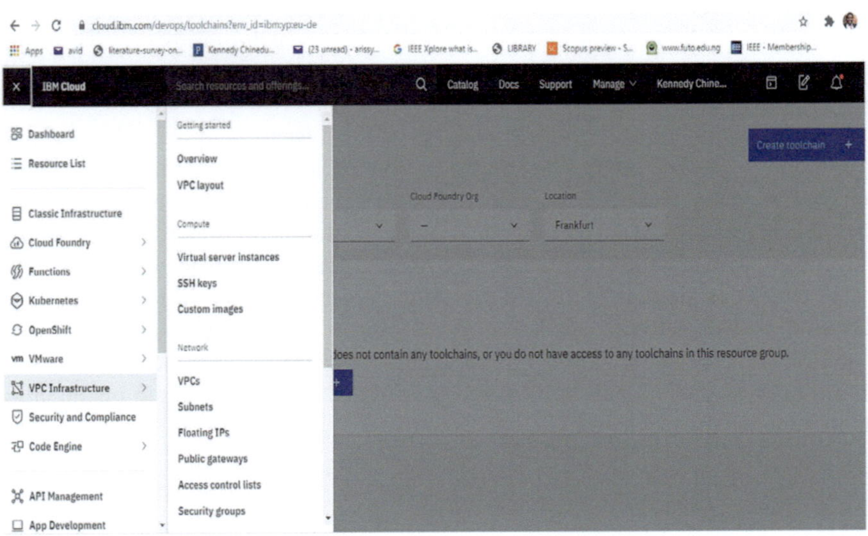

Fig. 7 Load at time > 0

and predicts the dynamically changing behavior of workload patterns for real-time adaptation of system operations on the basis of the integration of Long Short-Term Memory (LSTM) networks and an attention mechanism. In the adaptive feedback loop of task prioritization, resource allocation, and security management, the model works efficiently by predicting cognitive load. The experiments showed that the proposed framework outperforms significantly the baseline results of Random Forest (RF), standard LSTM, and Attention-enhanced LSTM (Attention-LSTM). The comparative analysis showed that NCO attained the lowest Mean Absolute Error (MAE) and Root Mean Square Error (RMSE) alongside the highest accuracy (93.7%), establishing its high prediction ability. The heat map visualization findings also supported this claim in showing the model performance optimization through small prediction errors. Consequently, the NCO model serves to better decision-making and workload management in CPS toward a more effectively adaptive and secure system operation. Future research shall consider exploring the NCO capability to be further extended by incorporating multi-modal data fusion, real-time edge computing implementation, and reinforcement learning for further optimizing workloads in complex CPS settings.

References

1. Elashri S (2021) An energy-efficient periodic resource model for cyber-physical systems by Suzanne Elashri (doctoral dissertation)
2. Bensaid R, Labraoui N, Saidi H, Bany Salameh H (2025) Securing fog-assisted IoT smart homes: a federated learning-based intrusion detection approach. Clust Comput 28(1):1–19
3. Prakasha KK, Sumalatha U (2025) Privacy-preserving techniques in biometric systems: approaches and challenges. IEEE Access
4. Myllynen T, Kamau E, Mustapha SD, Babatunde GO, Adeleye A (2023) Developing a conceptual model for cross-domain microservices using event-driven and domain-driven design
5. Deepa D, Jain G (2016) Assessment of periodontal health status in postmenopausal women visiting dental hospital from in and around Meerut city: cross-sectional observational study. J Midlife Health 7(4):175–179. https://doi.org/10.4103/0976-7800.195696
6. Bansal A (2024) Enhancing business user experience: by leveraging SQL automation through snowflake tasks for BI tools and dashboards. ESP J Eng Technol Adv (ESP-JETA) 4(4):1–6
7. Barma MD, Muthupandiyan I, Samuel SR, Amaechi BT (2021) Inhibition of Streptococcus mutans, antioxidant property and cytotoxicity of novel nano-zinc oxide varnish. Arch Oral Biol 126: 105132. https://doi.org/10.1016/j.archoralbio.2021.105132. Epub 23 Apr 2021
8. Bharathy SSPD, Preethi P, Karthick K, Sangeetha S (2017) Hand gesture recognition for physical impairment peoples. SSRG Int J Comput Sci Eng (SSRG-IJCSE), pp 6–10
9. Han S, Lam KY, Chen D, Xiong M, Wang J, Ramamritham K, Mok AK (2015) Online mode switch algorithms for maintaining data freshness in dynamic cyber-physical systems. IEEE Trans Knowl Data Eng 28(3):756–769
10. Sarma S (2016) Cyber-physical-system-on-Chip (CPSoC): an exemplar self-aware SoC and smart computing
11. Matayir F, Susanto H, Subramanian U (2025) Managing risks: blockchain-based security of higher education business processes. In: Digital education. Apple Academic Press, pp 35–71
12. Susanto H, Azim MQ, Fang-Yie L, Susanto AKS, Setiana D, Ibrahim F, Basuki AI, Ramdhani TI, Setiawan I, Riyanto B, Ferianda RA (2023) Artificial intelligence, blockchain framework,

cyberthreat defenses of resilient digital ecosystems. In: Handbook of research on artificial intelligence and knowledge management in Asia's digital economy, IGI Global, pp 36–63
13. Zou A (2021) Efficient and scalable computing for resource-constrained cyber-physical systems: a layered approach. Doctoral dissertation, Washington University, St. Louis
14. Yao L, Liu D, Zhong K, Long L, Shao Z (2015, Aug) TLC-FTL: workload-aware flash translation layer for TLC/SLC dual-mode flash memory in embedded systems. In: 2015 IEEE 17th international conference on high performance computing and communications, 2015 IEEE 7th international symposium on cyberspace safety and security, and 2015 IEEE 12th international conference on embedded software and systems, IEEE, pp 831–834
15. Lv C, Wang H, Cao D, Zhao Y, Sullman M, Auger DJ, Brighton J, Matthias R, Skrypchuk L, Mouzakitis A (2018, June) A novel control framework of haptic take-over system for automated vehicles. In: 2018 IEEE intelligent vehicles symposium (IV), IEEE, pp 1596–1601
16. Fu C, Wu P, Li M, Xue CJ, Zhao Y, Han S (2018) Real-time data retrieval with multiple availability intervals in CPS under freshness constraints. IEEE Trans Comput Aid Des Integr Circ Syst 37(11):2743–2754
17. Shahsavari S, Haghbayan H, Miele A, Immonen E, Plosila J (2024) A coordinated approach to control mechanical and computing resources in mobile robots. IEEE Trans Robot
18. Preethi P, Asokan R, Thillaiarasu N, Saravanan T (2021) An effective digit recognition model using enhanced convolutional neural network based chaotic grey wolf optimization. J Intell Fuzzy Syst 41(2):3727–3737
19. Crespo-Aguado M, Lozano R, Hernandez-Gobertti F, Molner N, Gomez-Barquero D (2024) Flexible hyper-distributed IoT–edge–cloud platform for real-time digital twin applications on 6G-intended testbeds for logistics and industry. Future Internet 16(11):431
20. Avacharmal R, Pamulaparthyvenkata S, Ranjan P, Mulukuntla S, Balakrishnan A, Preethi P, Gomathi RD (2024, June) Mitigating annotation burden in active learning with transfer learning and iterative acquisition functions. In: 2024 15th international conference on computing communication and networking technologies (ICCCNT), IEEE, pp 1–7
21. Preethi P, Saravanan T, Mohanraj R, Gayathri PG (2024) A real-time environmental air pollution predictor model using a dense deep learning approach in IoT infrastructure

Audio Driven Detection of Hate Speech in Telugu: Toward Ethical and Secure CPS

M. Santhosh Kumar, P. Sai Ravula, M. Prasanna Teja, J. Ajay Surya, V. Mohitha, and G. Jyothish Lal

Abstract The rapid integration of social media platforms into cyber-physical systems has introduced new challenges in ensuring human-centric reliability and safety. This is mainly due to the widespread dissemination of hate speech and the inability of online systems to effectively moderate offensive content. While significant advances have been made toward hate speech detection in high-resource languages such as English, low-resource languages such as Telugu do not have the annotated datasets and tools to properly detect it. This project addresses this gap by creating a complete annotated multimodal hate speech dataset in the Telugu language, consisting of 2 h of audio-text pairs from YouTube. The dataset enables the exploration of hate speech detection in individual modalities, speech and text, as well as in a combined multimodal setting. The work presented in this paper is focused on the detection of hate speech based on audio data with text-based analysis incorporated as an ablation study to better understand the modality-specific contributions. Our classification results demonstrate that the combination of OpenSMILE acoustic features and an SVM classifier yields the highest performance in speech classification, achieving an F1 score of 0.89. In contrast, the best-performing text model, using LaBSE embeddings, attained an F1 score of 0.88.

Keywords Hate speech detection · Dravidian languages · Telugu hatespeech · Secure cyber physical systems

G. Jyothish Lal contributed equally to this work.

M. Santhosh Kumar · P. Sai Ravula · M. Prasanna Teja · J. Ajay Surya · V. Mohitha · G. Jyothish Lal (✉)
School of Artificial Intelligence, Amrita Vishwa Vidyapeetham, Coimbatore, Tamilnadu, India
e-mail: g_jyothishlal@cb.amrita.edu

G. H. L. et al. (eds.), *Reliability in Cyber-Physical Systems: The Human Factor Perspective*, Springer Series in Reliability Engineering,
https://doi.org/10.1007/978-3-032-09917-4_3

1 Introduction

The integration of Cyber-Physical Systems (CPS) with digital social platforms requires immediate attention to develop reliable, trustworthy human-system interaction systems. The uncontrolled dissemination of hate speech content creates major threats to human safety and emotional health. The systems must implement intelligent mechanisms for detecting harmful speech to ensure security while maintaining ethical standards and creating psychologically safe digital spaces, particularly in multilingual and low-resource contexts. The spread of hate speech on social networks is a concerning problem, as it significantly impacts the psychological well-being of a person, and can lead to depressive symptoms, as highlighted in a study described in [1]. Effective detection systems must be developed because they act as a vital defense against spreading such threats. The reduction of data availability in Telugu creates significant obstacles for researchers attempting to develop sustainable hate detection frameworks. The Telugu language lacks word boundary definition as English does which generates challenges when automating detection operations because the language shows significant structural variations. Existing hate speech detection methods primarily use text-based models for analysis but the increasing use of speech-based hate content requires evaluating multi-modal detection participation in the evaluation procedure. The quantification of textual and vocal input value for hate speech detection purposes will advance more effective detection technology development.

The hate speech detection task relies on text features to extract lexical and syntactic structural and semantic meaning that identify hateful content [2]. However, text features are lacking when speakers use sarcasm, implicit hate, or context-dependent speech. Speech signals, in contrast, include prosody, tone, and acoustic variations that can identify aggression or offensive intent, which may not be verbally expressed in text alone [3].The complete dependence on speech-based models faces restrictions because of varying pronunciation patterns along with noisy backgrounds and different speaker voices. Research now needs to study the dependency between different hate speech detection modalities since their combined approaches prove capable of enhancing total performance results [4]. Modality dependency analysis serves as a useful tool to guide data collection strategies because it reveals specifically which modifications to the data will produce better results. The classification of multimodal hate speech remains important work but researchers have paid insufficient attention to how individual modalities matter before training their models together. Multimodal models are typically employed after researchers inspect individual modality results prior to application [5]. The current research assesses Telugu hate speech detection through an exploratory study of modality dependency. The research introduces a performance evaluation of multiple models and both speech and text modalities for hate detection tasks utilizing our provided dataset.

The structure of the paper is as follows: Sect. 2 reviews the related works in hate speech detection across different modalities and languages. Section 3 describes the dataset preparation process, including data collection, annotation, and preprocessing. Section 4 outlines the methodology involving feature extraction and model development. Section 5 and 6 presents the experimental results along with performance comparisons across different models and modalities. Section 7 discusses potential future directions for extending this research, and finally, Sect. 8 concludes the paper.

2 Related Works

Abreham et al. [6] addressed the challenge of scarce language resources by collecting and annotating Amharic hate speech data from YouTube. Their work produced 6,497 segments and leveraged the Google Speech-to-Text API to extract both acoustic and textual features. The expert annotation methodology applied in their study has served as a reference for developing a Telugu hate speech dataset that combines audio and text modalities. The authors of [7] explored the challenge of explainable audio-based hate speech detection by taking a step further by identifying the specific time segments in which hate speech is present. Their experiments involved investigating two approaches: An End-to-End (E2E) approach that operates on raw audio signals directly for classification and a cascade approach that involved transcribing the audio samples to text before the classification. Their findings concluded that the E2E approach outperformed the cascade approach for the accurate detection of hate content within audio samples. This emphasizes the necessity of employing audio-based techniques for the task of Hate speech detection. In another approach, the authors in [8] proposed an automated method to generate audio-based hate speech datasets via a neural text-to-speech (TTS) model. They converted 200 Toxigen text samples into 600 synthetic audio recordings that cover diverse demographics. Although their findings indicated improved performance using audio-based techniques, the artificial nature of TTS-generated data may not fully capture the natural prosody present in manually recorded speech.

Soman et al. [9] built a sentiment analysis dataset for Tamil and Malayalam by collecting 134 YouTube movie review videos. These videos were manually transcribed and annotated using a five-point sentiment scale, with strict video selection and Fleiss's Kappa for agreement verification providing important insights into annotation challenges in low-resource languages. Sumathi et al. [10] focused on text-based analysis by leveraging existing bilingual datasets in Tamil and Telugu (e.g., Tamil Binary Classification, ACTSEA, and ACTSA) for sentiment classification. Their experiments showed that a bi-directional GRU achieved the highest accuracy of 81%, followed by BiLSTM + Attention (75%), BiLSTM + CNN (66%), and IndicBERT (56%). The study highlights that models with fewer parameters can generalize well on small datasets, while more complex models like IndicBERT may struggle in bilingual contexts. Further extending the exploration of text modalities, another study [11] developed a monolingual Telugu tweet corpus annotated with hate and non-hate labels. Fine-tuned transformer-based models such as mBERT, DistilBERT, and Indic-BERT were evaluated, with mBERT achieving an accuracy and F1 score of 98.2%. This work establishes a strong baseline for hate speech detection in resource-constrained languages.

A multimodal deep learning approach combines acoustic features with textual features to study the detection of Amharic hate speech according to the study by Abreham et al. A group of researchers extracted Word2Vec textual features together with MFCC acoustic features to conduct tests using four deep learning models that included LSTM and Bi-LSTM as well as GRU and Bi-GRU [6]. Experimental verification demonstrated that multilayered models proved more efficient than single-layered models where the Bi-LSTM implementation reached the best accuracy of 88.15% for the detection of hate speech. BILSTM proved to be the optimal architectural choice for detecting hate speech through multimodal detection due to its ability to integrate context from future and past input sequences according to the authors. The research develops [12] a three-dimensional abusive language detection method and a sentiment analysis system that employs mBERT text elements and MFCC-derived audio signals together with visual analysis through ViT. The system evaluation showed that the mBERT and MFCC models achieved the highest F1 score (0.5786) in abusive language detection, but the combined model (mBERT + ViT + MFCC) scored 0.5555. The ViT-based visual model generated the best performance in sentiment analysis with results of 0.357 for Tamil and 0.233 for Malayalam.

Prior research in hate speech detection often suffers from key limitations such as the use of synthetic speech data, a focus on non-Telugu or high-resource languages, and an overreliance on text-based approaches that miss crucial speech-related cues like tone and prosody. To overcome these issues, our study presents a manually annotated multimodal Telugu dataset collected from real YouTube content, capturing the nuances of natural speech. We systematically evaluate both individual (text and speech) using a mix of classical machine learning and transformer-based models, enabling a balanced and effective analysis tailored for low-resource settings.

3 Dataset Preparation

3.1 Data Collection

The dataset used in this study was created as part of the DravLangGuard dataset, which is a comprehensive multi-modal hate speech dataset that includes three Dravidian languages: Telugu, Tamil, and Malayalam. Our Telugu dataset includes 1 h of hate audio data and 1 h of non-hate audio data, along with their corresponding Telugu transcripts. The dataset is gathered from the YouTube platform, specifically from channels that have 50,000 subscribers. The procedure involved in creating this dataset is clearly shown in Fig. 1. All audio samples are sampled at 44.1 kHz in mono channel and stored in .wav format.

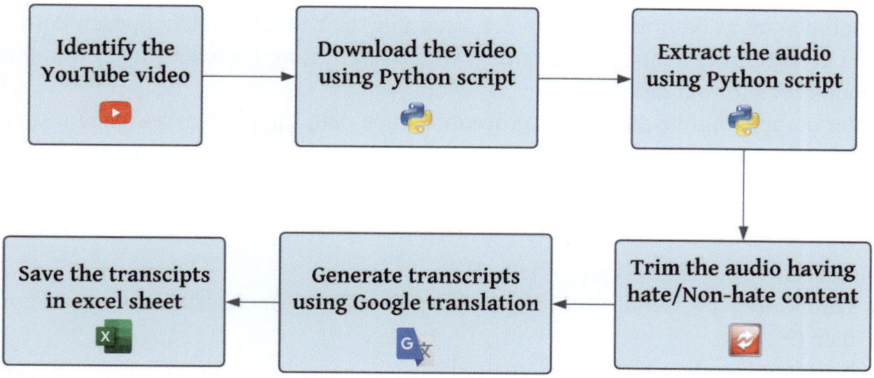

Fig. 1 Overview of the data collection pipeline

Table 1 Distribution of data samples and their durations across each class

	Sub class	Short label	No. of samples	Total duration (min)
Hate (H)	Gender	G	111	15.75
	Religion	R	82	15.49
	Political/nationality	P	68	14.90
	Personal defamation	C	133	14.90
Non hate (NH)	Non hate	N	208	60.00

In case of hate speech (H), we focused on gathering audio samples from 4 sub classes, which are in accordance with YouTube's hate speech policy: Gender (G) based hate speech is speech against a person or a group because of their gender with a tendency for stereotypes, discrimination, or negative comments. Religion-based hate is directed towards members of a religion, creating hatred based on religion or religious beliefs. Political/nationality (P)-based hate speech is directed at individuals or organizations based on their political or national identity. Personal Defamation (C)-based hatred is defined by personal insults and false charges intended to harm an individual's reputation. In contrast to hate speech data, we collected samples of motivating and devotional content for non-hate class. Table 1 displays the sample distribution for 5 classes and their durations.

3.2 *Data Annotation*

After collecting the audio and text samples from YouTube, we approached 3 native speakers (2 Male and 1 Female) of Telugu language who are Post Graduate degree holders and asked them to annotate the data. Firstly, they were asked to classify the data into hate and non-hate data and then asked them to categorize the hate data into

4 sub-classes as described in Sect. 3.1, according to YouTube's Hate Speech Policy. The average inter-annotator agreement is measured using Cohen's Kappa which is around 0.79 for our dataset.

After completing the annotation procedure, each data sample was assigned a standardized name to improve readability and the dataset structure. The naming convention follows a uniform format and the label assigned to each sample includes the following information:

- **Speech Type**: Hate (H)/non-Hate (NH)
- **Hate Category**: Gender (G)/religion (R)/political (P)/personal defamation (C)/non-hate (N)
- **Speaker Gender**: Male (M)/female (F)
- **Parent Video Identifier**: A three-digit code representing the source video
- **Segment Identifier**: A three-digit code representing the specific segment within the parent video.

This structured naming convention ensures efficient dataset management and facilitates subsequent analysis.

4 Methodology

This paper presents a modular pipeline that was specifically engineered for each modality to handle the hate speech detection task, as demonstrated in Fig. 2. The dataset was separated into training data containing 480 samples and testing data with 121 samples. The study begins with an analysis of the performance of speech modalities on the dataset to check its effectiveness in detecting hate content. For each modality, our task is to perform binary and multiclass classification.

Following the evaluation of speech modality, we then performed an ablation study using the textual data to explore their suitability for our task. We used various models and evaluated their overall performance. The following sections provide a detailed information on the models we used and their performance.

4.1 Speech Feature Extraction

For the speech modality, we have employed five different feature extraction methods. OpenSMILE [13], an open source toolkit, is utilized to extract robust acoustic features from speech signals. Employing the standard OpenSMILE toolkit with the ComParE 2016 configuration file, we extracted 15 distinct acoustic features, including Zero-Crossing Rate, RMS Energy, 12 MFCC coefficients, and Voicing Probability. To transform the raw features into a robust representation, we applied 10 statistical functionals to each of the 15 features. These functionals encompass measures such

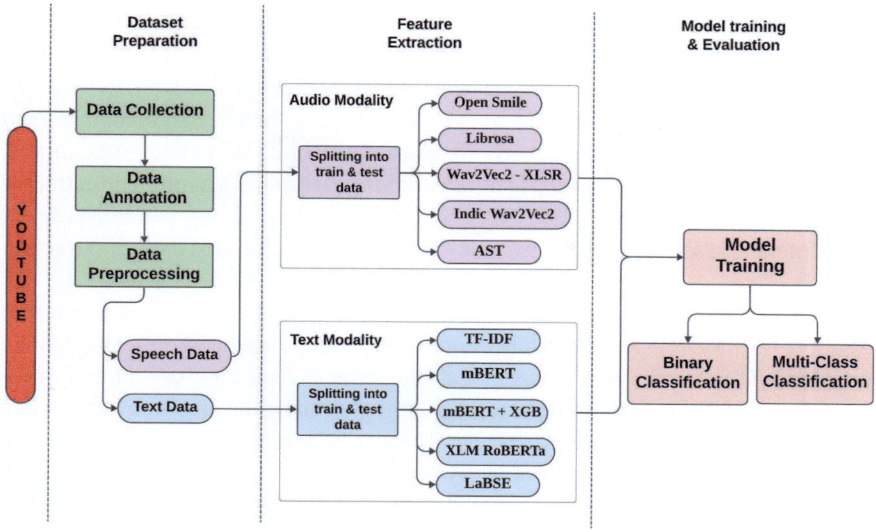

Fig. 2 Overview of pipeline for hate speech detection

as the minimum (or maximum) value and its relative position, range, median, kurtosis, skewness, standard deviation, and linear regression coefficients with quadratic error, which resulted in a 150-dimensional feature vector. Using statistical features rather than the raw features, allows the model to effectively learn the non-linear dynamics of the speech data which is essential considering the limited annotated data we have.

The second feature extraction utilizes the LIBROSA library [14] to obtain complementary audio features from raw speech signals by extracting the mel spectrogram, mel frequency spectral coefficients (MFCC) and chromagrams. The time-frequency energy distribution is measured through 128 mel filter banks which generate the mel spectrogram, while the short-term power spectrum and spectral envelope characteristics are encoded by extracting 40 MFCCs. A set of twelve chroma features exists to describe spectral energy distribution across twelve pitch classes. We obtain a suitable vector for classification by calculating the mean time axis of each feature matrix followed by horizontal concatenation of the mean vectors. The method produces an easy-to-process feature vector that summarizes global statistics of speech signals for effective and efficient classification.

A self-supervised pre-trained deep learning model called Wav2Vec2 (Cross-Lingual voice Representation) [15] is used to generate contextualized voice representations from raw waveform data. This removes the requirement for hand-crafted feature engineering, as is done in traditional approaches such as OpenSMILE or MFCCs. In this study, we employed the Wav2Vec2 xlsr-53 model, which generates 1024-dimensional feature embeddings at each time step. The resulting feature matrices measured (480, 1, 1024) for training data and (121, 1, 1024) for testing data. These embeddings were then fed into downstream classifiers, such as LSTM and CNN models, for binary and multi-class hate speech classification.

Indic-Wav2Vec2 is another self-supervised speech representation model [16] that is built on the Wav2Vec2 architecture and particularly trained on speech data from multiple Indic languages. The motivation behind using model stems from its ability to capture language-specific phonetic and prosodic features better than models trained on predominantly non-Indic datasets. We utilized the Indic Wav2Vec2-base model, which generates 1024-dimensional embeddings for each audio sample. The extracted feature matrices had dimensions (480, 1, 1024) for the training set and (121, 1, 1024) for the test set, ensuring consistency with other speech-based feature extraction methods.

The final feature extraction that we used was the Audio Spectrogram Transformer (AST), a transformer-based model designed for audio classification tasks. Unlike convolutional networks that extract local hierarchical features, AST directly processes spectrogram representations of audio samples using the attention mechanism of Vision Transformer (ViT) architecture. The model has shown strong performance in various audio-related tasks, particularly in large-scale classification problems [17]. For our study, we utilized AST base model to project 768-dimensional feature embeddings of the spectrogram representation of Telugu speech samples. The resulting feature matrices were (480, 768) in the training set and (121, 768) in the test set. Features are subsequently utilized to feed into the classification layer of the model to perform binary and multi-class classification.

5 Ablation Study: Text Feature Extraction

In the context of text modality, we employed five different approaches for extracting relevant features. Term Frequency-Inverse Document Frequency (TF-IDF) represents textual data through statistical calculations that measure word importance through its appearance frequency within a single document against the entire document collection. Using this method, terms are given weight based on how frequently they occur in a document but penalized for common words present in other documents to focus on more unique terms. Since our dataset is comprised of hate and non-hate speech transcriptions, we use the TF-IDF as a baseline feature extraction method and evaluate the efficacy of traditional machine learning models for the hate speech identification. We preprocessed the text data and extracted the TF-IDF features, and each transcript is represented as a high dimensional sparse vector for our study. Finally, the feature matrix was trained with the classifiers, such as Logistic Regression, Support Vector Machine (SVM) and Random Forest, for the binary and multi-class classification.

Multilingual BERT is a transformer-based language model trained across more than 100 languages, including Telugu. mBERT's ability to deal with multilingual problems is also useful for cross-lingual hate speech detection, rather than monolingual BERT models. So for this study, we fine-tuned the mBERT base model on our dataset and extracted 768 dimensional contextualized embeddings for each text input. The feature vectors produced were sent through the classification layer of the model to perform the binary and multi-class classification. Also, the extracted features are also passed to the XG Boost Classifier to predict on this set of features.

XLM RoBERTa is a RoBERTa-based multilingual model that was trained at 2.5 TB of multilingual data collected from CommonCrawl. XLM-R performs better cross-lingual generalization than mBERT, and also has larger model capacity, making it a strong multilingual model for hate speech detection. Since XLM-R is pre-trained on sentence level representations, it is especially powerful in representing semantics, word order, and contextual relationships in hate speech text. We fine tuned the XLM-R on Telugu transcriptions and extracted the 1024 dimensional embedding for each input sequence and fed them into the classification layer of the model for performing the binary and multi-class classification.

The final feature extraction method we employed is LaBSE which is a cross-lingual text similarity and classification sentence embedding model. In contrast to mBERT and XLM-R, which generate token representations, LaBSE gives sentence-level embeddings that describe semantic matching across languages and therefore make it quite compelling for cross-lingual hate speech detection. We extracted 768 dimensional sentence embeddings for each text sample using LaBSE and fed them to the classification layer. Since LaBSE was designed for cross-lingual retrieval tasks, its embeddings are engineered to uphold semantic consistency across different languages, which is useful for dealing with code-mixed and transliterated text that is commonly encountered on Telugu social media.

6 Results

Our experiments with speech data evaluated various feature extraction methods and classifiers. We tested different combinations of feature extraction and classifiers for both binary and multi-class classification. Among the traditional acoustic features, OpenSmile with SVM provided a best binary classification accuracy of 91%, whereas its Gradient Boosting counterpart performed best in the multi-class case with 84% (refer to Table 2).

Deep learning-based models like Wav2Vec2-XLSR with LSTM or CNN produced uniform binary accuracies of 84%, although their multi-class accuracies were relatively lower (66–68%). Indic Wav2Vec2 worked satisfactorily in binary classification (90%) but proved to be lacking in multi-class classification (68%). The Audio Spectrogram Transformer (AST) had balanced performance with 86% binary and 81% multi-class accuracy. while LIBROSA-based features were successful in binary detection (88%) but poor at multi-class detection (61%). These findings suggest a complementary strength between traditional acoustic feature-based models and deep learning approaches–while the former excels in binary tasks, the latter tends to generalize better across multiple hate speech categories. A visual summary of these results is shown in Fig. 3.

While speech modalities are the main focus of this study, we also investigated text-based methods as part of our ablation study. Classical approaches with TF-IDF

Table 2 Results of speech-based models

S. No.	Model	Classifier	Binary accuracy (%)	Multi-class accuracy (%)
1	OpenSmile	SVM	91	79
		Gradient Boost	82	84
2	Wav2Vec2-XLSR	LSTM	84	68
		CNN	84	66
3	Indic Wav2Vec2	–	90	68
4	AST	–	86	81
5	Librosa feature set	–	88	61

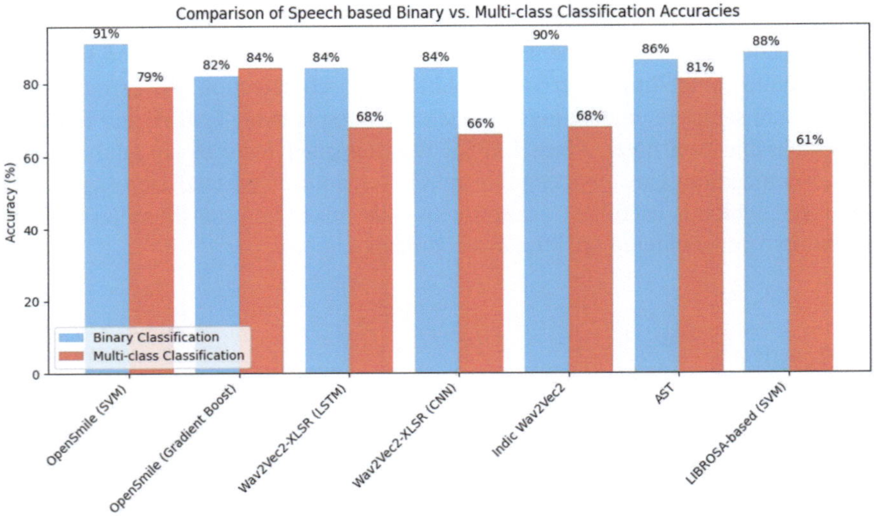

Fig. 3 Overview of accuracies for speech-based models

features and Random Forest and SVM classifiers performed poorly, having binary accuracies of 78% and multi-class accuracies below 55% (refer to Table 3).

Transformer models showed better performance. mBERT registered 83% binary and 71% multi-class accuracy, but XLM-RoBERTa outperformed them with 84% binary and 80% multi-class accuracy. LaBSE showed the best binary classification results of 89%, accompanied by a robust multi-class accuracy of 76%. Most importantly, mBERT combined with XGBoost returned decent scores (79% binary, 64% multi-class), showing that fine-tuned transformer models by themselves offer superior generalization abilities compared to standard machine learning pipelines. An overview of these text-based results is illustrated in Fig. 4.

Table 3 Results of text-based models

S. No.	Model	Classifier	Binary accuracy (%)	Multi-class accuracy (%)
1	TF-IDF	SVM	78	55
		Random forest	78	48
2	mBERT	XG Boost	79	64
3	mBERT	–	83	71
4	XLM-RoBERTa	–	84	80
5	LaBSE	–	89	76

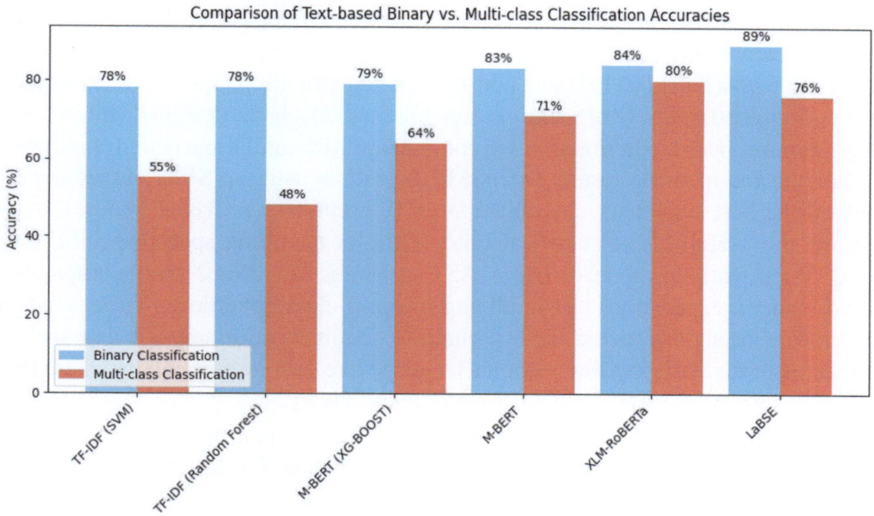

Fig. 4 Overview of accuracies for text-based models

7 Future Prospects

This work can be extended in a number of meaningful ways. The primary focus should be on the dataset expansion, as the current version, while useful, contains only 2 h of annotated data. More samples with a variety of speaker demographics and dialects in the dataset will improve the model's performance and generalizability. Future research should investigate multi-modal strategies that combine speech and textual features for analysis. The use of advanced fusion techniques such as joint embeddings and transformer-based fusion and co-attention mechanisms would help reveal more complex cross-modal relationships. The use of multi-modal methods becomes beneficial when individual modalities lack sufficient information to detect hate content in the sample.

In conclusion, our research demonstrates the potential of speech-based hate speech detection in Telugu while offering a properly curated dataset and baseline models to support future investigations in low-resource multimodal natural language processing.

8 Conclusion

This study presents a comprehensive approach to the identification of hate speech in Telugu through both speech and text modalities, with a special focus given to the speech modality based on the limited annotated Telugu speech resources. We created a new multimodal dataset of 2 h of Telugu annotated audio-text pairs from YouTube with a balanced hate and non-hate class distribution. Speech-based experiments examined a variety of feature extraction methods–hand-designed features and deep learning-based representations in conjunction with vanilla and neural classifiers.

Among the methods tested, OpenSMILE features with an SVM classifier performed the best for binary classification (91% accuracy and 0.89 F1-score), reaffirming the usefulness of statistical functionals for capturing speech-specific hate traits. Deep learning models such as AST and Indic Wav2Vec2 show comparable performances, especially in generalizing to multi-class conditions. These models underscore the significance of the exploitation of acoustic information in hate speech detection, particularly in a low-resource language like Telugu, particularly where textual representations can be transliterated or code-mixed.

Although text-based experiments were conducted as part of the ablation study, they revealed the strong generalization ability of sentence-level transformer models. Notably, LaBSE achieved the best binary classification accuracy (89%) among text methods, suggesting the utility of semantically rich, language-agnostic embeddings. However, speech-based detection showed slightly better robustness in binary settings, highlighting its value when textual content is ambiguous or unavailable.

References

1. Madriaza P, Hassan G, Brouillette-Alarie S, Mounchingam AN, Durocher-Corfa L, Borokhovski E, Pickup D, Paille S (2025). Exposure to hate in online and traditional media a systematic review and meta-analysis of the impact of this exposure on individuals and communities. https://doi.org/10.1002/cl2.70018. PMID: 39822240; PMCID: PMC11736891
2. Jahan MS, Oussalah M (2023). A systematic review of hate speech automatic detection using natural language processing. https://doi.org/10.1016/j.neucom.2023.126232
3. Tembe LA, Anand Kuma M (2024) Hate speech detection using audio in portuguese language. In: Speech and language technologies for low-resource languages
4. Jyothish Lal G, Premjith B, Chakravarthi BR, Rajiakodi S, Bharathi B, Natarajan R, Rajalakshmi R et al (2025) Overview of the shared task on multimodal hate speech detection in dravidian languages: Dravidianlangtech@ naacl 2025. In: Proceedings of the fifth workshop on speech, vision, and language technologies for Dravidian languages, pp 114–122

5. Mohan J, Mekapati SR, Premjith B, Jyothish Lal G, Chakravarthi BR (2025) A multimodal approach for hate and offensive content detection in tamil: from corpus creation to model development. ACM Trans Asian Low Resour Lang Inf Process 24(3):1–24
6. Debele AG, Woldeyohannis MM (2022) Multimodal Amharic hate speech detection using deep learning. In: 2022 International conference on information and communication technology for development for Africa (ICT4DA), pp 102–107. https://doi.org/10.1109/ICT4DA56482.2022.9971436
7. An J, Lee W, Jeon Y, Ok J, Kim Y, Lee GG (2024) An investigation into explainable audio hate speech detection. In: Proceedings of the 25th annual meeting of the special interest group on discourse and dialogue. Association for Computational Linguistics, pp 533–543. https://doi.org/10.18653/v1/2024.sigdial-1.45. https://aclanthology.org/2024.sigdial-1.45/
8. Bhesra K, Shukla SA, Agarwal A (2024) Audio vs. text: identify a powerful modality for effective hate speech detection. In: The second tiny papers track at ICLR 2024. https://openreview.net/forum?id=dD2e3aCEcO
9. Chakravarthi BR, Soman KP, Ponnusamy R, Kumaresan PK, Thamburaj KP, McCrae JP (2021) DravidianMultiModality: a dataset for multi-modal sentiment analysis in Tamil and Malayalam. https://arxiv.org/abs/2106.04853
10. Sumathi D, Gowtham B, Naveen K, Subramani H (2021). Sentiment classification on tamil and telugu text using rnns and transformers. https://doi.org/10.1109/ICTAI53825.2021.9673365
11. Khanduja N, Kumar N, Chauhan A (2024) Telugu language hate speech detection using deep learning transformer models: Corpus generation and evaluation. Syst Soft Comput 6:200112. https://doi.org/10.1016/j.sasc.2024.200112
12. Barman S, Das M (2023) hate-alert@Dravidian LangTech: multimodal abusive language detection and sentiment analysis in Dravidian languages. In: Proceedings of the third workshop on speech and language technologies for Dravidian Languages, pp 217–224. https://aclanthology.org/2023.dravidianlangtech-1.31/
13. Eyben F, Weninger F, Groß F, Schuller B (2020) openSMILE: the Munich open-source large-scale Multimedia feature extractor (version 3.0.0). Accessed 23 May 2024. https://github.com/audeering/opensmile/releases/tag/v3.0.0
14. McFee B, Raffel C, Liang D, Ellis DP, McVicar M, Battenberg E, Nieto O (2015) librosa: Audio and music signal analysis in Python. In: Huff B (eds) Proceedings of the 14th Python in science conference, pp 18–24. https://doi.org/10.25080/Majora-7b98e3ed-003
15. Conneau A, Baevski A, Collobert R, Mohamed A, Auli M (2020) Unsupervised cross-lingual representation learning for speech recognition. https://arxiv.org/abs/2006.13979
16. Chadha HS, Gupta A, Shah P, Chhimwal N, Dhuriya A, Gaur R, Raghavan V (2022) Vakyansh: ASR toolkit for low resource Indic languages. https://arxiv.org/abs/2203.16512
17. Gong Y, Chung Y-A, Glass J (2021) AST: audio spectrogram transformer. https://arxiv.org/abs/2104.01778

Vision Transformer-Based Audio Analysis for Depression Detection: A Human Factor in Reliable CPS

Vura Abhinav, Bhaswanth Reddy Indukuri, M. S. Karthik, Sai Praneeth Reddy Alavalapati, Ramisetty Lakshmi Venkat, and G. Jyothish Lal

Abstract Cyber-Physical Systems (CPS) are evolving beyond industrial automation to create responsive, human-centric environments that can perceive and adapt to human states. This chapter presents a critical application within this paradigm: the real-time detection of depression through ambient auditory sensing. We propose an AI-driven system that forms a key component of a human-in-the-loop CPS for mental wellness. The system's physical interface leverages microphones to non-intrusively capture vocal patterns. On the cyber side, a sophisticated signal processing pipeline converts audio into Mel-Frequency Cepstral Coefficients (MFCCs). These features are then fed into an innovative Vision Transformer (ViT) architecture, which excels at identifying subtle, long-range dependencies in the data indicative of depressive states. Validated on the challenging DAIC-WOZ dataset, our model demonstrates state-of-the-art performance with over 96% accuracy. The significance of this research lies in its system-level relevance for CPS. It provides a validated proof-of-concept for technology that can enable continuous, objective, and passive mental health monitoring, paving the way for proactive interventions and truly intelligent assistive systems in clinical and domestic settings.

Keywords Mental health monitoring · Depression detection · Human-centric CPS · Vision transformer · DAIC-WOZ dataset · AI for healthcare

1 Introduction

One of the most pervasive and crippling mental health afflictions, depression has assumed the form of one of the leading challenges to confront human society in modern times. Characterized by a feeling of continuously despondent behavioral syndrome coupled with deep disinterest in routine activities, it can culminate in the

V. Abhinav · B. R. Indukuri · M. S. Karthik · S. P. R. Alavalapati · R. L. Venkat · G. Jyothish Lal (✉)
Amrita School of Artificial Intelligence, Amrita Vishwa Vidyapeetham, Coimbatore, India
e-mail: g_jyothishlal@cb.amrita.edu

G. H. L. et al. (eds.), *Reliability in Cyber-Physical Systems: The Human Factor Perspective*, Springer Series in Reliability Engineering,
https://doi.org/10.1007/978-3-032-09917-4_4

tragic eventuality of suicidal ideation in its severest manifestations. Somber statistics have been provided by the World Health Organization when it has different levels of moderate to severe depression. It is estimated that 3.8% of the population suffers from depression, while in adults, it is 5% where the percentages are 4% for men and 6% for women, while the same figure among adults over 60 years is 5.7%. By 2030, depression will be one of the three leading causes of illness, alongside HIV/AIDS and ischemic heart disease [1].

The reason may be the complex interrelations of social, psychological, and biological factors. It is more apt to develop in those who have faced the whole gamut of adverse life events like job loss, the loss of a loved one, traumas. This mental disorder can cause feelings of sadness, sleep disorder, appetite fluctuations, difficulty participating in social activity, and in the most severe cases suicide [2, 3]. Previous researches report that the speech of depressed patients typically exhibits whispering, monotonous, slow, slurred, prolonged pauses and stuttering [4]. Fortunately, medical studies show that depression is curable, and early detection of depression is very important to control it at an initial stage [5, 6].

Against this background of the critical need for early detection and monitoring, mental health professionals turn to comprehensive clinical assessment tools supplemented by standardized questionnaires such as the PHQ-8/9 [7, 8] , Beck Depression Inventory [9] , Hamilton Depression Inventory [10]. These tools are complementary to professional depression assessment by measuring and managing disorders related to the mood, emotions, and relationships of an individual.

Recently, many works are being proposed to detect or claasify the impairments that can come to biomarkers like speech signals depending on the psychological or other cognitive conditions of the person [11, 12]. Also, there has been a growing and continuous effort towards addressing the problems and issues of depression detection. Hence major datasets have been created corresponding to surveys and patient responses. For instance, the Audio-Visual Emotion Recognition Challenge series [13–15] have given very key insights into multi modal data analysis for the assessment of depression. Further, audio recordings also help in assessing the mental states, specifically depression in a non-intrusive manner and with the preservation of privacy.

Keeping in view the scarcity of professional working in the field of mental health and the clinicians' subjective assessments, there is an increasing need for approaches that are effectual and scaly for the detection of depression. More than 75% of people in low and middle Nations have no treatment given for depression. Leaving aside the use of state-of-the-art technology and large-scale data sets, we aim to improve the early identification and treatment of depression, hence reducing its impact on society and contributing to the well-being of people all around the world. In this work, we propose a new approach where we have extracted MFCC features from the audio and fed them into Vision Transformer to classify depression from speech data.

The main contributions of this seminal work are as follows:

- This paper proposes a novel Transformer based approach in this paper for depression detection using MFCC features

- The proposed model's performance is evaluated using the DAIC-WOZ dataset. This dataset, which offers labeled audio samples classified as depressed and non-depressed, was created especially for the purpose of detecting depression. Our experimental results show that the suggested system outperforms current state-of-the-art methods by a wide margin.

Rest of the paper is organized as follows. Related literature is presented in Sect. 2. The proposed model is described in detail in Sect. 3, which includes Dataset Description,Data Pre-Processing, MFCC based feature extraction from audio,Experimental results present in Sect. 4 which also includes State of Art approaches. Section 5 concludes the paper by outlining the social impacts of this work and its future scope.

2 Related Works

The literature survey of transformer-based depression detection reviews some of the significant early contributions made by researchers between 2016. Xingchen Ma's [16] study introduced DepAudioNet, a model combining CNN and LSTM layers for audio-based depression classification using the DIAC dataset. This model effectively captures short-term and middle-term correlations using CNNs and long-term correlations with LSTMs. Concurrently, Valstar's [17] research on the RECOLA dataset tested multimodal fusion by using audio and video features with a random forest regressor for predicting levels of depression. Though these models remained unscalable and less robust to different data sets, they achieved early success in capturing speech features and employing a multimodality approach.

As of 2020, Huang and Epps [18] had developed a new approach based on VTC features that include formants, spectral correlation functions, and MFCCs. Epps, based on the DIAC and SH2FS datasets, established the FVTC-dilated CNN that succeeded in capturing long-range dependencies without drastically increasing model parameters. However, with noise conditions, this method posed significant challenges and warranted further improvements in feature extraction techniques. Building on these strides, the Hybrid CNN-SVM classifier [19], also in 2020, attempted to integrate CNNs for feature extraction with SVMs for classification, achieving 68% accuracy on the DAIC-WOZ dataset. However, the model required larger datasets and better parameter tuning for real-world applications.

During 2021, Rejabi [20] fine-tuned the MFCC-based RNN on the DIAC and RAVDESS datasets for depression detection. It further introduced advanced preprocessing techniques, including audio segmentation, speaker separation, and data augmentation techniques such as noise injection, pitch, and speed augmentation, which developed the robustness of the model. However, despite the progress, the research still pointed to the need for larger corpora and higher-level feature extraction. Similarly, a low-cost decision tree model [21] for voice data also improved the depression screening accuracy to 83.4%, while it continued to suffer due to small dataset sizes and simple feature extraction methods, calling for a more holistic approach.

By 2022, Sardari et al. [22] introduced CNN-based autoencoders that were specifically focused on latent feature extraction across datasets such as DIAC, German, Turkish, and Black Dog. This method efficiently distinguished between individuals with and without depression by creating concise latent feature representations. However, it underlined the need for additional data modalities and larger datasets to improve the model's performance. This opened avenues for more robust multimodal approaches, such as fusing audio with visual data for depression relapse prediction [23], which achieved above 80% accuracy. However, it still pointed to a need for improved robustness and data diversity.

In 2023, research explored different methodologies. This began with Gheorghe et al. [24], who used the DIAC and MODMA datasets to test CNN and MLP models. Additional features such as MFCC, MBF, and APF were incorporated to improve depression detection. Building on previous work, Du et al. [25] used LPC and MFCC features along with 1D-CNN and LSTM in his MSCDR model to capture both intra- and inter-segment depressive features. The High-Density EEG and Speech Signal-Based Deep Framework [26] further advanced multimodal feature extraction using vision transformers, achieving a precision-recall of 0.973 on the MODMA dataset. Also in 2023, the Transformer and CNN model [27] advanced temporal feature extraction, combining EEG and speech signals with transformers and LSTMs to detect mild depression with 93.5% accuracy, but it highlighted the need for better generalization across populations. However, the need for more efficient data fusion and improved transfer learning for mild depression detection remained. Meanwhile, Yang et al. [28] proposed the DALF model, which included time-domain filterbanks along with spectral attention modules on the NRAC dataset to extract features from specific frequency sub-bands, thus refining detection accuracy.

Finally, in 2024, many research efforts pushed depression detection to new heights. Sachi Gupta employed the DAttn-Conv 2D LSTM model [29], which combined convolutional layers, attention mechanisms, and LSTMs to capture long-term dependencies within speech data. Simultaneously, Yang et al. [30] integrated visual, audio, and text data using the RLKT-MDD framework, which improved feature extraction and fusion. Das et al. [31], working with the DAIC-WOZ and MODMA datasets, focused on spectrogram extraction through the Spectro-CNN model, which was integrated into the DND-Net framework for binary classification. Gao et al. [32] used a GRU, GAT, and GNN model to capture complex relationships within audio features for MLP prediction. The DepMSTAT model [33] addressed spatio-temporal feature integration, achieving 71.53% accuracy on the D-Vlog dataset. Despite this progress, challenges in real-world performance remained. The Speechformer-CTC model [34] also enhanced temporal speech pattern detection using Connectionist Temporal Classification, achieving an F1 score of 83.15%. While these models boosted detection accuracy, further refinement in pseudo-labeling techniques and the use of ASR data is necessary for future improvements.

The reviewed models have made significant strides in depression detection, from early CNN-LSTM approaches to recent multimodal frameworks that incorporate audio, video, and EEG data. Despite these advancements, challenges such as scalability, robustness, data diversity, and efficient feature extraction remain prevalent.

Vision Transformers (ViT) hold the potential to address these gaps by leveraging their capability to capture global and local dependencies efficiently in large datasets, improving both accuracy and robustness in depression detection tasks.

3 Materials and Methodology

3.1 Dataset Description

The Distress Analysis Interview Corpus (DAIC), [35] developed by J. Gratch and a multi-disciplinary team from USC Institute for Creative Technologies, supports the diagnosis of conditions like anxiety, depression, and PTSD. The DAIC-WOZ dataset includes four types of interviews: face-to-face, teleconference, Wizard-of-Oz (where Ellie is controlled by a human operator), and fully automated interviews. In our work, we utilized the Wizard-of-Oz interviews, where participants interacted with a virtual interviewer, Ellie, controlled by a human. These interviews, involving U.S. veterans and the general public, were recorded using a camera, a high-quality microphone, and a Kinect sensor. The sessions, typically lasting 5–20 min, started with neutral questions, progressed to mental health inquiries, and ended with a cool-down phase. Participants were classified using PHQ-8 scores and binary labels. Almost half of the participants are females, and one-third of the subjects are labeled as depressed. Among the depressed participants, almost half are females. The data is class-imbalanced, with non-depressed participants outnumbering depressed ones 3:1.

3.2 Proposed Model Overview

The proposed pipeline for depression detection from audio data integrates multiple stages, starting from raw data collection to final classification see Fig. 1. The process begins with the acquisition of audio recordings from DAIC-WOZ dataset, where each sample is associated with a unique participant ID and corresponding label indicating depressive or non-depressive states. These raw audio files often contain irrelevant speech, requiring a preprocessing stage to ensure high-quality input data. Preprocessing includes isolating the participant's voice from other speakers. Additionally, the audio is resampled to a consistent sampling rate of 16 kHz to standardize the data across different recording environments.

Once the audio is cleaned, data augmentation techniques such as noise augmentation are applied to enhance dataset diversity. This method simulates real-world variations in background noise, making the dataset more robust to different acoustic conditions. Following augmentation, the audio signals are segmented into smaller

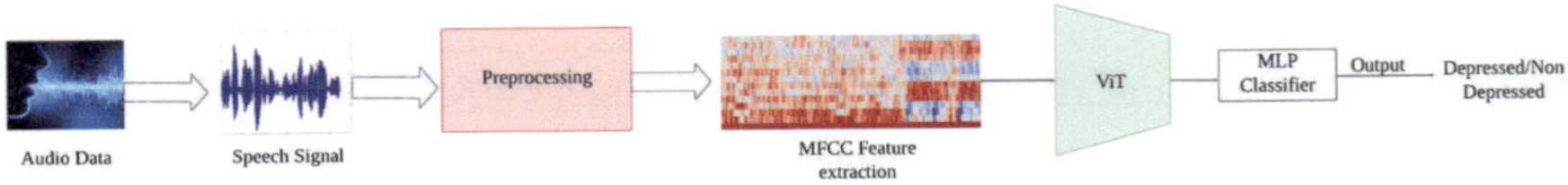

Fig. 1 Proposed methodology framework for depression detection

overlapping windows. This segmentation allows the pipeline to preserve temporal patterns in speech while dividing the data into manageable units for feature extraction.

The segmented audio signals are then processed to extract meaningful features. One key feature, MFCC, is computed from the audio. MFCCs are chosen for their strong ability to capture spectral and temporal patterns that are essential for detecting depressive states. The extracted MFCC features are transformed into visual representations and passed to a Vision Transformer (ViT) model [36]. The ViT processes the input MFCC image patches through its encoder, capturing both local and global dependencies. The output from the ViT is subsequently fed into a Multi-Layer Perceptron (MLP) classifier, which performs the final classification into depressed and non-depressed states. This end-to-end pipeline ensures that raw audio data is systematically processed, transformed, and analyzed, resulting in accurate and reliable depression state predictions.

3.3 Model Intution

The Model fetches MFCC images from the voice data given in the DAIC-WOZ dataset. Since ViT takes the sequence of 1D linear embeddings as input, it divides these 2D MFCC images into patches and then flattens the patches to ensure that it won't miss spatial information of the flattened patches with the inclusion of positional embeddings. These patches are then fed into a transformer encoder as an embedding vector sequence. ViT is composed of alternating layers of multi-head self-attention and MLP blocks with residual connections added after each block, with layer normalization applied before each layer; the final layer is using the softmax function to predict class probabilities. That manages to capture the local as well as the global dependencies within the image patches thereby effectively extracting the relevant features pertaining to classification.

3.4 Data Pre-processing

The raw audio data from the DAIC-WOZ dataset is collected for this study. The dataset is labeled according to depression states (depressed versus non-depressed) and presents unique challenges such as noise, variations in recording environments, and speech characteristics. A block diagram of the preprocessing steps is shown in Fig. 2.

The preprocessing pipeline begins with the removal of agent voices. Next, the cleaned audio is segmented based on speaker diarization, ensuring only the participant's voice is retained. Each segment is normalized and resampled to a fixed rate of 16 kHz to maintain consistency.

Data augmentation techniques, such as noise augmentation, enhance dataset diversity and improve model generalization. By adding background noise to the audio, this method simulates real-world acoustic variations, making the model more robust to diverse speech patterns and environments.

After applying the augmentation technique, the preprocessed audio is segmented into overlapping windows to facilitate feature extraction. This segmentation process ensures that temporal information crucial for depression detection is retained. By dividing the audio into smaller overlapping segments, the model can capture short-term variations in the signal, which are essential for identifying speech-related features indicative of depressive states.

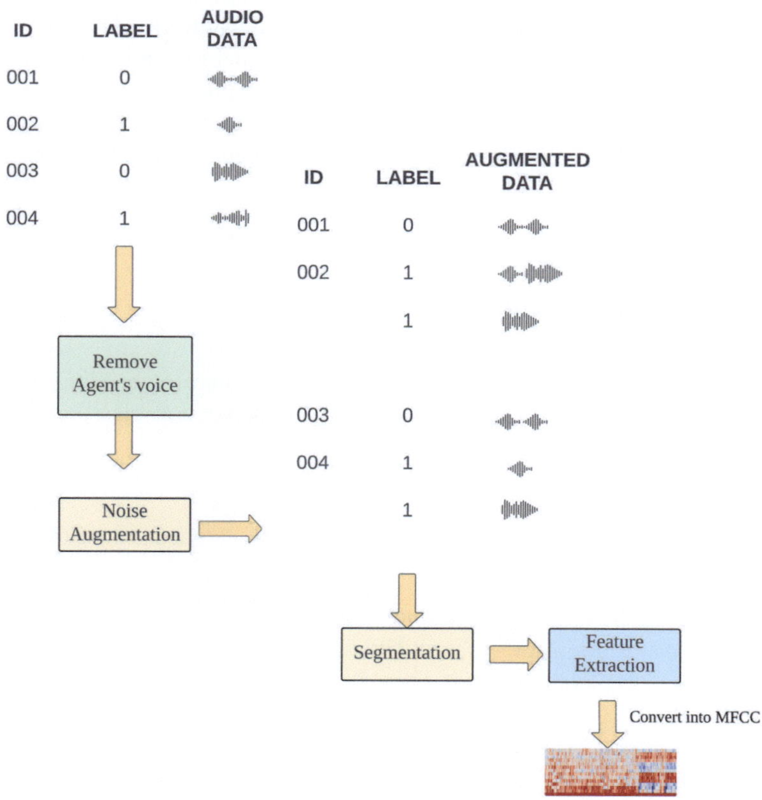

Fig. 2 Block diagram of data preprocessing pipeline. Block diagram of data preprocessing pipeline

3.5 MFCC Feature Extraction

MFCC is recognized as top-performing audio features in tasks such as speech and speaker recognition [37–40]. In this work, we extract MFCCs from raw audio signals as the primary feature representation. The MFCC features are generated by mapping the audio frequencies to the Mel scale, which mimics the human auditory system's sensitivity to different frequencies. This transformation effectively emphasizes low-frequency components that are more perceptually significant for detecting depressive traits in speech. The relationship between frequency in Hz and the Mel scale is given by:

$$Mel(f) = 2595 \cdot \log_{10}\left(1 + \frac{f}{700}\right)$$

The MFCC features capture both spectral and temporal information from the audio, making them suitable for identifying subtle variations in speech that are indicative of depression. These features are then converted into visual representations (MFCC plots) and fed into the Vision Transformer model for subject-level classification. This approach leverages the MFCCs' ability to represent depression-relevant acoustic patterns while preserving the temporal context of speech data extracted from interviews. MFCC plots for depressive and non-depressive patients are shown in Fig. 3.

3.6 Proposed Methodology

We propose a Vision Transformer (ViT) model for feature embedding, followed by a Multi-Layer Perceptron (MLP) classifier for depression classification. The architecture is shown in Fig. 4.

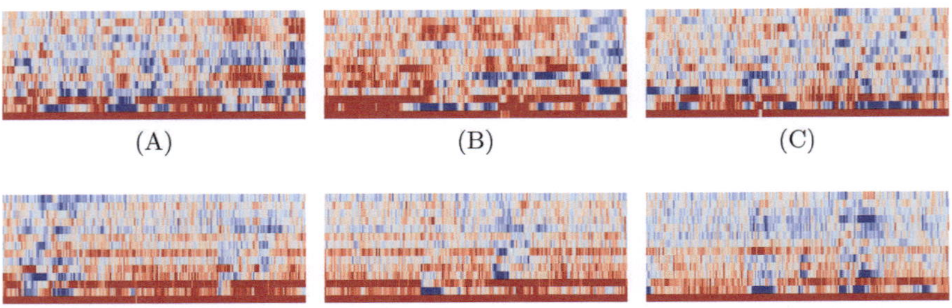

Fig. 3 Generated MFCC for DIAC-WOZ Samples. Depressed **a** id—319, **b** id—320, **c** id—321. Healthy **d** id—303, **e** id—304, **f** id—305

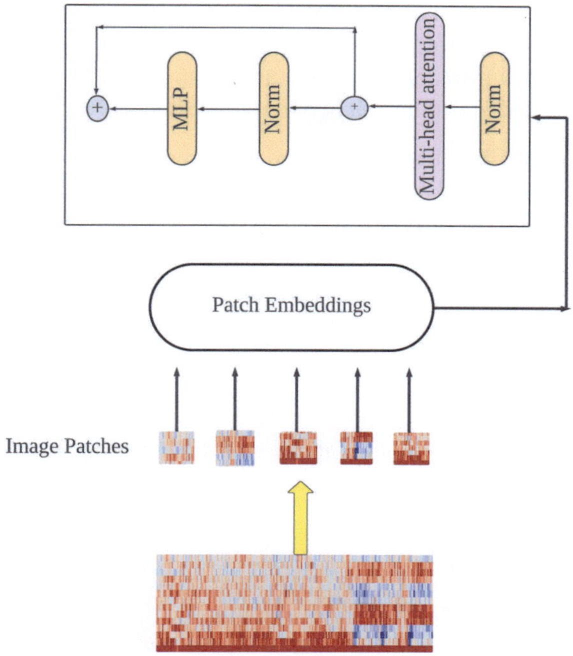

Fig. 4 Vision transformer (ViT) architecture

Vision Transformer (ViT) The Vision Transformer (ViT) model processes input spectrograms by dividing them into fixed-size patches, typically of size 16×16. Each patch is flattened into a one-dimensional vector and passed through a linear embedding layer. This embedding transforms the raw input patches into a suitable representation for the transformer architecture. The embedded patches are then concatenated with positional encodings to retain spatial information, ensuring that the sequential order of the patches is preserved throughout the model.

The core component of the ViT is the transformer encoder, which consists of multiple layers designed to extract meaningful features from the input embeddings. Within each encoder layer, the Multi-Head Attention mechanism plays a pivotal role in capturing both local and global dependencies among the patches. By attending to multiple regions of the spectrogram simultaneously, the Multi-Head Attention mechanism effectively models the interactions between different parts of the input.

Following the attention mechanism, a Feed-Forward Network (FFN) processes the outputs of the attention layer to enhance the feature representations. The FFN comprises two linear transformations separated by a non-linear activation function, allowing the model to learn complex patterns in the data. To ensure stability during training and to improve convergence, normalization layers are applied both before and after the attention and feed-forward operations. These layers normalize the intermediate feature maps, preventing gradient explosion or vanishing issues.

The output of the ViT model is a feature representation that captures both the global context and local patterns within the spectrogram. This representation is well-suited for downstream tasks, such as classification, as it encodes the relevant information needed to distinguish between depressed and non-depressed states.

MLP Classifier The features extracted by the ViT are passed through an MLP classifier. The classifier consists of fully connected layers with ReLU activation and dropout for regularization. The final layer applies the SoftMax function to produce probabilities for two classes:

$$SoftMax(X_i) = \frac{\exp(X_i)}{\sum_j \exp(X_j)}$$

The model is trained for 50 epochs with an initial learning rate of 10^{-5}. The Adam optimizer is used to minimize the cross-entropy loss.

3.7 Threshold

The nature of the interview data poses unique challenges for classification as even depressed subjects may not exhibit depressive traits throughout the interview. Similarly, non-depressed individuals might have moments that can be considered depressive. To address this, we introduced a threshold-based approach to classify the overall interview. A threshold duration of 20% was used to perform the classification of the subject. This value was selected based on extensive experimentation and optimization using the DAIC-WoZdataset.

4 Experimental Results and Discussion

4.1 Experimental Setup

An Intel i9-13900HX processor and a 4080 GPU were used to conduct the experiments.To manage memory limitations, the DAIC-WOZ dataset was split into 12 training batches, while separate batches were created for each speaker during testing. A batch size of 8, an Adam optimizer with a learning rate of 0.001, and a total of 50 epochs were used. To enable effective batch loading and model saving, a 10 epoch checkpoint period was used. To avoid overfitting and cut down on needless computational overhead, early halting with a 10 epoch patience was also used.

Table 1 Results—using threshold

Patient	Accuracy	F1 score	Recall	Precision
Depressed	0.92	0.98	0.91	0.97
Healthy	1.00	0.96	0.92	0.91

4.2 Evaluation Metrics

The performance of the proposed method is evaluated using standard measures such as accuracy, the confusion matrix, and the F1 score, consistent with previous research [41]. Mathematically, the measures are expressed as follows

$$\text{Accuracy} = \frac{TP + TN}{TP + FP + TN + FN} \tag{1}$$

$$\text{Precision} = \frac{TP}{TP + FP} \tag{2}$$

$$\text{Recall} = \frac{TP}{TP + FN} \tag{3}$$

$$F_1 \text{ Score} = \frac{2 \times \text{Precision} \times \text{Recall}}{\text{Precision} + \text{Recall}} \tag{4}$$

Here, TP and FP corresponds to True and false positive samples, where as TN or FN corresponds to True and False negative samples respectively.

4.3 Results of Performance Evaluation

As shown in Table 1, the results of the classification task using the DAIC dataset are promising. For the depressed patient class, the proposed method achieved an accuracy of 0.92, with a high F1 score of 0.98, indicating a strong balance between precision and recall. The recall for depressed patients was 0.91, meaning 91% of actual depressed cases were correctly identified. Moreover, the precision was 0.97, suggesting few false positives. In contrast, the healthy patient class achieved perfect accuracy, with an F1 score of 0.96. In this case, the recall was slightly higher at 0.92, indicating the model's ability to identify healthy cases effectively. However, the precision was observed to be 0.91, showing a marginal increase in false positives compared to the depressed class. In a nutshell, these results reflect a strong overall performance, with the system effectively distinguishing between the two classes. Further, these results confirm MFCC's power as a meaningful feature for understanding mental health through audio data. However, the higher accuracy and slightly better recall for healthy patients may point to class imbalance or feature dominance.

The Epochs versus Accuracy curve demonstrates the model's progressive learning behavior across training iterations see Fig. 5. It is observed that the application of a decision threshold during the testing phase yields consistently higher and more stable accuracy values, typically exceeding 90%. This emphasizes the role of thresholding in enhancing the model's generalization capability on unseen data, despite not being employed during training.

4.4 Performance Comparison with Other Methods

We compare the proposed method with eight state-of-the-art methods (discussed in Sect. 2) as shown in Table 2. The list of methods compared are given below.

- Multimodal fusion model for depression detection [17]
- Deep model for audio-based depression classification [16]
- Dilated CNN based vocal tract coordination exploitation for depression detection [18]
- MFCC based RNN for depression recognition (Rejaibi et al. 2022)
- Robust deep CNN based affect and depression recognition [20]
- Speaker embeddings and temporal context-based depression detection (Dumpala et al. 2021)
- Speech chain model fusing speech production and perception features [25]

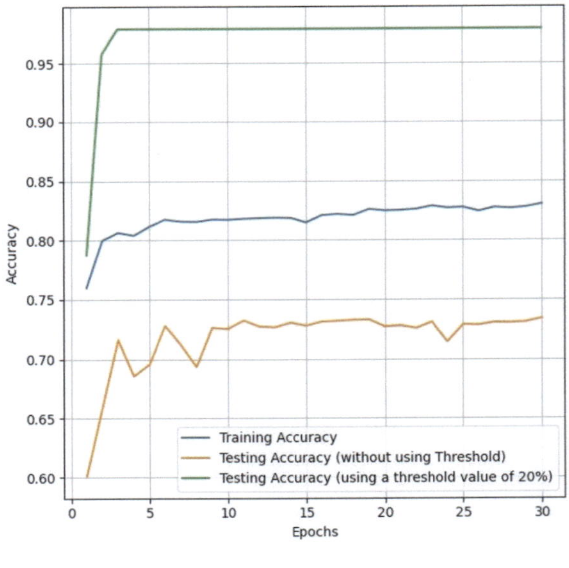

Fig. 5 Epoch versus accuracy

Table 2 Comparing with the state of art approaches

Model	Features	Related work	F1 score
SVM	F0, VUV, MFCC	Valstar et al. [17]	0.41/0.58
CNN-LSTM	Mel-Spec	Ma et al. [16]	0.52/0.70
FVTC-CNN	MFCC	Huang and Epps [18]	0.40/0.84
RNN	MFCC	Rejabi et al. (2022)	0.46/0.85
CNN	MFCC, Mel-Spec	Othmani et al. [20]	0.49/0.82
CNN-LSTM	LPC, MFCC	Du et al. [25]	0.67/0.83
MLP	MFCC	Gheorghe et al. [24]	0.40/0.84
CNN-ANN	MFCC, Mel-Spec	Arnab et al. (2024)	0.90/0.89
ViT	MFCC	Proposed methodology	0.98/0.96

- Depression screening and early warning based on voice data [21]
- Multilayer perceptron model or 1D CNN model from spectrogram [24]
- Graph neural network from audio-based signal (Chen et al. 2023).

The proposed ViT model achieves the highest F1 scores of 0.98 and 0.96, outperforming all the state-of-the-art methods. Confusion matrix for the proposed shown in Fig. 6. Traditional models like SVM [17], utilizing features such as F0, VUV, and MFCC, achieve lower F1 scores of 0.41/0.58. This shows the method's limited capacity for capturing complex dependencies in speech data. Among deep learning approaches, CNN-LSTM [25] with LPC and MFCC features attains an F1 score of 0.67/0.83, while CNN-ANN (Arnab et al. 2024) with MFCC and Mel-Spec features shows better performance (0.90/0.89) but still falls short of the proposed approach.

Fig. 6 Confusion matrix

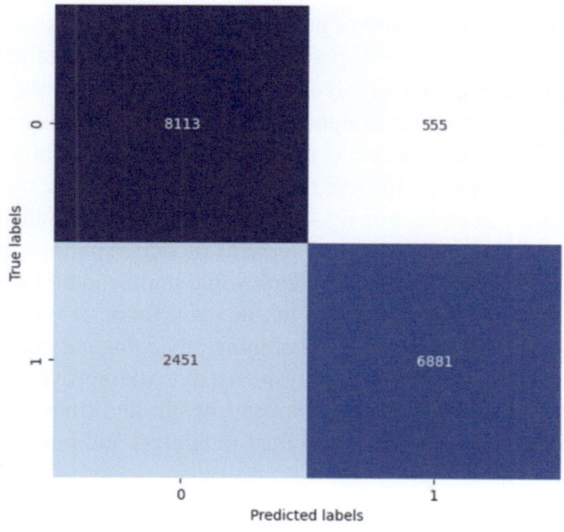

Table 3 Results—without threshold

Patient	Accuracy	F1 score	Recall	Precision
Depressed	0.77	0.77	1.00	0.63
Healthy	0.57	0.57	0.40	1.00

4.5 Results on DIAC Dataset Without Threshold

In the absence of threshold, the results of the DIAC dataset for subject-level classification present huge imbalances as shown in Table 3. For depressed subjects, it achieves a perfect recall of 1.00 but at the expense of a lower precision of 0.63, showing that there is a high false-positive rate. For healthy subjects, recall drops to 0.40 while precision is perfect at 1.00, implying that many healthy subjects are misclassified due to overprediction of depression in individual MFCC plots. The lack of a threshold mechanism exacerbates the model's failure to take into account the temporal nature of depressive traits over an entire interview. This underscores the importance of the threshold in capturing the intermittent nature of depressive behaviors and balance across classes for predictions.

5 Conclusion and Future Work

The advancement of human-centric Cyber-Physical Systems hinges on their ability to perceive and respond to nuanced human states. This study represents a significant step in this direction, offering a powerful new modality for passive mental state assessment. By innovatively repurposing a Vision Transformer to analyze MFCC spectrograms of human speech, we have shown it is possible to decode subtle, depression-indicative patterns from raw vocal data with remarkable accuracy. This achievement moves beyond the limitations of episodic clinical visits, presenting a pathway toward continuous and objective monitoring within everyday environments. The core contribution is not merely an improved algorithm, but a proof-of-concept for embedding affective computing into the fabric of CPS. This heralds a future of intelligent environments—from smart homes to telehealth platforms—that can sense emotional well-being and serve as proactive partners in promoting mental health.

The development of a deployable multi-modal Cyber-Physical System for mental health monitoring will be the main focus of future research. For thorough state modeling, this involves combining physiological inputs from wearables with video-based face expression analysis. Analytical improvements will use longitudinal analysis to monitor trends in mental health and transfer learning to increase generalization. Transparency and trust in clinical and human-in-the-loop applications will be made possible by prioritizing privacy-preserving strategies like federated learning and enhancing model interpretability through attention visualization in order to facilitate real-world deployment.

Acknowledgements This work was supported by Amrita Vishwa Vidyapeetham (AMRITA) Seed under Grant ASG2022118.

References

1. Mathers CD, Loncar D (2006) Projections of global mortality and burden of disease from 2002 to 2030. PLoS Med 3(11):e442
2. World Health Organization (2017) Depression and other common mental disorders: global health estimates, Technical documents
3. Jiang H, Hu B, Liu Z, Yan L, Wang T, Liu F, Kang H, Li X (2017) Investigation of different speech types and emotions for detecting depression using different classifiers. Speech Commun 90:39–46
4. Williamson JR, Godoy E, Cha M, Schwarzentruber A, Khorrami P, Gwon Y, Kung H-T, Dagli C, Quatieri TF (2016) Detecting depression using vocal, facial and semantic communication cues. In: Proceedings of the 6th International workshop on audio/visual emotion challenge, pp 11–18
5. Lejuez CW, Hopko DR, LePage JP, Hopko SD, McNeil DW (2001) A brief behavioral activation treatment for depression. Cogn Behav Pract 8(2):164–175
6. National Institute for Clinical Excellence et al (2009) Depression: the treatment and management of depression in adults (update). Clinical guidelines CG90
7. Kroenke K, Strine TW, Spitzer RL, Williams JB, Berry JT, Mokdad AH (2009) The phq-8 as a measure of current depression in the general population. J Affect Disord 114(1–3):163–173
8. Kroenke K, Spitzer RL (2002) The phq-9: a new depression diagnostic and severity measure, pp 509–515
9. Upton J (2013) Beck Depression Inventory (BDI). Springer, New York, pp 178–179 (Online). Available: https://doi.org/10.1007/978-1-4419-1005-9_441
10. Endicott J, Cohen J, Nee J, Fleiss J, Sarantakos S (1981) Hamilton depression rating scale: Extracted from regular and change versions of the schedule for affective disorders and schizophrenia. Arch Gen Psychiatry 38(1):98–103
11. Saran MDSR, Akshayaa B, Raghavendra RS, Poornima N, Lal GJ (2024) Accurate anomia severity detection in post-stroke aphasia patients using mobilenetv2. In: 2024 15th International conference on computing communication and networking technologies (ICCCNT). IEEE, USA, pp 1–7
12. Raj KD, Lal GJ, Gopalakrishnan E, Sowmya V, Orozco-Arroyave JR (2024) A visibility graph approach for multi-stage classification of Parkinson's disease using multimodal data. IEEE Access 12:87077–87096
13. Valstar M, Schuller B, Smith K, Eyben F, Jiang B, Bilakhia S, Schnieder S, Cowie R, Pantic M (2013) Avec 2013: the continuous audio/visual emotion and depression recognition challenge. In: Proceedings of the 3rd ACM international workshop on Audio/visual emotion challenge, pp 3–10
14. Ringeval F, Schuller B, Valstar M, Cummins N, Cowie R, Tavabi L, Schmitt M, Alisamir S, Amiriparian S, Messner E-M (2019) et al (2019) Avec 2019 workshop and challenge: state-of-mind, detecting depression with AI, and cross-cultural affect recognition. In: Proceedings of the 9th International on audio/visual emotion challenge and workshop, pp 3–12

15. Ringeval F, Schuller B, Valstar M, Gratch J, Cowie R, Scherer S, Mozgai S, Cummins N, Schmitt M, Pantic M (2017) Avec 2017: real-life depression, and affect recognition workshop and challenge. In: Proceedings of the 7th annual workshop on audio/visual emotion challenge, pp 3–9
16. Ma X, Yang H, Chen Q, Huang D, Wang Y (2016) Depaudionet: an efficient deep model for audio based depression classification. In: Proceedings of the 6th international workshop on audio/visual emotion challenge, pp 35–42
17. Valstar M, Gratch J, Schuller B, Ringeval F, Lalanne D, Torres Torres M, Scherer S, Stratou G, Cowie R, Pantic M (2016) Avec 2016: Depression, mood, and emotion recognition workshop and challenge. In: Proceedings of the 6th international workshop on audio/visual emotion challenge, pp 3–10
18. Huang Z, Epps J, Joachim D (2020) Exploiting vocal tract coordination using dilated CNNs for depression detection in naturalistic environments. In: ICASSP 2020-2020 IEEE International conference on acoustics, speech and signal processing (ICASSP). IEEE, USA, pp 6549–6553
19. Saidi A, Othman SB, Saoud SB (2020) (2020) Hybrid CNN-SVM classifier for efficient depression detection system. In: 2020 4th International conference on advanced systems and emergent technologies (IC_ASET). IEEE, USA, pp 229–234
20. Othmani A, Kadoch D, Bentounes K, Rejaibi E, Alfred R, Hadid A (2021) Towards robust deep neural networks for affect and depression recognition from speech. In: Recognition P (ed) ICPR International Workshops and Challenges: Virtual Event, 10–15 Jan 2021. Part II. Springer, Proceedings, pp 5–19
21. Chen X, Pan Z (2021) A convenient and low-cost model of depression screening and early warning based on voice data using for public mental health. Int J Environ Res Public Health 18(12):6441
22. Sardari S, Nakisa B, Rastgoo MN, Eklund P (2022) Audio based depression detection using convolutional autoencoder. Expert Syst Appl 189:116076
23. Othmani A, Zeghina AO (2022) A multimodal computer-aided diagnostic system for depression relapse prediction using audiovisual cues: a proof of concept. Healthc. Anal. 2:100090
24. Gheorghe M, Mihalache S, Burileanu D (2023) Using deep neural networks for detecting depression from speech. In: 31st European signal processing conference (EUSIPCO). IEEE, USA, pp 411–415
25. Du M, Liu S, Wang T, Zhang W, Ke Y, Chen L, Ming D (2023) Depression recognition using a proposed speech chain model fusing speech production and perception features. J Affect Disord 323:299–308
26. Qayyum A, Razzak I, Tanveer M, Mazher M, Alhaqbani B (2023) High-density electroencephalography and speech signal based deep framework for clinical depression diagnosis. IEEE/ACM Trans Comput Biol Bioinf 20(4):2587–2597
27. Yin F, Du J, Xu X, Zhao L (2023) Depression detection in speech using transformer and parallel convolutional neural networks. Electronics 12(2):328
28. Yang W, Liu J, Cao P, Zhu R, Wang Y, Liu JK, Wang F, Zhang X (2023) Attention guided learnable time-domain filterbanks for speech depression detection. Neural Netw 165:135–149
29. Gupta S, Agarwal G, Agarwal S, Pandey D (2024) Depression detection using cascaded attention based deep learning framework using speech data. Multimedia Tools Appl 1–39
30. Yang S, Cui L, Wang L, Wang T, You J (2024) Enhancing multimodal depression diagnosis through representation learning and knowledge transfer. Heliyon 10(4)
31. Das AK, Naskar R (2024) A deep learning model for depression detection based on mfcc and cnn generated spectrogram features. Biomed Signal Process Control 90:105898
32. Sun C, Jiang M, Gao L, Xin Y, Dong Y (2024) A novel study for depression detecting using audio signals based on graph neural network. Biomed Signal Process Control 88:105675
33. Tao Y, Yang M, Li H, Wu Y, Hu B (2024) Depmstat: Multimodal spatio-temporal attentional transformer for depression detection. IEEE Trans Knowl Data Eng
34. Wang J, Ravi V, Flint J, Alwan A (2024) Speechformer-ctc: Sequential modeling of depression detection with speech temporal classification. Speech Commun 163:103106

35. Gratch J, Artstein R, Lucas GM, Stratou G, Scherer S, Nazarian A, Wood R, Boberg J, DeVault D, Marsella S et al (2014) The distress analysis interview corpus of human and computer interviews. In: LREC, Reykjavik, pp 3123–3128
36. Alexey D (2020) An image is worth 16x16 words: transformers for image recognition at scale. arXiv preprint arXiv: 2010.11929
37. Cummins N, Epps J, Breakspear M, Goecke R (2011) An investigation of depressed speech detection: Features and normalization. In: INTERSPEECH 2011 12th Annual conference of the international speech communication association. Int Speech Commun Assoc 2997–3000
38. Low LSA, Maddage NC, Lech M, Allen N (2009) Mel frequency cepstral feature and gaussian mixtures for modeling clinical depression in adolescents. In: 2009 8th IEEE International conference on cognitive informatics. IEEE, USA, pp 346–350
39. Cutajar M, Gatt E, Grech I, Casha O, Micallef J (2013) Comparative study of automatic speech recognition techniques. IET Signal Proc 7(1):25–46
40. Lopez-Otero P, Dacia-Fernandez L, Garcia-Mateo C (2014) A study of acoustic features for depression detection. In: 2nd International workshop on biometrics and forensics. IEEE, USA, pp 1–6
41. Zhao Y, Liang Z, Du J, Zhang L, Liu C, Zhao L (2021) Multi-head attention-based long short-term memory for depression detection from speech. Front Neurorobot 15:684037

A Distributed Approach Based on Catboost, BlockChain and Edge Computing for IoT Security

Maryam Douiba, Chaimae Hazman, Azidine Guezzaz, Said Benkirane, Mourade Azrour, and B. J. Bipin Nair

Abstract Security has become a significant challenge, especially among the Internet of Things (IoT) that expanded rapidly in addition to generated vast amounts of data. Various connected devices, communication protocols, operating systems and hardware configurations increase the risk of cyber threats, which can compromise data confidentiality, integrity, and availability. Additionally, it makes attack detection more challenging, as cloud computing has to deal with the high latency of IoT applications which can make real-time processing difficult. Edge computing is the solution that solves these problems by processing data at local nodes and improving quality of service (QoS). As a result, real-time protection by integrating artificial intelligence (AI) and BlockChain technologies is essential to strengthening security because AI enhances attack detection through learning from data patterns. At the same time, BlockChain secures data exchanges and ensures integrity in a decentralized manner. In this context, we propose a distributed approach forsecuring IoT using BlockChain and Catboost in Edge computing; we explore how AI and BlockChain can secure IoT services in distributed Edge Computing.

Keywords IoT security · BlockChain · ML · Catboost · Edge computing · Ethereum · Smart contract

---

M. Douiba · C. Hazman · A. Guezzaz (✉) · S. Benkirane
SISAR Team, LaRTiD Laboratory, Technology Higher School Essaouira, Cadi Ayyad University, Marrakech, Morocco
e-mail: a.guzzaz@gmail.com

C. Hazman
e-mail: C.Hazman@emsi.ma

C. Hazman
Research Team LAMIGE, EMSI-Marrakech, Marrakech, Morocco

B. J. Bipin Nair
Department of Computer Science, Amrita School of Computing, Amrita Vishwa Vidyapeetham, Mysuru, India
e-mail: bj_bipinnair@my.amrita.edu

M. Azrour
Faculty of Sciences and Techniques, Moulay Ismail University of Meknes, Errachidia, Morocco

G. H. L. et al. (eds.), *Reliability in Cyber-Physical Systems: The Human Factor Perspective*, Springer Series in Reliability Engineering,
https://doi.org/10.1007/978-3-032-09917-4_5

1 Introduction

The IoT is predicated on a vast network of intelligent objects equipped with actuators, sensors and processors, interacting via various communication protocols. Choosing a particular protocol depends on the constraints imposed by the specific IoT application, particularly with regard to factors such as security, latency, bandwidth and energy consumption [1, 2]. Moreover, IoT has become a vital part of our modern world because it is key to collecting and analyzing data in real-time, offering significant benefits in efficiency, management, and decision-making. Its networked infrastructure connects a wide variety of smart devices across different sectors, including healthcare, industry, agriculture, and smart cities; this engenders an elevated probability of cyberattacks, which can impact the integrity, confidentiality and availability of sensitive data. This diversity poses significant challenges to network security, necessitating the development of innovative solutions adapted to the constraints of distributed environments for secure data exchanges especially when IoT devices are operating in an open and insecure environments. So, It is imperative to ensure the security of IoT networks to facilitate their effective operation [2, 3].

On the other hand, Edge computing is represented as a transformative technology in the field of IoT that offers services in real time through local data processing; This innovation involves directly processing data at its source, such as routers, surveillance cameras and devices to significantly reduce latency. Hence, The notable benefit of edge computing is its about the their potentials to enhance data security [1–3]. The transmission of data to remote servers is a practice that should be avoided in order to mitigate the risks of interception and information theft. Given the potentially catastrophic consequences of any data leakage the significance is especially pronounced in the context of critical infrastructures, including but not limited to power grids, hospitals and financial institutions [1–4].

To combat the challenges of IoT security, we propose an innovative approach to detecting and responding to cyberattacks on the Edge. This approach is built on a collaborative distributed intrusion detection system (IDS) deployed on the Edge, which uses Blockchain to guarantee reliability and trust between interconnected EdgeRaspberry Pi within a private network. Integration of blockchain in business and it is widely used in many other domain in recent days [5, 6]. The core of this model is an advanced IDS that employs binary intrusion classification implemented via the CatBoost library ensuring 99.2% accuracy and efficiency in detecting normal traffic and identifying any anomalous traffic as an attack. Furthermore, a hybrid smart contract is integrated into the BlockChain to secure the import of intrusion data, automate the execution of security functions, and reduce latency in decision-making. This study represents a substantial advancement by investigating the efficacy of collaborative IDS, the potential of BlockChain technology and the capabilities of machine learning algorithms in edge computing.

We structuredour paper as follows: Sect. 2 presents a related work about collaborative IDS using BlockChain and machine learning (ML) in IoT networks. The subsequent section, Sect. 3, outlines the proposed design, emphasizing its importance in addressing the pressing security challenges of IoT. The results of the experiment are discussed in Sect. 4.

2 Related Work

Because of the rise of complex cyber threats, researchers explored BlockChain as a potential solution in order to enhance security in collaborative IDS; BlockChain has led to numerous studies proposing innovative strategies that could significantly bolster data integrity, security and trust.

Selvarajan et al. [7, 8] created a BlockChain-based security model for industrial IoT systems. The model proposed by Selvarajan focuses on privacy protection and system efficiency using a simple proof-of-work consensus mechanism and a machine learning-based neural network for attack classification. Their system is secure, moreover it doesn't need a lot of computing power. Similarly, Truong et al. [2] introduced Meta Collaborative IDS; we talk here about a decentralized BlockChain-driven IDS that integrates two detection modules, first a multi-class classifier that learns from labeled-data with an attention mechanism then an anomaly detection module that uses a deep auto-encoder to find zero day attacks. The experiments of Truong et al. [2, 9] showed a detection accuracy between 96 and 99% across four different network intrusion datasets.

Alkadi et al. [10] developed a distributed intrusion detection system that uses BlockChain and smart contracts to strengthen IoT security and privacy. To enhance the accuracy of anomaly detection Alkadi et al. [10, 11] incorporates deep learning specifically a bi-directional LSTM algorithm. In a similar manner, Hu et al. [12] presented a decentralized multimodal gateway system (MMG) that boosts detection efficiency via BlockChain's consensus mechanisms and correlation-based anomaly detection techniques.

Aliyu et al. [1] proposed a federated learning BlockChain-based model for IDS in networks powered by SDN. Their system utilizes the Ethereum BlockChain along with Mininet simulation to enable manufacturers and end-users to engage in model training while safeguarding sensitive information. Storing model hashes on the BlockChain minimizes the risk of adversarial attacks and removes individual points of failure.

Madhusekhar et al. [13] investigated how effective BlockChain-integrated intrusion detection is during network blackout situations, utilizing Hyperledger Fabric to evaluate security performance. Their analysis focused on execution time, latency, throughput, and the efficiency of transaction processing using the UNSW NB15 dataset. Liang et al. [4, 14] investigated data clustering with efficiency and within BlockChain networks, they found that merely reducing clustering features did not adequately support dependable anomaly detection. To address this issue, they suggested a collaborative IDS that operates on the BlockChain, which systematically enhances attack detection patterns through ongoing competition among cluster nodes, ultimately producing weighted coefficients to improve classification.

3 Proposed Design

Developing innovative solutions for identifying and addressing cyber threats imperative faces to IoT increasing security challenges, a particularly encouraging approach involves the incorporation of technologies such as BlockChain, Edge and artificial intelligence. This integration has the potential to enhance the security of IoT infrastructures while optimizing system costs and responsiveness.

The proposed approach features a collaborative IDS deployed directly at the Edge, where devices like Raspberry Pi connect within a private network to work together in identifying suspicious activity. BlockChain technology is essential in this architecture because it provide dependable communication among devices and enhancing trust in the detection process. In intrusion detection, a new IDS model has been created, utilizing advanced binary classification techniques. We utilized the Gradient Boosted Decision Trees (GBDT) algorithm through the CatBoost open-source library to enables effective classification of normal traffic. Our model is enhanced by incorporating a hybrid smart contract that automatically uploads intrusion data to the BlockChain, performs security functions independently, and consequently minimizes decision-making delays. At present, the effectiveness of this approach is evaluated based on various criteria such as detection accuracy, response time, and energy efficiency, particularly with Raspberry Pi devices.

The proposed approach integrates collaborative IDS, BlockChain's reliability, and machine learning algorithms' potency to furnish a robust solution for securing IoT networks against emerging threats; This approach will improve ability to detect and respond to attacks, optimize system performance, and reduce the financial impact of cyberattacks on operational costs. The objective is to enhance the resilience of IoT infrastructures by ensuring proactive IDS, and rapid response to cyberattacks while guaranteeing optimal service continuity.

Integrating these technologies into an Edge environment has resulted in a highly effective and scalable approach to meeting the demands of contemporary IoT networks. Figure 1 offers a Visual representation of the process mentioned above:

4 Results and Discussions

We present an evaluation of the efficacy of our approach. The evaluation is conducted, and the results of implementing our system on six nodes are presented in this section.

The proposed model has been implemented on the private network of the Ethereum BlockChain. The implementation of our model necessitated the utilization of Geth,

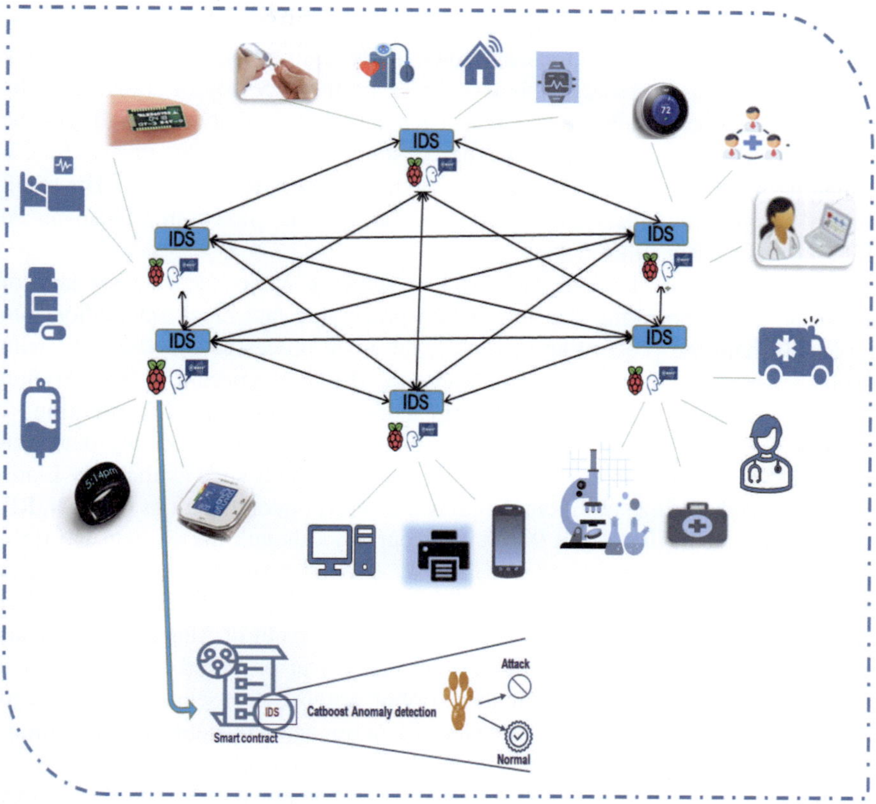

Fig. 1 Proposed smart and secure distributed intrusion detection architecture

Solidity, and Go, which are deemed essential components. This implementation was used to deploy the smart contract and operate the private BlockChain network. The private BlockChain network consists of six nodes, all of which are designated as proposing authorities. The model test was executed for a duration of 24 h, during which the following results were obtained on RPi: the various metrics generated by the nodes that were monitored using Grafana and matplotlib. Collaborative IDS, a BlockChain-based system, facilitates the establishment of connections and the exchange of data among nodes, including alerts. This is achieved through an automatic script where each node tracks its partners over a predefined period. The CICIOT2023 dataset was utilized for the model test, as illustrated in the subsequent Table 1.

Table 1 CICIOT2023 dataset description

Dataset	Attacks traffic samples (%)	Normal traffic samples (%)	Total samples
CIC-IoT-2023	61.7	38.3	963,773

Although the binary detection approach utilizing GBDT with CatBoost exhibited superior performance metrics, as shown in Fig. 2, imposing a minimum sample size constraint was necessary. This was driven by the necessity of adhering to the processing time constraints inherent to each transaction. Our tests highlight the impressive effectiveness of the model, demonstrating an average accuracy of 99.2% in identifying anomalies during each transaction on every Raspberry Pi; The result attained through the binary classification of transactions, showcasing its advanced ability to detect anomalous behavior.

Figure 3 is about the Geth dashboard. It shows a stable linear progression of block additions, indicating that transaction processing and block integration into the BlockChain are proceeding smoothly, the node processed block 26,712 with associated headers, receipts, and transactions. The node appears to be operating efficiently, continuously processing blocks and transactions without interruption.

Table 2 presents a comparison of machine learning-based anomaly detection approaches in blockchain environments. Previous methods such as those by Liang and Madhusekharutilize traditional models like SVM or combinations of CNN, RF, and XG-Boostachieving about 97.7% of accuracy, although the consensus mechanisms used are often unspecified. Hu et al. employ a simpler model One-class SVMwith a Delegated Proof of Stake consensus reaching 96.3% of accuracy. The ALIYU approach, based on Federated Forest and the Proof of Authority consensus on Ethereum, achieves a notable 98.1% accuracy. Our proposed approach which combines GBDT via CatBoost with the Proof of Authority consensus on Ethereumreaching 99.2% accuracy. This demonstrates superior anomaly detection capabilities and enhanced effectiveness in blockchain-based environments.

Finally we conclude that our proposal on Ethereum outperformed all other proposals and stands out by achieving an accuracy rate of 99.2%, as shown in Table 2, this superior accuracy can be attributed to the use of an advanced IDS based on GBDT, which is known for its ability to build highly accurate models, as we have previously demonstrated in our publications [3]. Moreover, choosing Ethereum with a proof-of-authority consensus for the BlockChain environment enhances model performance.

Fig. 2 Training anomaly detection results

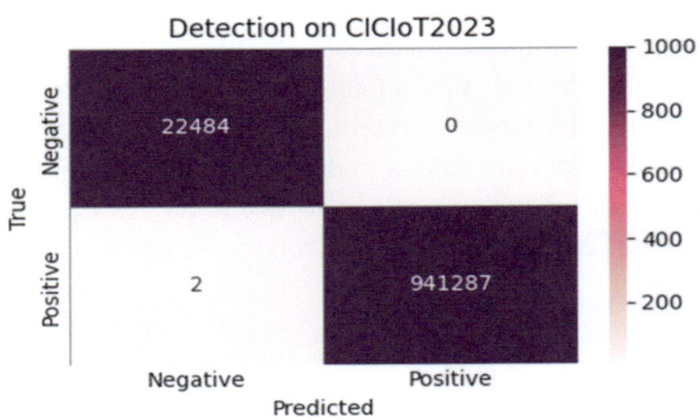

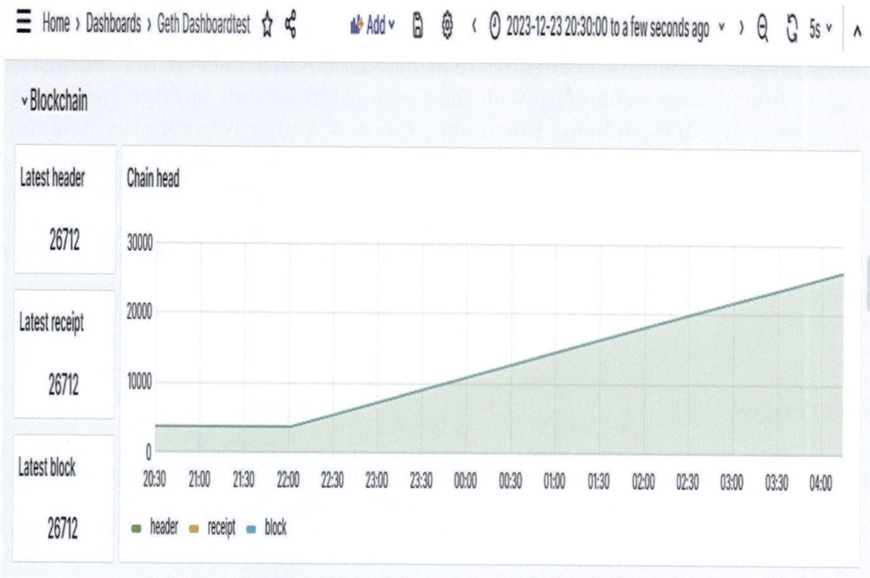

Fig. 3 An exposition on the functional dynamics of blockchain

Table 2 A comparison of distrusted collaborative IDS

	ML	Consensus	BlockChain	Accuracy %
Liang et al. [4]	SVM	Not specified	Hyperledger fabric	≅ 90
Hu et al. [12]	One-class SVM	Delegated proof of stake	Simulation	96.3
Madhusekhar et al. [13]	CNN, RF, and XG-boost	Not specified	Hyperledger fabric	97.7
Aliyu et al. [1]	Federated Forest	Proof of authority	Ethereum	98.1
Our approach	GBDT via Catboost	Proof of authority	Ethereum	99.2

To summarize, the effectiveness of the proposed model is showcased by its higher intrusion detection accuracy compared to all other models shown in this table, thereby confirming its robustness and efficiency in a BlockChain setting.

5 Conclusion and Future Work

This document introduced an effective distributed method to encrypt IoT settings by merging BlockChain technology with the CatBoost algorithm in Edge computing. Our collaborative IDS, utilizing Ethereum's Proof of Authority consensus, achieved an impressive accuracy of 99.2%, greatly surpassing current solutions. This success underscores the power of integrating BlockChain's dependability with cutting-edge machine learning techniques. Looking ahead, we intend to enhance our approach by addressing scalability challenges and assessing our model in broader, diverse IoT networks while evaluating its performance across different network conditions.

References

1. Huong TT, Bac TP, Long DM, Thang BD, Binh NT, Luong TD, Phuc TK (2020) LocKedge: low-complexity cyberattack detection in IoT edge computing
2. Spadaccino P, Cuomo F (2020) Intrusion detection systems for IoT: opportunities and challenges offered by edge computing and machine learning. arXiv preprint arXiv:2012.01174. https://arxiv.org/abs/2012.01174
3. Hosseininoorbin S, Layeghy S, Sarhan M, Jurdak R, Portmann M (2021) Exploring Edge TPU for network intrusion detection in IoT. arXiv preprint arXiv:2103.16295. https://arxiv.org/abs/2103.16295
4. Hafeez I, Antikainen M, Ding AY, Tarkoma S (2018) IoT-KEEPER: securing IoT communications in edge networks. arXiv preprint arXiv:1810.08415. https://arxiv.org/abs/1810.08415
5. Matayir F, Susanto H, Subramanian U (2025) Managing risks: Blockchain-based security of higher education business processes. In: Digital education. Apple Academic Press, pp 35–71
6. Subramanian U, Rahman NENA (2024) An exploratory study on the use of adopting blockchain technology in accounting. Int J Bus Innov Res 35(1):87–104
7. Douiba M, Benkirane S, Guezzaz A, Azrour M (2023) An improved anomaly detection model for IoT security using decision tree and gradient boosting. J Supercomput 79:3392–3411. https://doi.org/10.1007/s11227-022-04884-6
8. Selvarajan S, Srivastava G, Khadidos AO (2023) An artificial intelligence lightweight BlockChain security model for security and privacy in IIoT systems. J Cloud Comput 12(1):38. https://doi.org/10.1186/s13677-023-00378-3
9. Truong VT, Le LB (2023) MetaCIDS: privacy-preserving collaborative intrusion detection for metaverse based on BlockChain and online federated learning. IEEE Open J Comput Soc 4:253–266. https://doi.org/10.1109/OJCS.2023.3240329
10. Aliyu I, Feliciano MC, Van Engelenburg S, Kim DO, Lim CG (2021) A BlockChain-based federated forest for SDN-enabled in-vehicle network intrusion detection system. IEEE Access 9:102593–102608. https://doi.org/10.1109/ACCESS.2021.3096965
11. Alkadi O, Moustafa N, Turnbull B, Choo K-KR (2021) A deep BlockChain framework-enabled collaborative intrusion detection for protecting IoT and cloud networks. IEEE Int Things J 8(12):9463–9472. https://doi.org/10.1109/JIOT.2020.3049022
12. Madhusekhar Y, Priyanka S, Mandru DB, Srikanth T (2023) BlockChain: a safe way to transfer signatures in a distributed intrusion detection system. In: Manchuri AR, Marla D, Rao VV (eds) Intelligent manufacturing and energy sustainability, vol 334. Springer. https://doi.org/10.1007/978-981-19-8742-3_6

13. Hu B, Zhou C, Tian Y, Qin Y, Junping X (2019) A collaborative intrusion detection approach using BlockChain for multi-microgrid systems. IEEE Trans Syst Man Cybern Syst 49(8):1720–1730. https://doi.org/10.1109/TSMC.2018.2840321
14. Liang W, Xiao L, Zhang K, Tang M, He D, Li K-C (2021) Data fusion approach for collaborative anomaly intrusion detection in BlockChain-based systems. IEEE Int Things J. https://doi.org/10.1109/JIOT.2021.3082291

An Improved Anomaly Detection Based on Ensemble Learning and Deep Q-Network for Mobile Edge Computing Monitoring

Hafida Assmi, Said Jabbour, Azidine Guezzaz, B. J. Bipin Nair, Said Benkirane, and Mourade Azrour

Abstract Mobile edge computing (MEC) improves the performance of real-time services by bringing computing resources closer to users, thus reducing latency and optimizing network utilization. Task offloading plays a key role in this architecture, intelligently distributing loads between edge devices and servers, guaranteeing efficient use of resources and improved energy efficiency. This decentralization exposes edge infrastructures to increased cyber threats, making it essential to integrate intrusion detection systems (IDS) to guarantee the security and reliability of MEC environments. In this paper, we present an IDS model that uses the Deep Q-Network (DQN) algorithm with the ensemble learning method, specifically the voting classifier that combines three algorithms Random Forest (RF), Support Vector Machine (SVM) and Gradient Boosting (GB) on the two datasets Shanghai and Guangzhou. The model gave an accuracy between 67 and 99%.

Keywords Anomaly detection · Ensemble learning · Deep Q-Network · Mobile edge computing · Monitoring

---

H. Assmi · A. Guezzaz (✉) · S. Benkirane
SISAR Team, LaRTiD Lab, Technology Higher School Essaouira, Cadi Ayyad University, Marrakech, Morocco
e-mail: a.guzzaz@gmail.com

S. Jabbour
CRIL—CNRS, Artois University, Lens, France
e-mail: jabbour@cril.fr

B. J. Bipin Nair
Department of Computer Science, Amrita School of Computing, Amrita Vishwa Vidyapeetham, Mysuru, India
e-mail: bj_bipinnair@my.amrita.edu

M. Azrour
IDMS Team, Faculty of Sciences and Techniques, Moulay Ismail University of Meknes, Errachidia, Morocco

G. H. L. et al. (eds.), *Reliability in Cyber-Physical Systems: The Human Factor Perspective*, Springer Series in Reliability Engineering,
https://doi.org/10.1007/978-3-032-09917-4_6

1 Introduction

Mobile Edge Computing (MEC) is an emerging paradigm aimed at improving network performance by bringing computing capabilities closer to the edge. Faced with the increase in mobile devices and the need for real-time applications, traditional cloud architectures encounter latency and bandwidth limitations. MEC overcomes these challenges by deploying servers at the edge to offload heavy tasks and reduce latency. Task offloading is essential for optimizing the performance of MEC networks, by distributing tasks between mobile devices and peripheral servers according to factors such as network conditions, device capabilities and application needs. Intelligent offload strategies, based on machine learning and optimization techniques, aim to maximize efficiency by taking into account factors such as network congestion and energy consumption. The main challenge lies in accessing accurate, real-time information on network and device status to make optimal offloading decisions. This paper investigates task offloading decision-making in a Mobile Edge Computing (MEC) environment, where tasks can be processed either on the edge cloud or directly on a mobile device. To enhance this decision-making, a hybrid model is proposed that integrates a Deep Q-Network (DQN) neural network with the ensemble method (Voting classifier) and traditional machine learning techniques, including RF, GB, and SVM. The approach seeks to predict the most efficient offloading strategy based on key network parameters, including latency, downstream traffic, and duration, using datasets from Shanghai and Guangzhou. Two key contributions are presented:

- First, the DQN reinforcement algorithm is applied to optimize task offloading decisions in an edge computing environment (ECM).
- Secondly, the ensemble learning method is employed with three ML algorithms to optimize decision-making and enhance model performance, facilitating the prediction of the most effective load shedding strategy.

The paper is organized as follows: Sect. 2 offers a review of the background and pertinent studies on Mobile Edge Computing approaches. Section 3 explains the design of the framework proposed in this work. Section 4 examines the experimental assessment and the results achieved by the model. Finally, the study concludes by summarizing the key contributions and recommendations for future research.

2 Background and Related Works

This section analyzes the theoretical foundations and recent advancements in Intrusion Detection System (IDS) in the context of Mobile Edge Computing (MEC), emphasizing approaches aimed at strengthening the security of Internet of Things (IoT) devices. MEC offers storage, computational power, and application services through edge nodes, including autonomous vehicles, sensors, and more. It plays a key role in driving the progress of IoT with big data and artificial intelligence [1, 2].

In 2025, Tang et al. [3] proposed a dynamic offloading decision model for Mobile Edge Computing (MEC), utilizing a Deep Double Q-Network (DDQN) and a regular hexagonal server deployment structure. To assess its effectiveness, they conducted extensive experiments on both the system model and the proposed algorithm. Experimental results indicate that their approach significantly outperforms similar studies, offering good scalability and satisfactory responsiveness, even in complex scenarios where the load is influenced by the number of users or tasks. Meanwhile, Xiao et al. [4] focused on multi-objective task offloading and proposed a model based on a priori preference for decision-making. Their approach employs a neural combinatorial optimization method, combining an encoder-decoder model with reinforcement learning to simultaneously reduce energy consumption and latency. Experimental studies validate the efficiency of the proposed method by comparing it against three benchmark algorithms on instances of varying scales. In addition, To address the issue of task offloading in medical diagnostic services within MEC systems, Li et al. [5] developed a multi-objective, multi-agent reinforcement learning model. Their method incorporates a multi-agent tournament selection mechanism to identify historically significant preferences, enabling the system to consider the strategies of other agents while preserving the previously acquired policies of the current agent. Simulation results indicate that the proposed algorithm outperforms multiple baseline methods across a range of performance metrics. Assmi et al. [6] proposed an Intrusion Detection System (IDS) model leveraging ensemble learning, specifically a voting classifier that integrates three machine learning algorithms: Random Forest (RF), Support Vector Machine (SVM), and K-Nearest Neighbors (KNN). Their approach aims to detect normal or malicious activities on the NSL-KDD and Edge-IoT datasets. The obtained results show an accuracy ranging from 72 to 99%, a precision between 87 and 99%, and a recall and F1-score varying from 72 to 99%. To highlight the impact of feature selection integration in enhancing the security of Intrusion Detection Systems (IDS), Attou et al. [7] developed a model using a Radial Basis Function Neural Network (RBFNN) aimed at improving detection accuracy and supporting effective decision-making. They also incorporated the AdaBoost algorithm to refine feature selection and training on the NSL-KDD and CICIDS2017 datasets. A comparative analysis was conducted between the standalone RBFNN model and the RBFNN model combined with AdaBoost. The results demonstrate that the latter outperforms the former in terms of effectiveness.

As smart cities rapidly evolve, they also present new challenges in attack detection. Concerned with protecting citizens' privacy, In 2024, Hazman et al. [8] conducted a study to explore the significance of Intrusion Detection Systems (IDS) in these cities. Their research underscores the importance of IDS in detecting anomalies, preventing data breaches, and ensuring the security and reliability of urban electronic ecosystems. Furthermore, In the field of smart agriculture, Mohy-eddine et al. [9] developed an Intrusion Detection System (IDS) model based on Radial Basis Function Neural Networks (RBFNN), combined with several machine learning algorithms including Random Forest (RF), AdaBoost (ADA), Extra Trees (ET), LightGBM (LGBM), and

XGBoost (XGB), as well as a one-class Support Vector Machine (1-CSVM) to identify and remove outliers. When tested on the NF-Bot-IoT and NF-ToN-IoT datasets, the model achieved accuracy scores of 99.25% and 96.92%, respectively. Concomitantly, Amaouche et al. [10] concentrated on safeguarding AVENT networks from threats such as Denial of Service (DoS) attacks and false alarms. To tackle this challenge, they introduced an Intrusion Detection System (IDS) based on ensemble learning, utilizing the AdaBoost algorithm along with techniques such as SMOTE and Chi-Squared. Evaluated on the NSL-KDD, UNSW-NB15, and TON-IOT datasets, the model outperformed existing IDS solutions, achieving near-perfect results across all evaluation metrics, such as accuracy, precision, recall, and F1-score. In 2023, Zhang and Debroy [11] presented a comprehensive review of resource management issues in MEC, addressing the challenges related to resource allocation for applications with varying performance requirements. They examined existing solutions, including traditional optimization techniques and deep learning-based approaches, while focusing on popular MEC applications, especially in video analytics. Finally, they discussed the current research challenges and proposed potential avenues for future investigations in the field. Ning et al. [12] offer a comprehensive analysis of the technologies, applications and challenges involved in integrating MEC and ML into the Internet of Drones. It first presents the fundamental technologies, followed by the integration of these technologies and the problems specific to the Internet of Drones. The study also discusses applications of MEC and ML in urban, industrial and emergency scenarios, before summarizing current solutions and highlighting open challenges in this field. Simultaneously, Zhou et al. [13] analyzed the impact of task prioritization and mobile computing on MEC networks, where a UAV helps process user tasks across multiple access points. They optimized the system by selecting access points and offloading user tasks using a deep learning algorithm. The results are validated by simulations. Additionally, Wang et al. [14] introduced a WPMEC (Wireless Powered Mobile Edge Computing) model, merging MEC and WPT technologies. They analyze the main problems and techniques associated with this model, propose solutions for task offloading and resource allocation, and identify the challenges and avenues of research to be explored in this field.

In 2022, Feng et al. [15] studied computational offloading in MEC networks, exploring objectives such as delay reduction, energy management and utility maximization. They analyzed various approaches such as heuristic algorithms and reinforcement learning. The study highlights current challenges and future prospects, particularly concerning server selection and security. Chen et al. [16] introduced a hybrid energy provisioning model for task offloading in a dynamic MEC system, incorporating energy harvesting from IoT devices. It proposes the DTOME algorithm, which optimizes task offloading decisions by minimizing system costs and ensuring queue stability. Simulation results show that DTOME outperforms two baseline strategies in reducing system costs. In addition, Tang and Wong [17] proposed a distributed offloading algorithm based on deep reinforcement learning, integrating LSTM, dual DQN and double DQN. The results show that their approach optimizes the use of peripheral resources and effectively reduces delays and task abandonment rates compared with current approaches. In 2021, Naouri et al. [18] presented a DCC framework for task offloading, which comprises the device, cloudlet and cloud layers. It optimally allocates tasks, sending high-computation tasks to the cloud and low-cost tasks to the device to reduce communication delays. The framework uses a greedy optimization algorithm for task scheduling based on device capabilities. Experimental results confirm that the DCC framework delivers superior performance compared to current task offloading methods. Zhou et al. [19] focused on optimizing task offloading and resource allocation in a multi-user MEC system, aiming to reduce energy consumption while meeting delay constraints and handling uncertainty in resource demands. To achieve this, they introduced a Q-Learning-based approach and a Double Deep Q Network (DDQN) for improved efficiency. Simulation results demonstrate that the DDQN method significantly reduces energy consumption, achieving up to a 53% reduction compared to other methods.

In 2020, Shakarami et al. [20] proposed a taxonomy of ML-based computational offloading mechanisms in the MEC environment, divided into three categories: reinforcement learning, supervised and unsupervised. They analyze and compare these approaches in terms of their performance, advantages and limitations, before exploring the research challenges still to be met. Zhan et al. [21] addressed the optimization of task offloading and resource allocation for multiple users, balancing latency and power consumption. They proposed a heuristic algorithm (HMAOA) that takes mobility into account and decomposes the problem into local optimization sub-problems. The approach combines a numerical solution for the allocation of computations and a heuristic method for the load shedding decision. Simulations show that this algorithm outperforms other existing solutions in terms of performance. In 2019, Chen et al. [22] developed a dynamic algorithm, EEDOA, for the energy optimization of task shedding in EMCs, formulated as a stochastic optimization problem. The algorithm aims to minimize energy consumption while controlling queue length, despite the uncertainty of wireless channels and the dynamics of task arrivals. With polynomial complexity, EEDOA makes online decisions and demonstrates its effectiveness through experimental results. Huang et al. [23] presented a Deep Reinforcement learning-based Online Offloading (DROO) algorithm to optimize task offloading and resource allocation in wireless MEC networks. The algorithm bypasses complex combinatorial optimization problems, reducing complexity and automatically adjusting its parameters. Results show that this algorithm delivers near-optimal performance with latency of less than 0.1 s in a 30-user network, making real-time offloading viable even in fluctuating signal environments.

3 Proposed Model

This section provides an overview of our model's architecture, as well as the algorithms used for its design and implementation.

3.1 Model Architecture

In this section, we detail the process of developing a decision support model for task offloading in a MEC environment. The architecture of the proposed model is illustrated in Fig. 1.

- Dataset preprocessing: The first stage of this experiment involved rigorous preprocessing of the data, to ensure quality and protect sensitive information. This step is a fundamental prerequisite to any analysis and modeling, ensuring the reliability and relevance of the results obtained.
- Task offloading strategy: Our project adopts this strategy to enhance available information and ensure intelligent, dynamic decision-making for offloading tasks between mobile devices and the cloud at the edge. It integrates several key factors, including traffic load, network conditions, and user behavior, with the aim of maximizing Mobile Edge Computing (MEC) network performance.
- Evaluation metrics: These evaluation metrics enable us to assess the effectiveness of the proposed task offloading strategy, particularly in terms of latency reduction and decision accuracy. By analyzing these parameters, we can evaluate the performance of the dynamic task offloading strategy between mobile devices and the edge cloud.

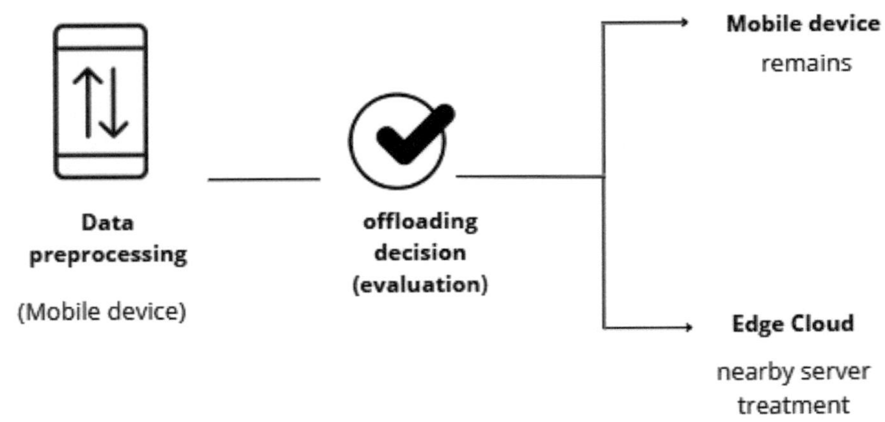

Fig. 1 Architecture for our model MEC

3.2 Description of Solutions

3.2.1 Dynamic Intelligent Edge Task Offloading Algorithm

The DITO algorithm is deployed to assess the effectiveness of offloading decisions between mobile devices and the edge cloud. Significantly improving application efficiency and responsiveness.

- Task Analysis: Assess the tasks to be performed by taking into account their resource requirements, complexity, and processing duration.
- Network Performance Evaluation: Assess the network conditions, such as latency, available bandwidth, and congestion, to pinpoint the most optimal offloading strategies.
- Resource Evaluation: Evaluate the capacity and availability of local devices and distant resources including Edge servers or cloud to perform task execution.
- Dynamic Decision Making: Leverage ML algorithms or heuristics to identify the most suitable location (local or remote) for task execution, informed by prior assessments.
- Task Offloading: Transfer tasks to the appropriate server Edge or Cloud, when it optimizes performance, ensuring that latency and resource limitations are fulfilled.
- Surveillance and Adjustment: Continuously track task performance and network conditions, adjusting offloading decisions in real-time whenever necessary.

3.2.2 Deep Q-Network (DQN)

A deep Q network (DQN) is a model that integrates Q learning with a deep convolutional neural network (CNN) to train a network to approximately estimate the value of the Q function, associating state-action pairs with their expected discounted performance.

- Using a neural network to estimate the Q function

The neural network $Q(s, a, \theta)$ takes an input condition s and generates Q as output values for each possible action a. The goal is to enable this network to identify the ideal policy that maximizes the long-term cumulative reward.

- Leveraging Experience Replay

Recorded transitions (s, a, r, s') are stored in a buffer.

To mitigate temporal correlations and promote training stability, transitions are randomly sampled in mini-batches in the training phase

- Target Network (Q′) integration

To determine the learning target, a target network $Q'(s, a; \theta^{-})$ is utilized.

The network replicates the principal network but updates less often to avoid unstable oscillations.

- Adjusting the Q function based on the Bellman equation:

The Q value is adjusted during training to minimize the mean square error between current and target values, as defined by:

$$y_i = \begin{cases} r_j \; si \; s_{j+1} \; _{un\ e'tat\ de\ fin} \\ r_j + \gamma \max_{a'} Q'\left(s_{j+1}, a'; \theta^-\right) \end{cases} \tag{1}$$

Stochastic gradient descent (SGD) is used to adjust the network's weights θ.

- ϵ-greedy strategy

The DQN employs an ϵ-greedy policy to promote exploration of the environment, selecting an action randomly with a probability of ϵ.

The optimal action, arg $max_a Q(s, a)$, is chosen with a probability of $1 - \epsilon$.

4 Experimental Evaluation and Results

This section provides an in-depth analysis of the data used and the results obtained from the model we developed.

4.1 Preprocessing of Dataset

In our system, the models leverage crucial variables such as 'Downlink traffic,' 'Duration,' and 'Latency' as input features, while the 'Offloading Decision' column representing the target variable. The selection of these fields is decisive for our model, providing crucial information on network status and device performance.

The selection of these fields is decisive for our model, providing crucial information on network status and device performance.

- The 'Start Time' and 'End Time' columns are changed to date time format to improve processing efficiency
- By adding a new column, 'Offloading Decision,' which specifies if a task is offloaded to the edge cloud or carried out locally on a mobile device. In this situation, are transferred to the Edge Cloud when download traffic exceeds upload traffic; if not, they are handled on the mobile device.
- By incorporating a new column, 'Latency,' which measures latency according to the offloading decision. if the job is transferred to the peripheral cloud, an additional 5 s are added to the job duration. Otherwise, the task duration is used as is.
- A new 'Throughput' column is added to calculate each task's throughput by splitting the combined traffic (upload + download) by its respective latency.

- The percentage of edge cloud utilization is obtained by computing the total latency of offloaded to the cloud relative to the total latency of all tasks, and then multiplying by 100.

4.2 Experimental Finding and Analysis

The efficiency of the four models is assessed based on their accuracy. As depicted in Fig. 2, The Random Forest (RF) model demonstrates excellent performance across both offloading classes. While the Gradient Boosting model trails slightly, its performance is nearly equivalent to RF. In comparison, the SVM model produces less favorable results, particularly because of its underperformance in the Edge Cloud class. The ensemble method (Voting Classifier) combines the strengths of the three base algorithms, resulting in an impressive accuracy of 98%. In contrast, the DQN model exhibits relatively low accuracy, suggesting that it requires additional refinement to better align with the specific requirements of this task.

Figure 3 demonstrates the accuracy differences of our classifiers (RF, SVM, GB, DQN, and Voting Classifier) based on precision for the two classes: Edge Cloud and Mobile Device. In the Edge Cloud class, DQN exhibits a precision of 0%, indicating no correct predictions. On the other hand, Random Forest and Gradient Boosting display substantial improvements, each achieving a peak precision of 99%. In comparison, the SVM model attains 80% precision, reflecting a decline in performance for this class. In the Mobile Device class, DQN attains a precision of 67%, significantly trailing behind the other models. With an outstanding precision of 99%, Random Forest and Gradient Boosting highlight their effectiveness in managing this

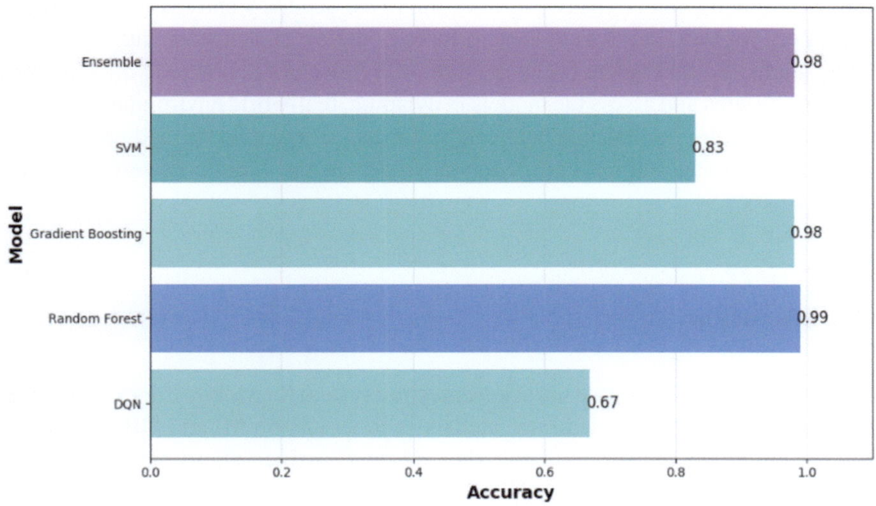

Fig. 2 Accuracy metric evaluation of the intrusion detection model

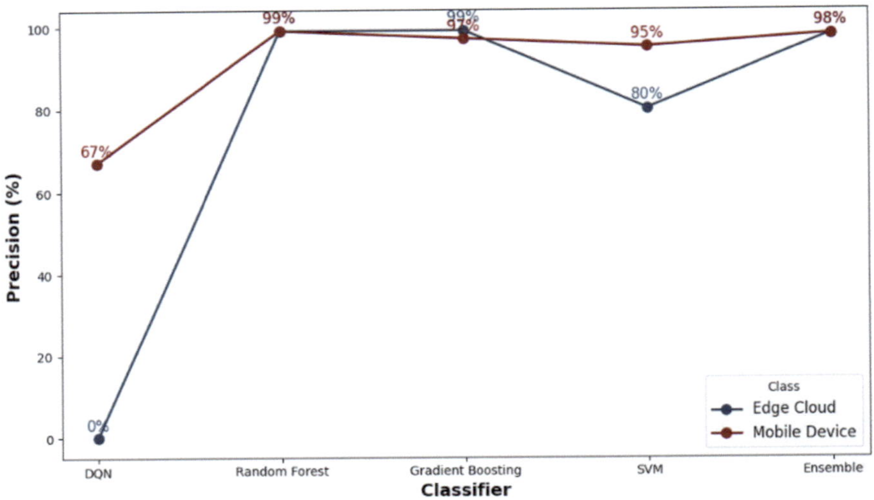

Fig. 3 Precision metric evaluation of the intrusion detection model

class. While SVM is slightly lower, it still delivers a strong precision of 95%. Lastly, the Voting Classifier algorithm delivers an impressive accuracy of 98% for both classes. In general, Random Forest and Gradient Boosting distinguish themselves with consistent, strong performance in both classes, whereas DQN reveals notable limitations, particularly in the Edge Cloud class.

While the models used produce satisfactory results, some limitations persist. SVM and DQN show inefficiencies, as SVM is impacted by long execution times and class imbalance, whereas DQN demonstrates consistently subpar performance. In contrast, RF and Gradient Boosting offer improved accuracy and reasonable execution times, yet they may struggle with more complex or dynamic data scenarios. Future studies could explore the integration of lightweight neural networks tailored for MEC systems and assess the models in dynamic, real-world environments. Moreover, investigating the balance between performance, complexity, and energy efficiency could broaden the scope of this approach in the Mobile Edge Computing (MEC) sector.

5 Conclusion

We have created an innovative intrusion detection system for Mobile Edge Computing (MEC), utilizing machine learning models such as RF, SVM, and GB, combined with an ensemble method utilizing the Voting Classifier algorithm, and neural network models such as Deep Q-Network (DQN). The system was assessed using the Shanghai and Guangzhou datasets, with the results confirming its capability to support task offloading decisions in MEC, demonstrating the effectiveness of our approach. Future work will focus on integrating neural networks optimized for MEC and evaluating the system in real-world, dynamic environments to refine its performance. Furthermore, analyzing the balance between performance, complexity, and energy efficiency will play a key role in improving the applicability of this approach to MEC.

Acknowledgements We are very grateful to CNRST for the invaluable support it has given us through the PASS (PhD-Associate Scholarship) program.

References

1. Susanto H, Azim MQ, Fang-Yie L, Susanto AKS, Setiana D, Ibrahim F, Basuki AI, Ramdhani TI, Setiawan I, Riyanto B, Ferianda RA (2023) Artificial intelligence, Blockchain framework, Cyberthreat defenses of resilient digital ecosystems. In: Handbook of research on artificial intelligence and knowledge management in Asia's digital economy. IGI Global, pp 36–63
2. Isa H, Subramanian U (2024) The impact of big data in auditing. Proc Comput Sci 238:841–848
3. Tang X, Tang T, Shen Z, Zheng H, Ding W (2025) Double deep Q-network-based dynamic offloading decision-making for mobile edge computing with regular hexagonal deployment structure of servers. Appl Soft Comput 169:112594. https://doi.org/10.1016/j.asoc.2024.112594
4. Xiao X-J, Wang Y, Huang P-Q, Wang K (2025) Neural combinatorial optimization for multi objective task offloading in mobile edge computing. IEEE Trans Veh Technol:1–12. https://doi.org/10.1109/TVT.2025.3546914
5. Li S, et al. (2025) Dynamic weight reinforcement learning method considering multiple factors in mobile edge computing system. Neurocomputing 620:129194. https://doi.org/10.1016/j.neucom.2024.129194
6. Assmi H, et al. A robust security detection strategy for next generation IoT networks. EBSCOhost [Online]. Available https://openurl.ebsco.com/contentitem/doi:10.32604%2Fcmc.2024.059047?sid=ebsco:plink:crawler&id=ebsco:doi:10.32604%2Fcmc.2024.059047. Accessed 15 Mar 2025
7. Attou H, Guezzaz A, Benkirane S, Azrour M (2025) A new secure model for cloud environments using RBFNN and AdaBoost. SN Comput Sci 6(2):188. https://doi.org/10.1007/s42979-025-03691-1
8. Hazman C, Guezzaz A, Benkirane S, Azrour M (2024) A smart model integrating LSTM and XGBoost for improving IoT-enabled smart cities security. Clust Comput 28(1):70. https://doi.org/10.1007/s10586-024-04780-1
9. Mohy-eddine M, Guezzaz A, Benkirane S, Azrour M (2024) Malicious detection model with artificial neural network in IoT-based smart farming security. Clust Comput 27(6):7307–7322. https://doi.org/10.1007/s10586-024-04334-5
10. Amaouche S, Guezzaz A, Benkirane S, Azrour M (2024) A robust model for predicting abnormal behavior in vehicular networks using AdaBoost and Chi-Square. Wirel Pers Commun 138(4):2583–2611. https://doi.org/10.1007/s11277-024-11615-0
11. Zhang X, Debroy S (2023) Resource management in mobile edge computing: a comprehensive survey. ACM Comput Surv 55(13s): 291:1–291:37. https://doi.org/10.1145/3589639
12. Ning Z, et al. (2023) Mobile edge computing and machine learning in the internet of unmanned aerial vehicles: a survey. ACM Comput Surv 56(1): 13:1–13:31. https://doi.org/10.1145/3604933
13. Zhou W, et al. (2023) Priority-aware resource scheduling for UAV-mounted mobile edge computing networks. IEEE Trans Veh Technol 72(7):9682–9687. https://doi.org/10.1109/TVT.2023.3247431

14. Wang X, et al. (2023) Wireless powered mobile edge computing networks: a survey. ACM Comput Surv 55(13s): 263:1–263:37. https://doi.org/10.1145/3579992
15. Feng C, Han P, Zhang X, Yang B, Liu Y, Guo L (2022) Computation offloading in mobile edge computing networks: a survey. J Netw Comput Appl 202:103366. https://doi.org/10.1016/j.jnca.2022.103366
16. Chen Y, Zhao F, Lu Y, Chen X (2023) Dynamic task offloading for Mobile edge computing with hybrid energy supply. Tsinghua Sci Technol 28(3):421–432. https://doi.org/10.26599/TST.2021.9010050
17. Tang M, Wong VWS (2022) Deep reinforcement learning for task offloading in Mobile edge computing systems. IEEE Trans Mob Comput 21(6):1985–1997. https://doi.org/10.1109/TMC.2020.3036871
18. Naouri A, Wu H, Nouri NA, Dhelim S, Ning H (2021) A novel framework for mobile-edge computing by optimizing task offloading. IEEE Internet Things J 8(16):13065–13076. https://doi.org/10.1109/JIOT.2021.3064225
19. Zhou H, Jiang K, Liu X, Li X, Leung VCM (2022) Deep reinforcement learning for energy-efficient computation offloading in mobile-edge computing. IEEE Internet Things J 9(2):1517–1530. https://doi.org/10.1109/JIOT.2021.3091142
20. Shakarami A, Ghobaei-Arani M, Shahidinejad A (2020) A survey on the computation offloading approaches in mobile edge computing: a machine learning-based perspective. Comput Netw 182:107496. https://doi.org/10.1016/j.comnet.2020.107496
21. Zhan W, Luo C, Min G, Wang C, Zhu Q, Duan H (2020) Mobility-aware multi-user offloading optimization for mobile edge computing. IEEE Trans Veh Technol 69(3):3341–3356. https://doi.org/10.1109/TVT.2020.2966500
22. Chen Y, Zhang N, Zhang Y, Chen X, Wu W, Shen X (2021) Energy efficient dynamic offloading in mobile edge computing for internet of things. IEEE Trans Cloud Comput 9(3):1050–1060. https://doi.org/10.1109/TCC.2019.2898657
23. Huang L, Bi S, Zhang Y-JA (2020) Deep reinforcement learning for online computation offloading in wireless powered Mobile-edge computing networks. IEEE Trans Mob Comput 19(11):2581–2593. https://doi.org/10.1109/TMC.2019.2928811

Efficient Anomaly Detection for Cyber-Physical Leveraging Knowledge Distillation and Model Quantization

Kamal Bella, Azidine Guezzaz, Ravikumar Patchmuthu, Said Benkirane, Mouaad Mohy-Eddine, and Mourade Azrour

Abstract Cyber-Physical Systems (CPS) are fundamental to modern technological advancement, they integrate physical and computational processes enabling real time interactions, and are often under constant security threats. Intrusion Detection Systems are essential in this context, they continuously monitor network traffic for anomalies that could indicate cyber-attacks. However, CPS are usually resource constrained environments and IDS can be computationally demanding, which makes real-time intrusion challenging. In this work, we address these limitations by providing a lightweight intrusion detection combining feature selection, knowledge distillation and quantization, aiming to drastically reduce the intrusion detection model size and inference time. In this setup, we experimented with 10, 20, 30 and 40 features from the NSL-KDD dataset, and developed a teacher model, a student model and a quantized model for each set of features. As a result, we found that the 30 features configuration set achieves the best balance between performance and efficiency, where the teacher model achieves 98.91% accuracy at 627 KB, the distilled model with 98.78% accuracy at 39.9 KB and finally, the quantized model achieves 98.80% accuracy at only 7.96 KB. These results showcase the abilities of combining these techniques in achieving low-cost intrusion detection without compromising performance.

K. Bella · A. Guezzaz (✉) · S. Benkirane
SISAR Team, LaRTiD Laboratory, Higher School of Technology, Cadi Ayyad University, Marrakesh, Morocco
e-mail: a.guzzaz@gmail.com

R. Patchmuthu
Universiti Teknology Brunei, Bandar Seri Begawan, Brunei Darussalam
e-mail: ravi.patchmuthu@utb.edu.bn

M. Mohy-Eddine
DELTA Lab, National School of Arts and Crafts, Hassan II University, Casablanca, Morocco

M. Azrour
Faculty of Sciences and Technics, Moulay Ismail University, Meknes, Morocco

G. H. L. et al. (eds.), *Reliability in Cyber-Physical Systems: The Human Factor Perspective*, Springer Series in Reliability Engineering,
https://doi.org/10.1007/978-3-032-09917-4_7

Keywords Intrusion detection · Cyber-physical systems · Knowledge distillation · Post-training quantization · Feature selection · Resource constraints

1 Introduction

Cyber-Physical Systems (CPS) seamlessly blend the physical world with digital computing and networking, reshaping everything from manufacturing and transportation to energy and healthcare [1, 2]. These systems use sensors, actuators, and smart software to perform complex tasks that were once highly manual [3, 4]. Yet, this close integration means that a vulnerability in one part of the system can quickly lead to significant operational disruptions and financial losses [5].

CPS play a vital role in many aspects of our daily lives, so keeping them secure is not just a technical issue, it's essential for our safety. Cyber-attacks can disrupt real-world operations, steal sensitive information, or even put lives at risk. The traditional security measures we know today need a fresh approach to counter fast-moving and unpredictable threats, including persistent, advanced attacks and unforeseen vulnerabilities. On top of that, many CPS elements, like edge devices and embedded controllers are limited by their small size and computing power. This makes it clear that we need security solutions that are both lightweight and tough enough to defend these critical systems [6–8].

Intrusion Detection Systems (IDS) serve as an early warning mechanism, catching potentially harmful activity before it can inflict lasting damage. They monitor network traffic, user actions, and system events, and then flag any irregular patterns that might signal an attack [9–12]. This capability is especially crucial for CPS, where even a slight delay in threat detection could lead to physical harm or extended downtime. Although IDS were originally designed for enterprise IT networks, their use in CPS environments is on the rise [13]. The real challenge now is tailoring these systems to the specific performance and resource limitations of CPS without sacrificing speed or accuracy.

To address these challenges, we leverage feature selection, knowledge distillation and quantization, to maintain intrusion detection accuracy and the ability to operate within the resource constraints of CPS, feature selection helps identify and eliminate low to no impact features, while knowledge distillation transfers the insights of larger and complex teacher model into a simpler and smaller student model, and finally, quantization reduces the precision of the model's weights in aim to reduce its size and computational requirements. Together, these methods help produce an IDS that balances high detection performance with efficient resource usage, making it suitable for deployment in diverse CPS settings [14–16].

The remainder of this paper is presented as follows: Sect. 2 reviews related work on IDS in resource-limited CPS settings. Section 3 covers our methodology, with Sect. 3.2 describing the proposed solutions and Sect. 3.1 detailing the model architecture. Section 4 introduces our experimental study: the NSL-KDD dataset (4.1), feature selection (4.2), and evaluation metrics (4.3). Section 5 presents and discusses the results, and Sect. 6 concludes the paper.

2 Related Works

Wang et al. [17] proposed an efficient intrusion detection model combining Spiking Neural Networks (SNNs) and Convolutional Neural Networks (CNNs) to create a lightweight, high-performance system for resource-constrained IoT environments. It introduces time-value encoding, a custom loss function, and dynamic thresholding to improve efficiency and adaptability. The model significantly reduces energy consumption (up to 90%), processes 5333 samples/sec, and maintains high accuracy (98.82% on CSE-CIC-IDS2018, 99.86% on CIC-DDoS2019), making it a promising solution for real-time, low-power intrusion detection. Otokwala et al. [18] proposed OCFSDA, a lightweight IoT intrusion detection model using feature selection, LSTM-autoencoder compression, pruning, and quantization. Deployed on Raspberry Pi 4, it achieves 99 and 97% accuracy on MQTT-IoT-IDS2020 and CIC-IDS2017, with 2 KB memory usage and 0.12 s inference time, ensuring efficiency and robustness against adversarial attacks. Tiwari et al. [19] proposed a high-accuracy IIoT intrusion detection using PSO, PCA, and MARS, achieving 100% accuracy on WUSTL-IIOT-2021. Quantized (4-bit and 8-bit) and deployed on Azure IoT Edge, it reduces latency by 25%, enabling real-time cybersecurity. Bouayad et al. [20] introduced Lightweight-Fed-NIDS, a federated learning-based NIDS with zero-shot pruning, reducing model size by 90% while maintaining 99% accuracy. It achieves a 3× training speedup, enhancing efficiency and scalability for privacy-preserving, distributed NIDS. Putrada et al. [21] presented QUIDS, an edge-based IoT botnet detection using KNN, quantization, random sampling, and feature selection. It reduces features from 115 to 30, achieving a 175× compression ratio while maintaining detection accuracy, enabling efficient real-time security. Mittal [22] presented CyberSentinel, a high-speed anomaly detection system for programmable switches, using knowledge distillation from autoencoders to iForest. Deployed on Intel Tofino, it boosts throughput by 66.47% and cuts latency by 50%, enabling real-time, low-latency intrusion detection. Abbasi et al. [23] explored knowledge distillation for IoT traffic classification, compressing deep models into lightweight student models for resource-constrained devices. The student model maintains high accuracy (98.3% on Cambridge, 92.7% on CIC-IDS2017) while reducing model size and computational cost. It generalizes well with limited data, outperforming pruning and traditional ML models, making it a scalable, efficient solution for real-time IoT traffic analysis.

Existing intrusion detection models optimize efficiency for IoT and IIoT (as depicted in Table 1) but often overlook the extreme resource constraints and real-time demands of CPS. Many prioritize accuracy over deployability, lacking the compact design needed for CPS environments. While knowledge distillation reduces complexity, few combine it with post-training quantization for further efficiency. A

Table 1 Related works

	Year	Methods	Feature selection	Datasets
Wang et al. [17]	2024	SNN + CNN, time-value encoding, custom loss, dynamic thresholding	Yes (SelectKBest)	CSE-CIC-IDS2018, CIC-DDoS2019
Otokwala et al. [18]	2024	OCFSDA: LSTM-autoencoder, pruning, quantization	Yes (OCFS)	MQTT-IoT-IDS2020, CIC-IDS2017
Tiwari et al. [19]	2024	PSO, PCA, MARS, quantization (4-bit and 8-bit)	Yes (PSO, PCA)	WUSTL-IIOT-2021
Bouayad et al. [20]	2024	Federated learning, zero-shot pruning	Yes (CNN)	UNSW-NB15, USTC-TFC2016, and CIC-IDS-2017
Putrada et al. [21]	2024	KNN, quantization, random sampling	Yes (random forest)	Kaggle botnet attack dataset
Mittal [22]	2024	Knowledge distillation (autoencoders to iForest)	No	UNSW
Abbasi et al. [23]	2021	Knowledge distillation, model compression	No	Cambridge, CIC-IDS2017

solution is needed that ensures high detection accuracy with minimal size and latency, making real-world CPS deployment feasible.

3 Methodology

In this solution, our aim is to provide a framework designed to build an efficient IDS solution ready for deployment in resource limited CPS environments.

3.1 Model Architecture

As depicted in Fig. 1, we begin with a data source that provides the raw information—often network logs or a similar dataset. Before any modeling takes place, the data goes through a preprocessing phase: we remove incomplete (NaN) values, drop any duplicates that could skew the results, and encode categorical features into a format suitable for machine learning.

Once the data is clean, we use feature selection to pick out the attributes that matter most for our prediction goals. This step helps strip away noise, so our model can focus on the strongest signals. We then feed these selected features into a prediction

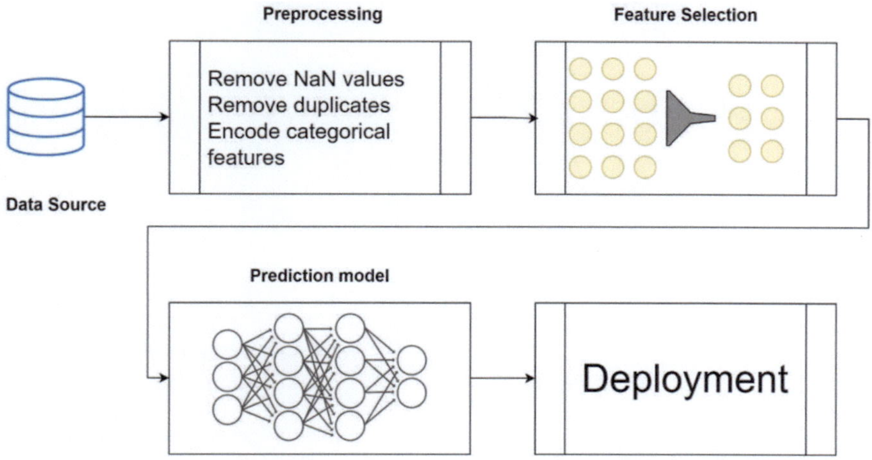

Fig. 1 Proposed model

model, often a neural network, that learns to identify patterns or anomalies in the data.

Finally, the trained model is handed off to a deployment stage, where it's integrated into a real-world environment ready to handle new inputs, flag unusual behavior, or generate predictive insights on live data. This pipeline ensures that only the most relevant data is used, the model is well-trained, and the final system can run efficiently in practice.

3.2 Description of Solutions

The student model itself contains two hidden layers of 64 and 32 units, also using ReLU activations. The architecture details are shown in Fig. 2. Because it has fewer parameters, it runs faster and takes up less memory—advantages that are especially valuable for real-time or embedded applications. However, such a compact model might not match the performance of the teacher model if it were trained on the same dataset in isolation. This is where knowledge distillation comes in: the student learns not only from the true labels but also from the teacher's "soft" predictions, which provide insights into the teacher's decision-making process. By imitating the teacher's output distribution, the student can capture much of the teacher's expertise without growing to the teacher's size.

After the student model converges, the final step is to apply post-training quantization. This process compresses the student model's weights—often down to 8-bit integers—dramatically reducing both the model's footprint and its computational requirements. Quantization is especially important for devices with limited resources,

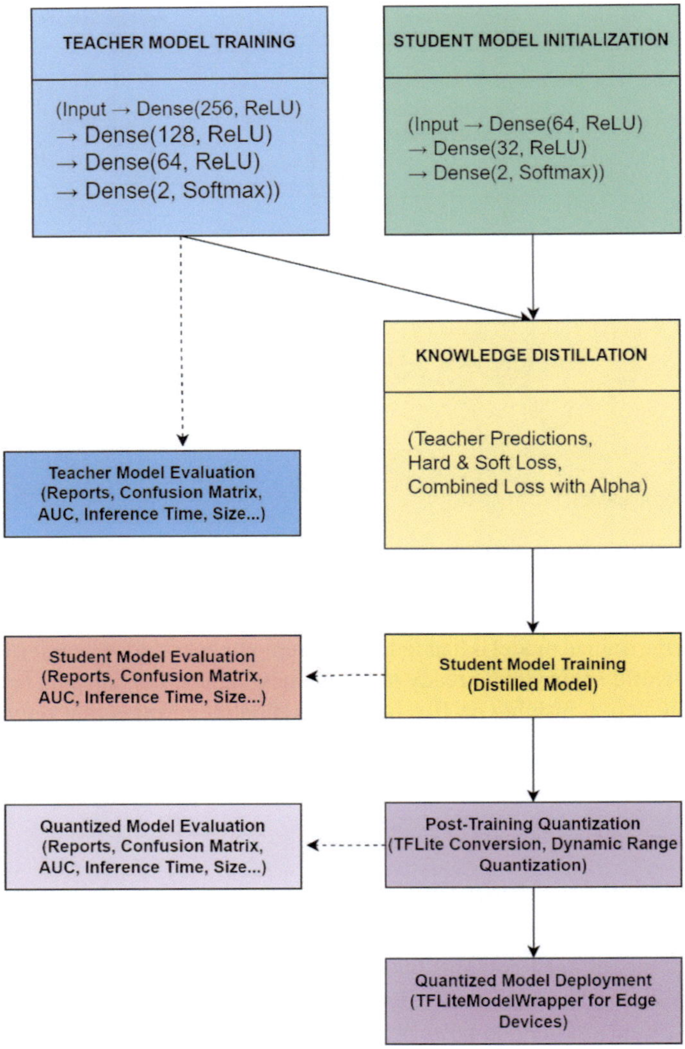

Fig. 2 Architecture solution

as it preserves most of the model's accuracy while making inference far more efficient. At the end of this pipeline, we have a small, accurate, and resource-friendly model that is ready for real-world deployment.

4 Experimental Study

This section describes the experimental study performed to assess the proposed IDS for IoT and CPS environments. The proposed solution integrates a feature selection approach with a knowledge distillation-based model to attain high detection accuracy, computational efficiency, and adaptability to resource-constrained settings. In this experiment, we used the NSL-KDD dataset and evaluated feature subsets of 10, 20, 30, and 40 features to find the optimal balance for the IDS. The dataset was pre-cleaned, so we only needed to encode the categorical features and target variables.

4.1 NSL-KDD Dataset

NSL-KDD is a refined version of the original KDD Cup 99 dataset and is a common choice for IDS research. It offers realistic network traffic with fewer redundant records, featuring over 140,000 entries and 41 features that cover normal activity as well as attacks like DoS, Probe, R2L, and U2R. Its diverse attack patterns make it an excellent benchmark, which is why we chose it to evaluate our IDS model.

4.2 Feature Selection

Choosing the right features is key for an effective IDS in CPS. Using the NSL-KDD dataset's 41 features, we applied a random forest to rank their importance and then trained models with 40, 30, 20, and 10 features. This streamlined approach cut noise, improved accuracy, and reduced computational costs—crucial for deploying IDS on resource-limited edge devices in IIoT environments.

4.3 Evaluation Metrics

To assess the IDS model's performance, we utilized different metrics, such as accuracy, precision (PR), recall (or true positive rate TPR), and F1-score. These metrics provided a comprehensive assessment of the model's effectiveness and efficiency in detecting intrusions.

- Accuracy: $ACC = \frac{TP+TN}{TP+TN+FP+FN}$
- Precision: $PR = \frac{TP}{TP+FP}$
- Recall: $TPR = \frac{TP}{TP+FN}$
- F1 score: $F1\ Score = 2 \times \frac{PR \times TPR}{PR+TPR}$

Table 2 Confusion matrix

	Predicted as positive	Predicted as negative
Labeled as positive	TP	FN
Labeled as negative	FP	TN

Here, TP, TN, FP, and FN represent true positives, true negatives, false positives, and false negatives, respectively as shown in the confusion matrix in Table 2.

In addition to these metrics, we evaluate our model using model size in bytes, and inference time in milliseconds, which is the time it takes the model to classify one instance of traffic.

5 Results and Discussion

These experiments compared the performance of three models—Base (Teacher), Distilled, and Quantized—across four different feature subsets (10, 20, 30, and 40). Table 3 summarizes the key metrics, while Figs. 3, 4, and 5 offer a visual overview of inference times, accuracies, and model sizes, respectively.

Figure 3 presents the inference times for each model. The Base (Teacher) and Distilled models both take around 0.05 ms per sample, whereas the Quantized model operates at about 0.008–0.01 ms. Even with an increasing number of features, the Quantized model remains consistently faster. This speed advantage makes it especially appealing for real-time scenarios or edge devices with limited computational resources.

Figure 4 illustrates the accuracy trends (and aligns with other performance metrics such as AUC-ROC, Precision, Recall, and F1-Score). All three models achieve very high accuracies—well above 96%—even with only 10 features. As the number of features rises, the models converge around 98–99% accuracy, demonstrating that the Distilled and Quantized models can perform almost on par with the larger Teacher model. AUC-ROC values follow a similar pattern, remaining above 0.99 once the feature set reaches 20 or more.

Figure 5 shows the substantial differences in model sizes. The Teacher model ranges from about 566 KB at 10 features to roughly 658 KB at 40 features. In contrast, the Distilled model is under 50 KB across all feature sets, while the Quantized model is even smaller—under 9 KB for 40 features. This reduction is significant for memory-constrained environments and directly complements the fast inference times seen in Fig. 3.

Overall, these results highlight the trade-offs among the three approaches. The Base (Teacher) model offers a slightly higher top-end performance but comes with a larger footprint and slower inference speed. The Distilled model preserves most of the Teacher's accuracy while being much smaller, and the Quantized model goes a

Table 3 Evaluation of all models across different numbers of features

		Number of features			
Model	Metric	10	20	30	40
Base model (teacher)	Model size (in bytes)	566,080	596,800	627,520	658,240
	Inference time (in ms)	0.05320988349	0.05232449044	0.05189269789	0.05161634617
	AUC-ROC	0.9968	0.9987	0.9992	0.9991
	ACC	0.9738755723	0.9839078912	0.9890923781	0.9882844061
	PR	0.9745152714	0.9870833062	0.9886636004	0.9903001107
	Recall	0.9751476218	0.9818311596	0.9903315813	0.9870871455
	F1-score	0.9748313441	0.9844502277	0.989496888	0.9886910178
Distilled model	Model size (in bytes)	34,752	37,312	39,872	42,432
	Inference time (in ms)	0.05001397169	0.05099471932	0.05021298025	0.04998390449
	AUC-ROC	0.9955	0.9987	0.9992	0.9991
	ACC	0.9666711554	0.9836385672	0.9877794236	0.9881497441
	PR	0.9664876755	0.9809861424	0.9876215165	0.9895325401
	Recall	0.9693725261	0.9876062553	0.9888391409	0.9876062553
	F1-score	0.9679279513	0.9842850676	0.9882299536	0.9885684593
Quantized model	Model size (in bytes)	7800	7320	7960	8600
	Inference time (in ms)	0.008254713554	0.009748706463	0.009882780795	0.008780753038
	AUC-ROC	0.9956	0.9987	0.9992	0.9991
	ACC	0.9667721519	0.9827295987	0.9880487476	0.9882844061
	PR	0.966977467	0.9795287756	0.9880705394	0.9892172783
	Recall	0.9690480825	0.9873467004	0.9889040296	0.9881902537
	F1-score	0.9680116675	0.9834222007	0.9884871088	0.9887034993

step further by reducing both size and inference time dramatically. For many practical scenarios, especially those running on limited hardware, the Quantized model delivers an excellent balance of accuracy, speed, and resource usage.

Table 4 summarizes the metrics of the three models using 30 features, this variation yielded the best performance to size tradeoff. With 30 features, the Base (Teacher) model continues to show top-tier accuracy and AUC-ROC, but its size and inference time are significantly higher. The Distilled model is smaller and nearly matches the Teacher's performance. Meanwhile, the Quantized model remains the most efficient

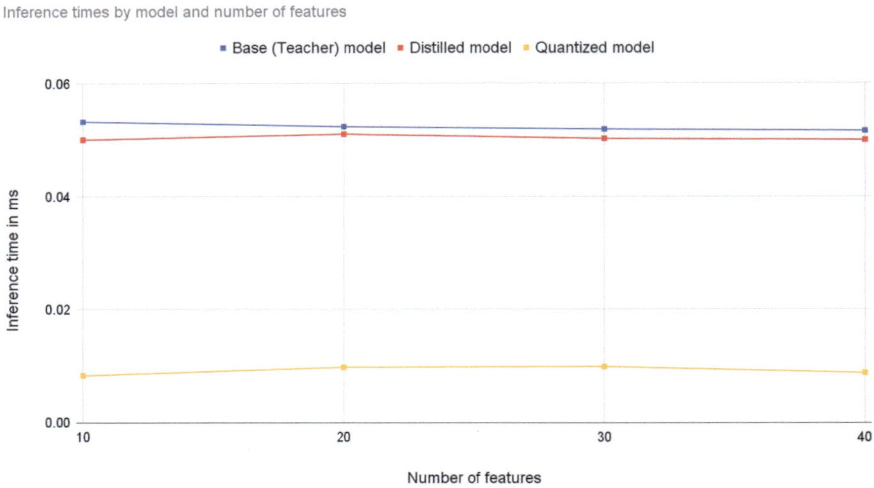

Fig. 3 Inference time by model across different numbers of features

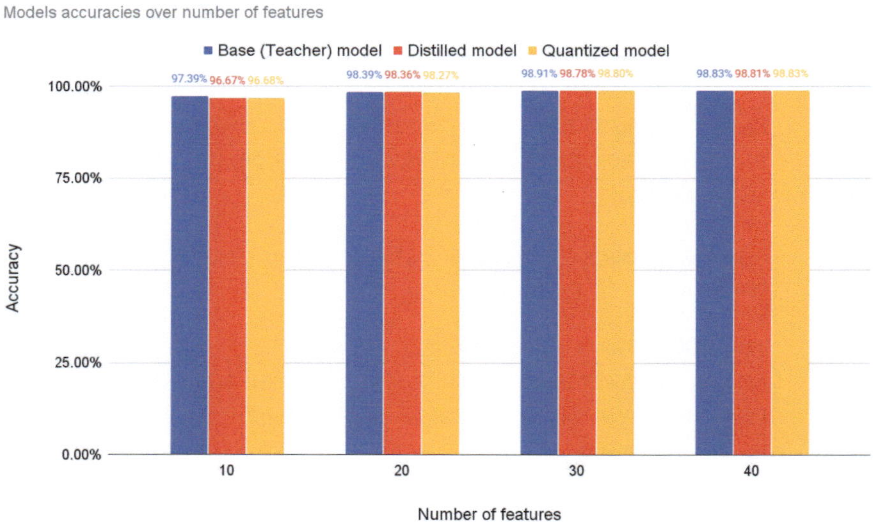

Fig. 4 Accuracies over different numbers of features (10, 20, 30, 40) for three compared models

in terms of both memory footprint and speed, and it still achieves excellent classification metrics. Overall, the 30-feature configuration strikes a good balance: it preserves high accuracy while reducing resource requirements. Figure 6 displays the ROC curve for the Quantized model using 30 features, with an AUC of 0.9992. The curve's proximity to the upper-left corner highlights its strong discriminative ability, indicating that it can accurately distinguish between normal and anomalous traffic

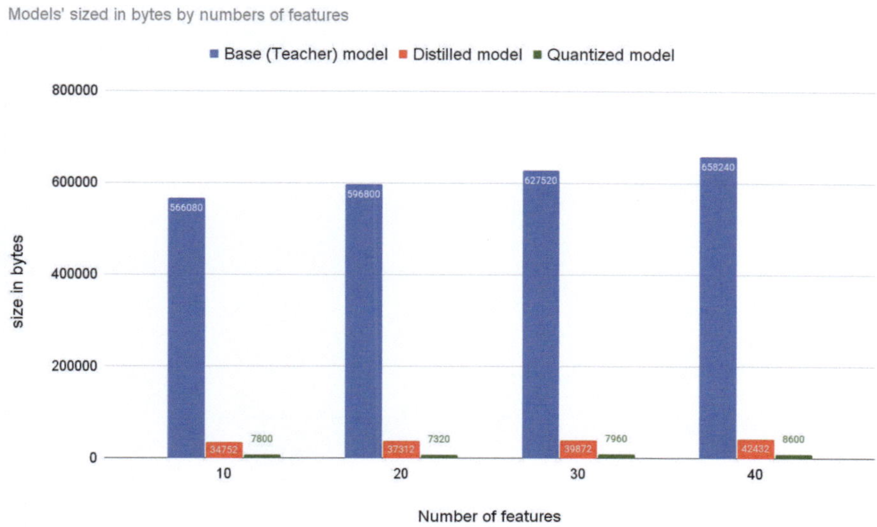

Fig. 5 Models sizes over different numbers of features

Table 4 Models performances using 30 features

Metric	Base (teacher)	Distilled model	Quantized model
Model size (bytes)	627,520	39,872	7960
Inference time (ms)	0.0519	0.0502	0.0099
AUC-ROC	0.9992	0.9992	0.9992
Accuracy	0.9891	0.9878	0.9880
Precision	0.9887	0.9876	0.9881
Recall	0.9903	0.9888	0.9889
F1-score	0.9895	0.9882	0.9885

most of the time. This near-perfect AUC confirms that reducing model size through quantization does not compromise the crucial task of reliable detection.

6 Conclusion and Future Works

This work presents a lightweight intrusion detection model tailored for Cyber-Physical Systems (CPS) by integrating feature selection, knowledge distillation, and post-training quantization. Our approach significantly reduces model size and inference time while maintaining high detection accuracy. Experimental results demonstrate that the quantized model achieves 98.80% accuracy with a minimal 7.96 KB footprint, making it highly efficient for deployment in resource-constrained CPS

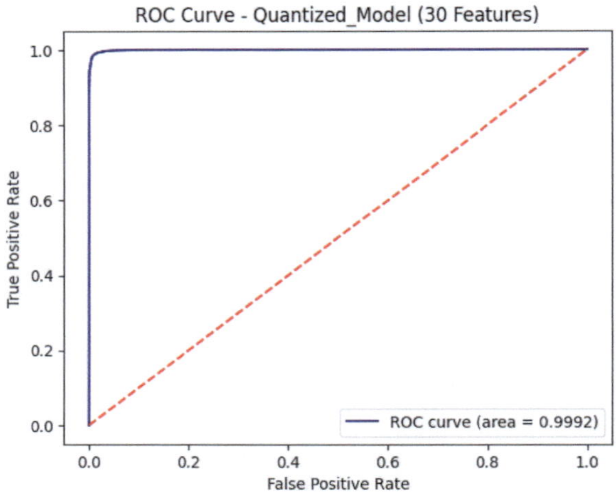

Fig. 6 ROC curve for the quantized model with 30 features

environments. For future work, we aim to extend this approach to real-time CPS deployments and explore adaptive learning techniques to enhance model resilience against evolving cyber threats. Additionally, we will investigate the integration of differential privacy and secure model aggregation to further improve the security and privacy of CPS-based intrusion detection systems.

References

1. Shi J, Wan J, Yan H, Suo H (2011) A survey of cyber-physical systems. In: 2011 international conference on wireless communications and signal processing (WCSP), IEEE, Nanjing, China, Nov 2011, pp 1–6. https://doi.org/10.1109/WCSP.2011.6096958
2. Bella K, Guezzaz A, Benkirane S, Azrour M, Mohy-eddine M (2024) Blockchain and machine learning applications in overcoming security challenges for CPS and IoT Systems. In: Farhaoui Y, Hussain A, Saba T, Taherdoost H, Verma A (eds) Artificial intelligence, data science and applications, vol 838. Lecture notes in networks and systems, vol 838. Springer Nature Switzerland, Cham, pp 208–213. https://doi.org/10.1007/978-3-031-48573-2_30
3. Volkan G, Steffen P, Tony G, Frank V (2014) A survey on concepts, applications, and challenges in cyber-physical systems. KSII Trans Internet Inf Syst 8(12). https://doi.org/10.1142/S2424862217500129
4. Chen H (2017) Applications of cyber-physical system: a literature review. J Ind Integr Manag 02(03):1750012. https://doi.org/10.1142/S2424862217500129
5. Gao Y, et al. (2013) Analysis of security threats and vulnerability for cyber-physical systems. In: Proceedings of 2013 3rd international conference on computer science and network technology. IEEE, Dalian, Oct 2013, pp 50–55. https://doi.org/10.1109/ICCSNT.2013.6967062
6. Cao L, Jiang X, Zhao Y, Wang S, You D, Xu X (2020) A survey of network attacks on cyber-physical systems. IEEE Acc 8:44219–44227. https://doi.org/10.1109/ACCESS.2020.2977423
7. Duo W, Zhou M, Abusorrah A (2022) A survey of cyber attacks on cyber physical systems: recent advances and challenges. IEEECAA J Autom Sin 9(5):784–800. https://doi.org/10.1109/JAS.2022.105548
8. Alguliyev R, Imamverdiyev Y, Sukhostat L (2018) Cyber-physical systems and their security issues. Comput Ind 100:212–223. https://doi.org/10.1016/j.compind.2018.04.017

9. Khraisat A, Gondal I, Vamplew P, Kamruzzaman J (2019) Survey of intrusion detection systems: techniques, datasets and challenges. Cybersecurity 2(1):20. https://doi.org/10.1186/s42400-019-0038-7
10. Mohy-eddine M, Guezzaz A, Benkirane S, Azrour M, Bella K (2024) A whale optimization algorithm feature selection model for IoT detecting intrusion in environments. In: Farhaoui Y, Hussain A, Saba T, Taherdoost H, Verma A (eds) Artificial intelligence, data science and applications, vol 838. Lecture notes in networks and systems, vol 838. Springer Nature Switzerland, Cham, pp 413–419. https://doi.org/10.1007/978-3-031-48573-2_59
11. Bella K, et al. (2024) An efficient intrusion detection system for IoT security using CNN decision forest. Peer J Comput Sci 10:e2290. https://doi.org/10.7717/peerj-cs.2290
12. Mohy-Eddine M, Bella K, Guezzaz A, Benkirane S, Azrour M (2023) Design and implementation of intrusion detection model with machine learning techniques for IoT security. In: Blockchain and machine learning for IoT security, 1st edn. Chapman and Hall/CRC, New York, pp 135–148. https://doi.org/10.1201/9781003438779-9
13. Mitchell R, Chen I-R (2014) A survey of intrusion detection techniques for cyber-physical systems. ACM Comput Surv 46(4):1–29. https://doi.org/10.1145/2542049
14. Zhao R, Chen Y, Wang Y, Shi Y, Xue Z (2021) An efficient and lightweight approach for intrusion detection based on knowledge distillation. In: ICC 2021—IEEE international conference on communications, IEEE, Montreal, Jun 2021, pp 1–6. https://doi.org/10.1109/ICC42927.2021.9500574
15. Ali T, Eleyan A, Bejaoui T, Al-Khalidi M (2024) Lightweight intrusion detection system with GAN-based knowledge distillation. In: 2024 international conference on smart applications, communications and networking (SmartNets), IEEE, Harrisonburg, May 2024, pp 1–7. https://doi.org/10.1109/SmartNets61466.2024.10577682
16. Sharmila BS, Nagapadma R (2023) QAE-IDS: DDoS anomaly detection in IoT devices using post-quantization training. Smart Sci 11(4):774–789. https://doi.org/10.1080/23080477.2023.2260023
17. Wang Z, Ghaleb FA, Zainal A, Siraj MM, Lu X (2024) An efficient intrusion detection model based on convolutional spiking neural network. Sci Rep 14(1):7054. https://doi.org/10.1038/s41598-024-57691-x
18. Otokwala U, Petrovski A, Kalutarage H (2024) Optimized common features selection and deep-autoencoder (OCFSDA) for lightweight intrusion detection in internet of things. Int J Inf Secur 23(4):2559–2581. https://doi.org/10.1007/s10207-024-00855-7
19. Tiwari RS, Lakshmi D, Das TK, Tripathy AK, Li K-C (2024) A lightweight optimized intrusion detection system using machine learning for edge-based IIoT security. Telecommun Syst 87(3):605–624. https://doi.org/10.1007/s11235-024-01200-y
20. Bouayad A, Alami H, Janati Idrissi M, Berrada I (2024) Lightweight federated learning for efficient network intrusion detection. IEEE Acc 12:172027–172045. https://doi.org/10.1109/ACCESS.2024.3494057
21. Putrada AG, Alamsyah N, Fauzan MN, Prabowo S, Oktaviani ID (2024) QUIDS: a novel edge-based botnet detection with quantization for IoT device pairing. Indones J Comput Indo-JC, Jan 2024, pp 29–41. https://doi.org/10.34818/INDOJC.2023.8.3.878
22. Mittal S (2024) CyberSentinel: efficient anomaly detection in programmable switch using knowledge distillation. arXiv. https://doi.org/10.48550/ARXIV.2412.16693
23. Abbasi M, Shahraki A, Prieto J, Arrieta AG, Corchado JM (2024) Unleashing the potential of knowledge distillation for IoT traffic classification. IEEE Trans Mach Learn Commun Netw 2:221–239. https://doi.org/10.1109/TMLCN.2024.3360915

Interpretable Anomaly Detection for Cyber-Physical System Risk Mitigation Using CNN and SHAP

Kamal Bella, Azidine Guezzaz, Ravikumar Patchmuthu, Said Benkirane, Mouaad Mohy-Eddine, and Mourade Azrour

Abstract Cyber-physical systems (CPS) increasingly underlie critical infrastructures, and ensuring they remain free of anomalies is paramount, although many existing detection techniques either struggle to adapt to novel threats or offer little insight into how decisions are reached. In this paper, we introduce a convolutional neural network (CNN) combined with SHAP (SHapley Additive exPlanations) that not only achieves an accuracy of 98.89% but also provides transparent justifications for each anomaly prediction. By carefully preprocessing the NSL-KDD dataset—merging and standardizing features, as well as encoding categorical fields—we create a robust foundation for our CNN, which employs one-dimensional convolutions, dropout, and global average pooling to capture subtle patterns while mitigating overfitting risks. Empirical evaluations show that precision, recall, and F1-score all exceed 98%, with the ROC AUC reaching 99.93%, yet the true novelty lies in how SHAP clarifies feature-level contributions, empowering system operators and security professionals to understand precisely why particular instances are flagged as malicious or benign.

Keywords Shapley values · Anomaly detection · XAI · CNN · NSL-KDD · IDS · CPS

K. Bella · A. Guezzaz (✉) · S. Benkirane
SISAR Team, LaRTiD Laboratory, Higher School of Technology, Cadi Ayyad University, Marrakech, Morocco
e-mail: a.guzzaz@gmail.com

R. Patchmuthu
Universiti Teknology Brunei, Bandar Seri Begawan, Brunei Darussalam
e-mail: ravi.patchmuthu@utb.edu.bn

M. Mohy-Eddine
DELTA Lab, National School of Arts and Crafts, Hassan II University, Casablanca, Morocco

M. Azrour
Faculty of Sciences and Technics, Moulay Ismail University, Meknes, Morocco

G. H. L. et al. (eds.), *Reliability in Cyber-Physical Systems: The Human Factor Perspective*, Springer Series in Reliability Engineering,
https://doi.org/10.1007/978-3-032-09917-4_8

1 Introduction

In the evolving domain of cyber-physical systems (CPS), the delicate interplay of digital networks and physical processes underpins a wide array of critical infrastructures—ranging from power grids and industrial control systems to advanced healthcare devices and transportation networks [1, 2]. These environments face substantial risks when anomalies or intrusions go undetected, as the consequences can be financial, reputational, and, in some cases, life-threatening [3]. Traditional anomaly detection strategies [4, 5]—which are often based on signature-based approaches or overly simplistic statistical methods—frequently struggle to adapt when malicious actors refine their tactics [6]. In light of this persistent challenge, one might ask, how can we secure these intricate systems, while at the same time preserving trust in automated detection, especially considering the dynamic and evolving nature of modern cyber threats? [7].

Development of automated tools to anomaly detection plays an important role [8]. Recent advances in deep learning offer a promising solution [9], even if it must be acknowledged that many of these methods operate as "black boxes" which makes their interpretation quite challenging and at times results in an opacity that renders them unsuitable for real-world deployments where transparency is absolutely crucial [10]. Convolutional neural networks (CNNs), which are recognized for their remarkable ability to capture local feature interactions, show a distinct proficiency at identifying subtle patterns in network traffic and sensor data—patterns that are likely to be overlooked by classical approaches [11]. It remains of utmost importance that these learned representations receive a detailed explanation, a requirement that cannot be overstated [12] especially in safety–critical scenarios where operators and engineers need to clearly grasp why an alarm was raised, ensuring that the system's decisions are both transparent and justifiable. SHAP (SHapley Additive exPlanations), a state-of-the-art interpretability framework, provides insight by attributing each feature's contribution to the final model output, thereby bridging the gap between high-performance detection and the need for accountability [13].

This paper addresses the dual objectives of achieving robust anomaly detection in CPS while ensuring interpretability, which is crucial for risk mitigation and stakeholder confidence. By integrating CNNs with SHAP, we create a pipeline that not only identifies malicious behavior with high accuracy but also reveals which features most strongly influence each prediction. After discussing related work in anomaly detection and model interpretability, we present our methodology—detailing data preprocessing, model architecture, and the SHAP-based explanation layer—followed by comprehensive experiments on the NSL-KDD dataset to validate performance. Ultimately, we argue that this framework elevates CPS security by providing both the deep learning power needed to spot complex attacks and the transparency needed to justify critical decisions in real-world settings.

The remainder of this paper is organized as follows: in Sect. 2 we present related works focusing on IDS for CPS security and explainability, Sect. 3 is our proposed methodology and solution, and in Sect. 4 we present our experimental setup, our environment, dataset and evaluation metrics, we then present the obtained results and discussion in Sect. 5 highlighting and interpreting the model's performance, and finally we conclude in Sect. 6 and outline the direction of our future research.

2 Related Works

Rahim et al. [14] proposed a host-based IDS for CPS security using a mixture-of-experts model and a context-aware CPS-SNORT ruleset for inspecting Gcode instructions. Testing on a new CPS dataset showed that a random tree classifier achieved over 99% accuracy for known attacks, logistic regression reached 85% in a semi-supervised setting, and an LSTM for anomaly detection achieved 99.9% accuracy. Catillo et al. [15] introduced CPS-GUARD, an intrusion detection approach for CPSs using a semi-supervised autoencoder with an outlier-aware thresholding technique. Experiments across six datasets yielded recall scores between 0.949 and 1.000, precision between 0.961 and 0.999, and false positive rates of 0.006–0.027, showing competitive performance compared to other methods. Akintade et al. [16] explored reducing the feature space for IDS models using permutation importance in XGB alongside SHAP and LIME for explainability. For NSL-KDD, features were reduced from 42 to 20, improving AUC from 0.8530 to 0.8751 with a 60% reduction in training time; for CCID-V1, features dropped from 82 to 22, achieving an AUC of 0.9999 with a 46% improvement in computing time. Ahmed et al. [17] introduced a hybrid ensemble method that integrates Explainable AI—using Shapley Additive explanations for feature selection—with a novel approach called Hybrid Bagging-Boosting and Boosting on Residuals. The multi-step process evaluates both binary and multiclass classification using standard metrics, achieving a peak accuracy of 98.47% and an F1 score of 96.19%, thereby significantly enhancing model performance, robustness, and transparency in Intrusion Detection and Prevention Systems. Bhatt et al. [18] enhanced industrial control security by using a Measurement Intrusion Detection System (MIDS) that analyzes measurement data directly, supported by a hardware-in-the-loop system to generate labeled attack data. Machine learning models, including random forest, detect anomalies with high accuracy (ROC of 0.9999 and 97.95% accuracy), while explainable AI techniques provide decision-making transparency. Sivamohan1 And Sridhar [19] introduced BiLSTM-XAI, an intrusion detection system for Industry 4.0 that combines data cleaning, normalization, and feature selection via the Krill Herd Optimization algorithm with a Bidirectional LSTM model. Integrating SHAP and LIME for explainability, the method was evaluated on Honeypot and NSL-KDD datasets using MATLAB 2016, achieving a classification accuracy of 98.2%. Attique et al. [20] proposed a Federated Learning-based IDS for Industrial IoT, using a deep neural network (EX-DFL) to detect anomalies while preserving data privacy. By integrating SHAP-based Explainable AI for better interpretability, the approach achieved over 99% accuracy on the CICIDS2017 dataset. Existing research in anomaly detection for cyber-physical systems often focuses solely on achieving high accuracy while operating as opaque "black boxes," leaving a critical gap in interpretability. This lack of transparency hinders trust and limits practical deployment in safety–critical environments, as operators are left without clear insights into why anomalies are flagged. Our paper addresses this gap by integrating a CNN with SHAP-based explanations, providing both robust detection performance and transparent, feature-level insights for effective risk mitigation.

3 Methodology

Our methodology, as depicted in Fig. 1, which is intended to provide interpretable anomaly detection in cyber-physical systems, commences with a thorough data preprocessing phase that involves merging the NSL-KDD training and testing files into a single Data Frame, removing irrelevant columns (if present), and labeling any non-normal entry as "anomaly," which ensures a consistent definition of attack behavior across the entire dataset. We handle missing values by either imputing or discarding them, depending on their distribution, and then proceed to identify non-numeric columns that are converted into numerical form via label encoding so the CNN can process them effectively. Once label encoding is complete, we standardize the data using a Standard Scaler to mitigate discrepancies in feature magnitude, which is vital for stable model convergence, and we finally split the data into training and test sets in a stratified manner, thereby preserving class balance and promoting a fair evaluation of the proposed approach.

Following this preprocessing, we reshape each sample into a three-dimensional array to accommodate one-dimensional convolutions, then feed these arrays into a CNN architecture that can include up to three convolutional layers with ReLU activation, each designed to uncover latent relationships among features that might otherwise remain hidden, particularly in complex CPS environments. We employ global average pooling to reduce the dimensionality of feature maps and thus curb overfitting, after which a dense layer of 16 units processes the aggregated information for enhanced representation learning, and a dropout layer introduces additional regularization. To conclude the forward pass, a single neuron with a sigmoid activation outputs the probability of an instance being anomalous or normal, while the Adam optimizer and binary cross-entropy loss function guide the learning process, and a series of hyper parameter tuning steps—covering filter sizes, regularization parameters, and convolutional depth—aim to identify an optimal balance between model capacity and generalization performance, with accuracy, confusion matrices, classification metrics, and the ROC AUC score serving as the principal indicators of success.

Finally, we incorporate SHAP to illuminate the inner workings of the trained model, which is particularly important when addressing sensitive or safety–critical situations in CPS, because operators, engineers, and researchers alike require justifications for the system's decisions rather than mere numerical predictions. By leveraging Gradient Explainer, we compute Shapley values that highlight the contribution of each feature in steering the model's output toward normal or anomalous

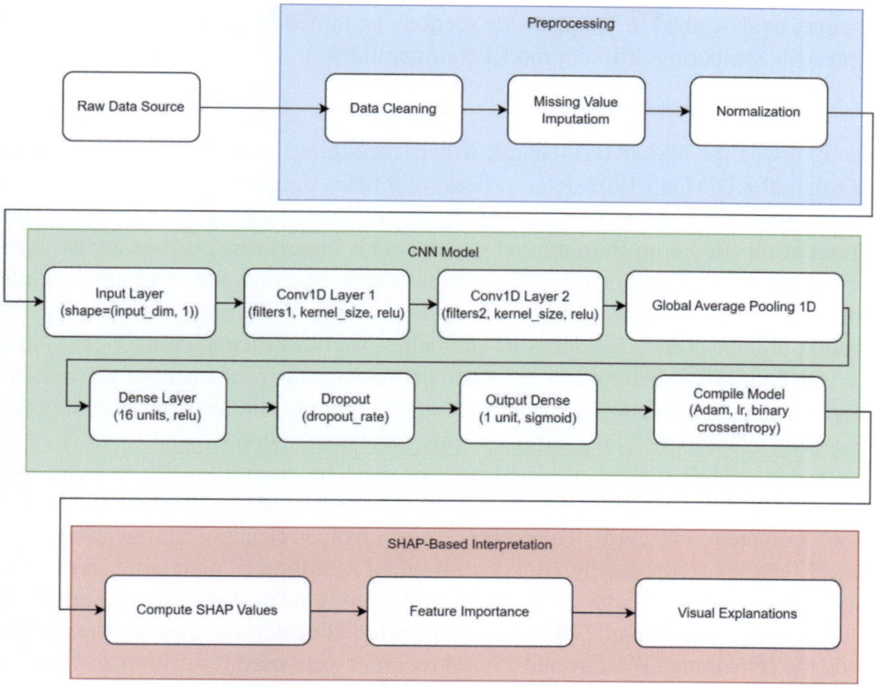

Fig. 1 Proposed solution of interpretable anomaly detection for cyber-physical system

classifications, and we visualize these values using summary plots that reveal overall feature importance trends, as well as force plots that break down the exact influences in individual cases. This interpretability layer, combined with the CNN's ability to capture subtle interactions and the inclusion of regularization methods like global average pooling and dropout, results in a model that not only demonstrates high accuracy but also provides transparent insights into how anomalies are detected, thereby facilitating trust, enabling faster root cause analysis, and promoting adaptability to other industrial or network-based scenarios where both performance and clarity are paramount.

4 Experimental Setup

In this section, we describe the environment and methodology used to train, validate, and test our anomaly detection model, while sometimes relying on a Kaggle-hosted notebook with a P100 GPU for handling the heavier computational tasks. All hyperparameter tuning, including adjustments to the number of convolutional layers, kernel sizes, and regularization parameters, was performed through a systematic search over predefined parameter grids, and by applying our data preprocessing pipeline, as described in the previous section, we aimed to ensure consistency and fairness in comparing different model configurations.

1. NSL-KDD Dataset

We employed the NSL-KDD dataset, widely recognized as an enhanced version of the original KDD Cup 1999 dataset, because it helps mitigate issues like redundant records, thus offering a more balanced distribution of normal and attack traffic. This dataset comprises both training and test sets that incorporate various attack categories, and each record includes 41 features—some of which are continuous, while others are categorical—encompassing crucial details about network behavior. By labeling all non-normal instances as anomalies, we constructed a binary classification task focused on distinguishing normal network traffic from potentially malicious activities, and this structure is particularly suitable for examining how effectively our model can adapt to the complexities of cyber-physical systems.

The NSL-KDD dataset was first preprocessed by combining the training and testing CSV files, resulting in an aggregated dataset of approximately 148,517 network connection records. The preprocessing involved cleaning and encoding categorical features, followed by applying stratified sampling to maintain class distribution. The combined dataset was then partitioned into two subsets using an 80/20 split: roughly 80% (about 118,813 records) were allocated for training the model, while the remaining 20% (around 29,704 records) were used for testing and validation during model development. This stratified splitting ensured that both training and evaluation phases effectively represented the inherent class imbalances present in the original NSL-KDD dataset.

2. Evaluation Metrics

We assessed our model using multiple metrics, such as accuracy, the confusion matrix, precision, recall, F1-score, and the ROC AUC, because no single measure can fully capture the challenges inherent in anomaly detection. While accuracy is straightforward, it can prove misleading if the dataset is highly imbalanced, which prompted us to incorporate precision (the proportion of predicted anomalies that are actually anomalous) and recall (the fraction of true anomalies correctly identified) to gain deeper insights into the model's behavior. The F1-score, which combines precision and recall, gives a balanced view of performance while the ROC AUC demonstrates the model's ability to distinguish normal from anomalous instances by integrating true and false positive rates across different thresholds.

5 Results and Discussion

Table 1 shows that our CNN-based anomaly detection approach achieves 98.89% accuracy, 99.62% precision, 98.23% recall, and an F1-score of 98.92%, clearly indicating that the model can detect anomalies with high accuracy while keeping false alarms to a minimum—a critical factor in cyber-physical systems where operational stability and security must be maintained.

The confusion matrix in Fig. 2 illustrates the model's performance by showing that 99.59% of normal samples are correctly classified as normal and only 0.41% of these instances are misclassified as anomalies and this result offers clear evidence of the model's capability to accurately distinguish between normal and anomalous cases. At the same time, 98.23% of anomalous samples are correctly identified, leaving just 1.77% of attacks undetected. Finally the ROC curve in Fig. 3 shows an AUC of 99.93%, confirming the model's strong discriminative power and its effective balance of sensitivity and specificity across classification thresholds.

Figure 4 displays the SHAP summary plot generated by the GradientExplainer, offering both global and local insights into how each feature contributes to the final model output. The features are sorted by their mean absolute SHAP values, with "logged_in" emerging as the most influential predictor across the test samples. In practical terms, this implies that whether or not a user session is logged in can strongly shift the model's decision boundary toward identifying a particular record as either normal or anomalous. Following "logged_in," "protocol_type," "flag," "count," and "same_srv_rate" also exhibit substantial impact, suggesting that variations in network protocol usage, packet flags, or the frequency of same-service requests can play a decisive role in anomaly detection. The color gradient in the summary

Table 1 Model performance

ACC	PR	Recall	F1-score
0.9888567196	0.9961832061	0.9822853806	0.9891854804

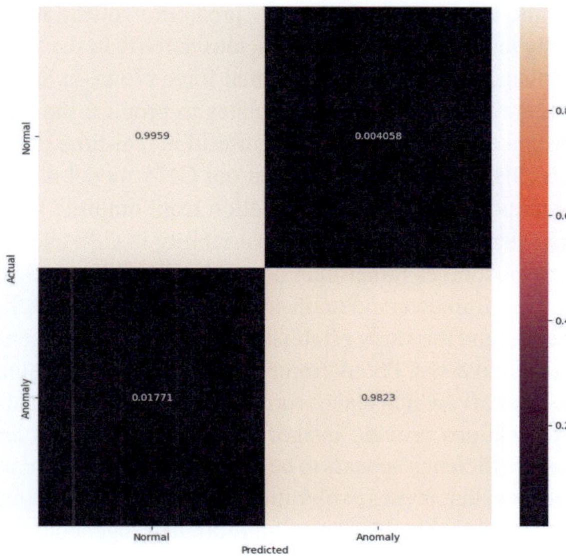

Fig. 2 The confusion matrix to illustrate the model's performance

Fig. 3 Roc curve

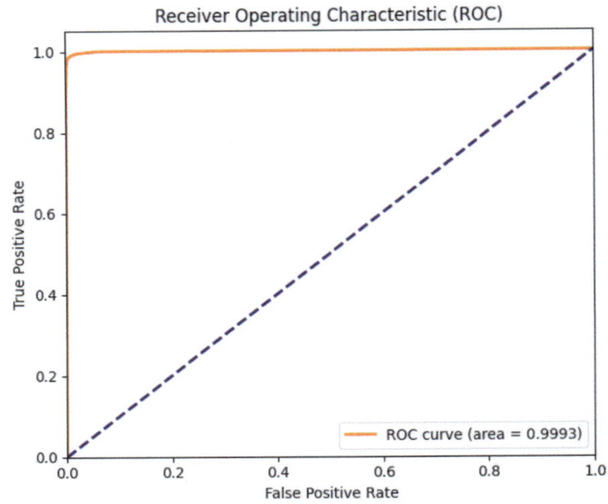

plot indicates the feature values from low (blue) to high (red), and the corresponding SHAP values on the x-axis reflect how strongly each feature value pushes the model's prediction toward normal or anomaly. Thus, a strongly positive SHAP value for a given feature indicates that this feature-value combination increases the likelihood of an anomaly, whereas a negative SHAP value has the opposite effect.

The underlying SHAP analysis offers a granular understanding of the CNN's decision-making process, highlighting not just which features matter, but also how they matter. For instance, certain protocols or flags may consistently be associated with higher SHAP values, signifying that these network conditions are more frequently linked to anomalous behavior. Conversely, features such as "logged_in" might, in many cases, steer the prediction toward normal, if being logged in is typically less correlated with malicious activity in the training data. By examining local explanations—such as individual force plots—it becomes possible to see how each feature's contribution accumulates to produce the final classification for a specific instance, which is particularly useful in real-time investigation or incident response. The SHAP results confirm that our CNN model does not rely on a single dominant predictor; it integrates information from multiple features, which indicates a detection strategy that handles the variability in CPS anomalies by drawing on data from several sources rather than depending solely on one factor.

Our framework demonstrates accuracy and interpretability on the NSL-KDD dataset and the study establishes a basis that might benefit from additional validation across diverse CPS environments given that the evaluation is conducted on a single benchmark, which leaves room to question its applicability in varied settings; SHAP explanations provide insight into model decisions and optimizing their computational efficiency appears to be a promising avenue to enhance real-time applicability, a prospect that invites exploration in practical implementations. Future work is planned to extend the evaluation to multiple heterogeneous datasets, refine the explanation

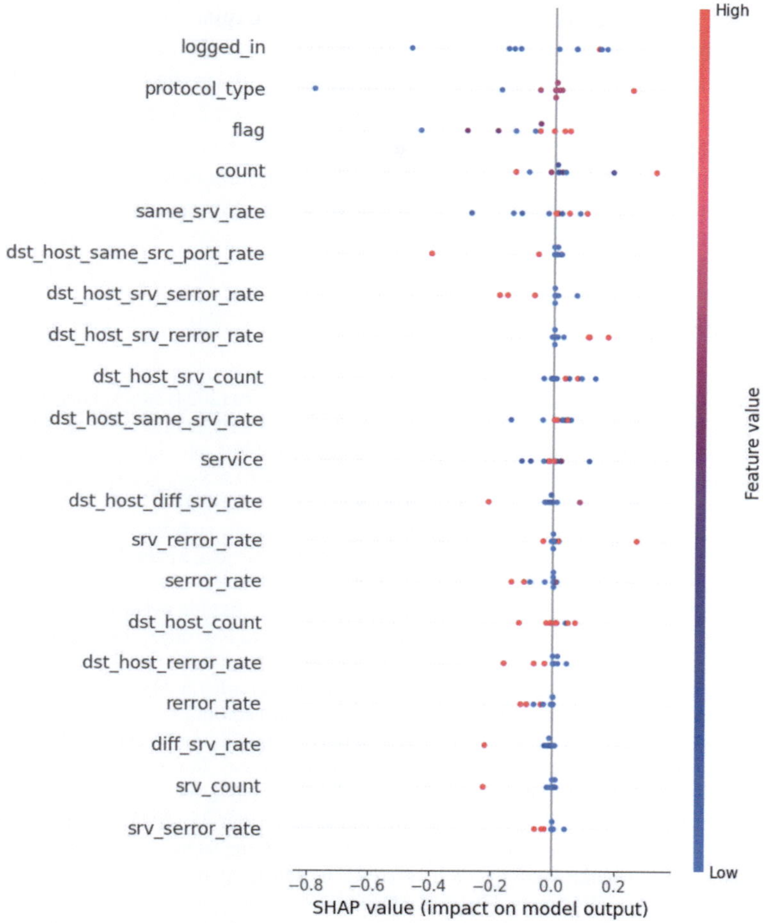

Fig. 4 SHAP summary

process, and explore complementary architectures such as attention mechanisms with an aim to improve scalability and robustness in operational settings.

6 Conclusion

In cyber-physical systems, where uptime and reliability are required to maintain operations, intrusion detection systems (IDS) serve a role in countering malicious incursions, many existing solutions function as opaque "black boxes" that complicate understanding how anomalies are flagged. Our work addresses this gap by introducing a CNN combined with SHAP, a strategy that ensures both detection accuracy and model interpretability, and experiments on the NSL-KDD dataset show that our model achieves an accuracy of 98.89% while offering explanations for the impact of each feature, a result that supports operator trust and speeds incident response. Our framework shows excellent accuracy and interpretability on the NSL-KDD dataset, yet broader evaluation on diverse CPS environments could further confirm its generalizability. Refining the computational efficiency of SHAP explanations may also enhance real-time applicability, while exploring complementary architectures, such as attention mechanisms, could further improve scalability and robustness.

References

1. Bella K, Guezzaz A, Benkirane S, Azrour M, Mohy-Eddine M (2024) Blockchain and machine learning applications in overcoming security challenges for CPS and IoT systems. In: Farhaoui Y, Hussain A, Saba T, Taherdoost H, Verma A (eds) Artificial intelligence, data science and applications. Lecture notes in networks and systems, vol 838. Springer Nature Switzerland, Cham, pp 208–213. https://doi.org/10.1007/978-3-031-48573-2_30
2. Ahmed SH, Kim G, Kim D (2013) Cyber physical system: architecture, applications and research challenges. In: 2013 IFIP wireless days (WD), Valencia, Spain, Nov 2013. IEEE, pp 1–5. https://doi.org/10.1109/WD.2013.6686528
3. Zahid M, Inayat I, Daneva M, Mehmood Z (2021) Security risks in cyber physical systems—a systematic mapping study. J Softw Evol Process 33(9):e2346. https://doi.org/10.1002/smr.2346
4. Mohy-Eddine M, Guezzaz A, Benkirane S, Azrour M, Bella K (2024) A whale optimization algorithm feature selection model for IoT detecting intrusion in environments. In: Farhaoui Y, Hussain A, Saba T, Taherdoost H, Verma A (eds) Artificial intelligence, data science and applications. Lecture notes in networks and systems, vol 838. Springer Nature Switzerland, Cham, pp 413–419. https://doi.org/10.1007/978-3-031-48573-2_59
5. Mohy-Eddine M, Bella K, Guezzaz A, Benkirane S, Azrour M (2023) Design and implementation of intrusion detection model with machine learning techniques for IoT security. In: Blockchain and machine learning for IoT security, 1st edn. Chapman and Hall/CRC, New York, pp 135–148. https://doi.org/10.1201/9781003438779-9
6. Issa MM, Aljanabi M, Muhialdeen HM (2024) Systematic literature review on intrusion detection systems: research trends, algorithms, methods, datasets, and limitations. J Intell Syst 33(1):20230248. https://doi.org/10.1515/jisys-2023-0248
7. Susanto H, Azim MQ, Fang-Yie L, Susanto AKS, Setiana D, Ibrahim F, Basuki AI, Ramdhani TI, Setiawan I, Riyanto B, Ferianda RA, Haris Subramanian U (2023) Artificial intelligence, blockchain framework, cyberthreat defenses of resilient digital ecosystems. In: Handbook of research on artificial intelligence and knowledge management in Asia's digital economy. IGI Global, pp 36–63
8. Isa H, Rahim MEA, Ariffin NAM, Embran RA, Han SHMR, Subramanian U, Kawi F, Abdullah N (2022) Study on the different types of accounting fraud and tools to detect and prevent fraud. In: 2022 7th International conference on business and industrial research (ICBIR). IEEE, pp 627–631
9. Bella K et al (2024) An efficient intrusion detection system for IoT security using CNN decision forest. PeerJ Comput. Sci. 10:e2290. https://doi.org/10.7717/peerj-cs.2290
10. Patrício C, Neves JC, Teixeira LF (2024) Explainable deep learning methods in medical image classification: a survey. ACM Comput Surv 56(4):1–41. https://doi.org/10.1145/3625287
11. Mohammadpour L, Ling TC, Liew CS, Aryanfar A (2022) A survey of CNN-based network intrusion detection. Appl Sci 12(16):8162. https://doi.org/10.3390/app12168162
12. Ibrahim R, Shafiq MO (2023) Explainable convolutional neural networks: a taxonomy, review, and future directions. ACM Comput Surv 55(10):1–37. https://doi.org/10.1145/3563691

13. Lundberg SM, Lee S-I (2017) A unified approach to interpreting model predictions. In: Guyon I, Luxburg UV, Bengio S, Wallach H, Fergus R, Vishwanathan S, Garnett R (eds) Advances in neural information processing systems. Curran Associates, Inc. [Online]. Available https://proceedings.neurips.cc/paper_files/paper/2017/file/8a20a8621978632d76c43dfd28b67767-Paper.pdf
14. Rahim K, Nasir ZUI, Ikram N, Qureshi HK (2025) Integrating contextual intelligence with mixture of experts for signature and anomaly-based intrusion detection in CPS security. Neural Comput Appl 37(8):5991–6007. https://doi.org/10.1007/s00521-024-10967-9
15. Catillo M, Pecchia A, Villano U (2023) CPS-GUARD: Intrusion detection for cyber-physical systems and IoT devices using outlier-aware deep autoencoders. Comput Secur 129:103210. https://doi.org/10.1016/j.cose.2023.103210
16. Akintade S, Kim S, Roy K (2023) Explaining machine learning-based feature selection of IDS for IoT and CPS devices. In: Maglogiannis I, Iliadis L, MacIntyre J, Dominguez M (eds) Artificial intelligence applications and innovations. IFIP advances in information and communication technology, vol 676. Springer Nature Switzerland, Cham, pp 69–80. https://doi.org/10.1007/978-3-031-34107-6_6
17. Ahmed U et al (2024) Hybrid bagging and boosting with SHAP based feature selection for enhanced predictive modeling in intrusion detection systems. Sci Rep 14(1):30532. https://doi.org/10.1038/s41598-024-81151-1
18. Bhatt DG, Kyada PU, Rathore RS, Nallakaruppan MK, Jhaveri RH (2025) Enhancing anomaly detection in industrial control systems through supervised learning and explainable artificial intelligence. J Cybersecurity Inf Manag 15(1). https://doi.org/10.54216/JCIM.150125
19. Sivamohan S, Sridhar SS (2023) An optimized model for network intrusion detection systems in industry 4.0 using XAI based Bi-LSTM framework. Neural Comput Appl 35(15):11459–11475. https://doi.org/10.1007/s00521-023-08319-0
20. Attique D, Hao W, Ping W, Javeed D, Adil M (2024) EX-DFL: an explainable deep federated-based intrusion detection system for industrial IoT. In: 2024 21st International joint conference on computer science and software engineering (JCSSE), Phuket, Thailand. IEEE, June 2024, pp 358–364. https://doi.org/10.1109/JCSSE61278.2024.10613665

Intrusion Detection Approaches in Healthcare Systems: An Overview

Chaimae Hazman, Maryam Douiba, Azidine Guezzaz, Vinayakumar Ravi, Mourade Azrour, and Said Benkirane

Abstract Since healthcare facilities been more reliant on digital technology, safeguarding critical customer data and medical devices against cyber attacks has become critical. Intrusion Detection Systems (IDS) are critical defense mechanisms for identifying and reacting to breaches of security. This study investigates the importance of IDS for medical security, focusing on major techniques for detection including based on signatures, anomaly-based, and hybrid approaches. It also handles issues like as false positives, limited resource availability, and attacker flexibility. In addition, we investigate how advances in artificial intelligence and cloud computing might increase IDS efficacy. Our findings highlight the significance of IDS in improving cybersecurity in health care organizations and protecting the security of vital health data.

Keywords Healthcare · Security · Intrusion detection · Artficial intelligence

C. Hazman
Research Team LAMIGEP, EMSI-Marrakech, Marrakech, Morocco
e-mail: C.Hazman@emsi.ma

C. Hazman · M. Douiba · A. Guezzaz (✉) · S. Benkirane
SISAR Team, LaRTiD Laboratory, Technology Higher School Essaouira, Cadi AyyadUniversity, Marrakech, Morocco
e-mail: a.guzzaz@gmail.com

V. Ravi
Center for Artificial Intelligence, Prince Mohammad Bin Fahd University, Khobar, Saudi Arabia
e-mail: vravi@pmu.edu.sa

M. Azrour
Faculty of Sciences and Techniques, Moulay Ismail University of Meknes, Errachidia, Morocco

G. H. L. et al. (eds.), *Reliability in Cyber-Physical Systems: The Human Factor Perspective*, Springer Series in Reliability Engineering,
https://doi.org/10.1007/978-3-032-09917-4_9

1 Introduction

The increased use of computer technology in healthcare has improved patient care, data administration, and treatments. Still, this transition has created substantial cybersecurity threats, such as hacking, attacks involving ransomware, and unauthorised system access. Safeguarding critical patient information and ensuring the security of medical systems have become major considerations [1–5]. Intrusion Detection Systems (IDS) assist in mitigating these risks by constantly tracking network traffic and system activity for unusual activity. To identify and respond to possible threats, these technologies employ a variety of detection approaches, including signature-based, anomaly-based, and hybrid strategies. IDS with signatures uses predetermined patterns to detect known attacks, whereas anomaly-based IDS looks for anomalies from typical activity, which renders them useful against new and developing threats. Hybrid IDS combines the two techniques to improve detection capability. Notwithstanding its advantages, adopting IDS in healthcare contexts offers obstacles. High levels of false positives can overload security personnel with superfluous alarms, making it difficult to discern between genuine threats and typical system activity. Furthermore, because healthcare systems function in real time, intrusion detection systems (IDS) must detect and react to intrusions without disturbing vital medical operations [6–8]. The intricate nature of medical networks, that frequently contain older systems and networked devices, hinders IDS installation. Innovations in artificially intelligent (AI) and cloud computing have the potential to improve IDS precision and productivity. AI-powered IDS uses machine learning algorithms to evaluate large volumes of data, find trends, and eliminate false positives. Cloud-based IDS offer scalable security solutions, allowing for real-time surveillance of threats and central oversight across numerous healthcare institutions. These innovations enhance IDS performance, making them more adaptable to modern cyber threats. As cyberattacks targeting healthcare systems become more sophisticated, organizations must continuously update and optimize their security measures [9].

IDS play a crucial role in strengthening healthcare cybersecurity by detecting, analyzing, and mitigating potential threats. To summarize, the uniqueness of this study lies in our approach that combines efficient feature selection with advanced machine learning techniques, all optimized for real-time applications in IoT environments. Feature selection plays a crucial role in enhancing both the accuracy of intrusion detection and the speed of predictions. By focusing on the most relevant features, we were able to simplify the model, improving its performance and making it much faster, ideal for real-time use. The results demonstrate that our approach outperforms existing methods across all key metrics, including accuracy and precision. The ability to process data rapidly while maintaining high detection accuracy is essential for securing IoT systems. Our model provides an effective, reliable solution, highlighting how the combination of smart feature selection and machine learning can deliver both robust security and fast performance. This makes it a practical tool for protecting IoT-based systems in dynamic and critical environments. Figure 1 shows a smart healthcare environment.

Fig. 1 Smart helthcare environment

2 Intrusion Detection System Overview

Intrusion detection research commenced in 1972, after James Anderson, a US Air Force the researcher, issued a study emphasizing the necessity to identify vulnerabilities in IT infrastructure [10–12]. Administrators are used to do detection of attacks (ID) by sitting near the edge of a terminal and watching what users were doing. They may identify breaches by seeing, for instances, that a heading member has checked in locally or if a rarely utilized printing device is abnormally busy. Because of the inaccuracies it created, this early version of ID was useless. In the beginning of the 1980s, James Anderson released an article titled "Computer security threat monitoring and surveillance" [13–17], which introduced the concept of audit logs including the vital data, which may be useful in tracing abuse and comprehending user behaviors. As a result, the following stage in the detection of intrusions included reviewing audit logs for signs of anomalous or hostile activities. Until the 1980s, supervisors produced audit reports on fan folded paper, that were frequently piled up to five feet deep at the end of a normal week. Naturally, looking across large an abundance took an extended period of time. Given so much data and limited time for analysis, controllers mostly employed the audit logs as a kind of forensic instrument to establish the root of a particular safety issue following it had occurred. There had been little chance of capturing a live attempt. As storing grew more affordable, records of auditing were put on the internet and academics created tools to evaluate the data. Yet, since research was delayed and frequently physically costly, intrusion detection algorithms were typically performed at least as the computer's user traffic was low. As a result, the majority of breaches continued to be discovered after they happened. As a consequence of Denning's Security Monitoring System, scientists built real-time IDS that analyzed auditor information as it was generated in the beginning of the 1990s. This facilitated the identification of assaults and planned assaults as they happened, allowing for real-time reaction and, in certain situations, attack preemptive. IDS solutions were originally launched in the the middle of the 1990 by two organizations: Internet Security System Inc (ISS) and Wheelgroup. They created network-based intrusion detection systems dubbed RealSecure and Netranger, respectively. The original release of RealSecure 1.0 for Windows NT 4.0 was released by ISS Inc. RealSecure employed a repository of information for matching signatures, but it was unsuccessful against new assaults, thus was a big failure. Back in 1995, Wheelgroup's Netranger was a well-known networking based intrusion detection system that worked by examining network activity. Wheelgroup was purchased by Cisco in the second quarter of 1998 and is now an integral element of Cisco's safety. Nowadays security detection activities are focused on creating solutions that consumers can successfully implement in big networks. Considering rising safety issues, myriad new attack strategies, and regular shifts in the computer surroundings, this is no simple feat. IDS is a system designed to identify suspicious or odd behavior on the object being monitored. As a result, it is feasible to take measures to guard against the dangers of attack. In simple terms, it is equipment, software, or a mix of each of them that has the goal of tracking just one computer or system for harmful activity including to track and manage safeguarding the network against policy violations and cyber assaults. For example, safeguarding a network of IoT devices by deploying IDSs at various levels is a novel approach to dealing with privacy risks and weaknesses. In fact, integrating IDSs within IoT system components is necessary to improve the security of IoT devices. Furthermore, intrusion detection systems (IDS) are commonly employed to identify network anomalies and attacks. Devices can be embedded in clothing and the heart of people to track and regulate how well the body is functioning. Detecting is a type of classification that produces a pair of categorie (Normal and Attack). AI approaches have to be equipped to distinguish between aberrant and regular network actions in order to identify infiltration. Several AI algorithms may be utilized to identify intrusions in IoT systems to achieve the greatest outcomes. Its primary goal is to safeguard consumers from these kinds of assaults by utilizing AI approaches. Classification can be used to identify assaults. For example, if we accept a request, we first determine if it is regular or threat, and then determine the type of the discovered assault [18]. An IDS constitutes a risk-prevention technique designed to identify aberrant network or host activity. Both of the main notions are FP and FNR. The first refers to an IDS warning that is not related to a real assault (i.e., regular traffic is deemed an attack), while the other refers to a true incursion that remained unnoticed. Recognition of outliers or anomaly identification aids in reducing TN and FPR, hence reducing AI incorrect projections and improving outcomes. It is also important in safeguarding IoT devices or any type of safety infrastructure. As a result, intrusion detection systems (IDSs) are widely employed to monitor malicious activities on the network or on one computer. Its goal is to find or discover weaknesses and advise the system manager or provide timely notifications of any possible threats or assaults. DL and ML approaches can be applied in recognizing IoT assaults. This part describes two primary IDS categories: detection-based and data-based IDSs.

3 IDS Categories

Intrusion Detection Systems (IDS) use several methods to observe and examine network activity for possible threats. As dipicted in Fig. 2. One popular approach is signature-based detection, which identifies assaults through the comparison of inbound traffic patterns to a database of known risk identifiers. This method works well for recognizing previously known cyber threats, but it fails with new or changing assaults. Another way is anomaly-based identification, which creates an initial representation of typical network activity and alerts to variations that may signal a security violation. While based on signatures IDS, anomaly-based systems may identify before undiscovered threats, such as zero-day attacks, although they may produce false positives. Integrating both of these methods allows firms to enhance both threat detection and effective safety profile [16].

1. **Detection-Based IDS Methods**
    - **Signature intrusion detection systems (SIDS)**

This strategy is additionally referred to as information-based identification or misuse detection. A signature-based intrusion detection system (IDS) examines network packets and compares them to a database of signatures or characteristics of past attacks. If there is an appropriate match, when the network behavior succeeds, it is classified as an attack, and alert data is produced. For recognizing illegal traffic,

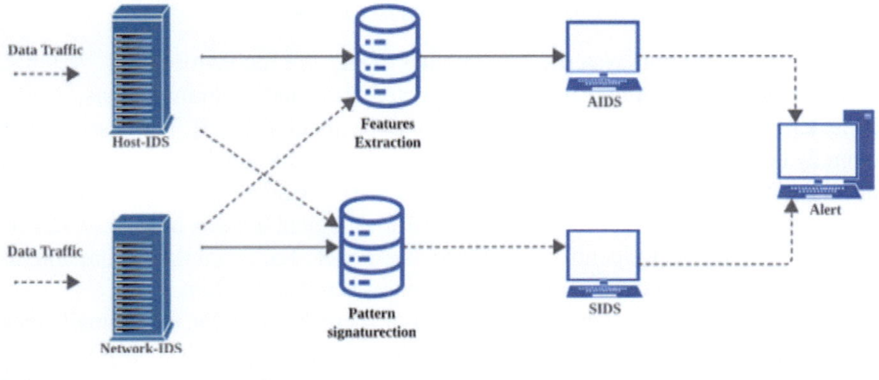

Fig. 2 Description of main elements integrating in intrusion detection system

signature provide a mix of the packet's header and packet content analysis criteria. A packet preamble rule is a 5-pairs filtering on packages that contains the origin and target IP addresses and ports. The packet content verification criteria is made up of a string or recognized structure that must be checked with the packet contents. On the other hand, the packet's header matched necessitates methods of classification which can be accomplished by employing TCAM-Ternary Content Accessible Stories, whereas patterns that match necessitate deep packet inspection, which entails scanning every piece of information in a packet [19].

Patterns are usually expressed as precisely match statements. Obviously in part to their widespread acceptance and use, various high-speed and effective character matching approaches have lately been presented. Such often involve string matching methods like AhoCorasick, Commentz-Walter, and Wu-Manber, which employ a beforehand processed data-structure to provide exceptionally well matching. Between these instances, Aho-Corasick has seen the most widespread adoption [20]. Misuse IDS works similar to a malware scanner, looking for a recognized identity or signature for every particular intrusion occurrence. It may be deployed on a system to monitor network vulnerabilities or on an individual machine. As a result, the frequency of false positive alerts is lower with signature-based intrusion detection systems. This approach works particularly effectively assuming the system in question has a current repository of signatures that indicate suspected assaults. Additional benefits of this system include the simplicity of maintaining and upgrading the predefined/preconfigured rules, the fact that new identities may be introduced to the database without affecting current ones, and the fact that it is fast and accurate. Quad net evaluation, artificial neural networks, systems of experts, transitional state evaluation, and feature analysis are examples of common abuse detection approaches. It should be mentioned that a variety of business applications incorporate abuse detection methods. The fundamental problem of abuse detection techniques is the fact that they can only identify known assaults for which a specified signature exists; therefore, they cannot identify zero-day attacks or new or undetected attacks. Engineers have to simulate and add new attacks to the signature repository when they are identified.

- **Anomaly-based intrusion detection system (AIDS)**

Anomaly detection systems are another name for behavior-based systems. Anomaly detection employs models of expected users and program actions, identifying departures from this "normal" behavior as an attack [21–23]. The basic idea of anomaly detection is that assaults deviate from regular activity, and a breach creates unusual use of assets or manifests unexpected behavior on a component of the user in question. As a result, several ways to learning typical actions have been developed for the purpose to identify any major divergence. For example, we are able to accurately model particular users' daily activities (kind and volume).

Assume a certain user signs in around ten o'clock in the morning, reads email, does database interactions, gets an absence around midday and 1 p.m., has only a few file access failures, and furthermore. If the operating system observes an identical user logging in at 3 a.m., starting to utilize compilers and tools for debugging, and having frequent access to files failures, it will mark this behavior as suspect. Actually, this method compares actual user actions toward stored profiles of people or networks in order to identify anomalous or unexpected activity that could indicate breaches. The profile information might be continuous or unchanging, and they can match with anticipated or actual usage patterns. Typical behaviors of users, connections to networks, or hosts are monitored for a specified amount of time termed the learning phase [21] to construct a user profile. Profiles are created utilizing various information such as unsuccessful attempts at logging in, the variety of instances a file is read by a certain user during a specific time period, CPU utilization, and so on.

The threshold being identified, based on rules measurements, measurements based on statistics, algorithmic learning, and data mining technologies are used in the anomaly detection approach. The first strategy uses tallies to express various aspects of both system and user behavior. The last one is then compared to a sensitivity threshold. The next approach attempts to provide a set of principles that may be employed to determine the question of whether or not a given behavior is normal. Statistical measures are used to study the spatial distribution of network activity features. They can be parametric or non-parametric; the former is expected to match a specific pattern, while the latter is learnt from a collection of previous data points.

The final machine learning and data mining approach learns from a collection of training data and builds a model capable of classifying fresh network activity as genuine or malicious. The fundamental benefit of detection of anomalies systems is the fact that they may identify known, unknown, and even zero-day assaults. This validity is especially important because novel kinds of attacks and vulnerabilities are continually emerging. This detection technique can also be signatures for abuse detectors are defined using this method. But new valid activity might be mistakenly detected as an assault, culminating in a false positive as a result, the approach has a somewhat high false positive rate. Table 2 summarizes the key disadvantages and advantages regarding behavioral and signatures techniques.

- **Hybrid intrusion detection system (AIDS)**

The efficiency of IDS can be considerably increased by integrating signature-based and anomaly-based approaches, an approach known as hybrid detection. By doing so, we may take benefit of each strategy while eliminating or mitigating its weaknesses. As a consequence, the hybrid IDS developed will be capable of detecting simultaneously known and unknown threats with high accuracy and a low FPR lately novel hybrid IDS have been created with substantial efficacy.

- **Collaborative Intrusion Detection Systems**

A Collaborative Intrusion Detection System (CIDS) is a specialized type of IDS in which multiple entities work together to detect and mitigate security threats. These systems can be categorized based on the type of information shared and the level of cooperation among participating entities. The collaborative model classifies CIDS into three types: centralized, distributed, and hybrid [20, 21, 24]. In a centralized CIDS, all security-related data is sent to a central server, which processes the information and generates alerts if threats are detected. This model is easy to manage but introduces a single point of failure if the central server is compromised. In contrast, a distributed CIDS consists of multiple independent intrusion detection systems that share threat intelligence, making the system more resilient to failures but more complex to coordinate. A hybrid CIDS combines features of both centralized and distributed models, where some entities report to a central server while others share data directly with each other. This approach enhances security and reliability but requires careful implementation to balance efficiency and complexity.

2. **Data-Based IDS Methods**

Data-based IDS techniques are divided into three categories [24]: host-based IDS, network-based IDS, and hybrid-based IDS.

- **Network-based intrusion detection system (NIDS)**

NIDS has evolved into a vital safety net for networked computers a comprehensive defense. It is a significant preventative measure that is commonly used in network topologies to monitor traffic for irregularities, raise alarms, and ultimately react to any kind of illegal or suspicious behavior in the network [24]. In the simplest terms, NIDS is a security layer that identifies actual time violent conduct, illicit activity, or abnormalities in the network's traffic and performs remedial action to prevent or reduce the incidence of assaults. As a consequence, the confidentiality of information will be protected from assaults, networking safety and reliability will remain intact, the ability to manage of the infrastructure administrator's protection will be enhanced, and the system's overall operational efficiency will be maximized. NIDS typically detects intrusions by analyzing the IP and transport-level flags of each packet. To identify intrusions, NIDS employs atypical behavior or signature-based detection technique. The signature-based detection technique searches for the association of collected packets with signatures that indicate previous attacks, whereas the anomaly recognition method compares the monitoring information to the typical background and issues an alarm if aberrant behavior occurs. Deploying NIDS at the cloud's the user interface aids in the identification of outside network threats and breaches. Fortunately, this is insufficient since the infrastructure could potentially be incapable of detecting inside attacks [25]. To address this problem, placing NIDS on analyzing systems at the cloud's back-end aids in the detection of breaches at the cloud's internal networks through tracking the network's virtualization and the passage of information coming or going to the analyzing servers on the physical network.

- **Host-based intrusion detection system (HIDS)**

HIDS is a piece of computer programs that is put on an environment to monitor and analyze the information acquired from the system in question in order to identify illegal and invasive occurrences. The key tenet of HIDS success is that intruders often leave a trail showing their activity. For example, they could put software on a computer that they obtained control. In general, a HIDS will keep a record of system items that need to be monitored. The database holds critical details on these computer items, such as properties, update duration, terms of size, and so on. Furthermore, the database's contents may include a verification number or hashing of the system's objects. These items can then be compared to the existing system components. If an item proves contradictory at any particular, the entire system could send out an alert. HIDSs are often employed to safeguard software security. HIDS's efficacy can be increased by selecting attributes that supply it with additional data for identification. Still, additional storage is required to facilitate analysis. In the context of a cloud-based computing network, HIDS can be deployed on the hypervisor, virtual machine, or host to examine log files, user authentication information, or enforce access regulations and identify intrusion activities. The cloud user has accountability to monitor HIDS implemented on a VM, whereas the cloud supplier is required for deploying HIDS on a hypervisor that is used. HIDS can analyze encrypted data and requires not any additional equipment because it can be deployed on current host server infrastructure, making it affordable for a small- and medium-network. The primary disadvantages of HIDS are that it cannot detect assaults originating from the computer when it is installed, and it is vulnerable to DoS attacks and can even be deactivated [25].

4 Description of a Proposed Model

In this section, we demonstrate an example of an intrusion detection system based on ensemble learning. For implement this IDS, we provide numerous options to test our intrusion detection technique for IoT security.

1. **Our Proposed Model**

The suggested detection framework includes four different phases:

For increase data quality, our suggested approach concentrates on preparation and normalization. Preparation is a beneficial process that aims to remove noise and cleanse data. In addition, the data are normalized in the interval [0, 1] using the min max approach to minimize the unfavorable effect of characteristics with a heavy weight of entries. At the implementation level, we recommend extracting dataset examples to minimize various difficulties such as preparation and a large number of data. There are multiple ways for reducing the amount of characteristics prior to using a dataset for training and validation of a classification algorithm. In this scenario, we apply PCA to do this assignment. The PCA is a mathematical approach that aims to generate a reduced dimensional representation of the source data by lowering large and complicated complexity while retaining all sensitive details. The reduction procedure reduces learning curve and computing costs without impacting outcomes effectiveness.

The primary goal is to uncover and identify relevant data (features) across huge network traffic datasets, to ensure that the proposed ensemble-based intrusion detection framework achieves reduced adaptation errors, fewer FAR, and accurate results and malware detection. As a result, they minimize cost while also enhancing the efficiency of the intrusion detection framework. As a result, we employ the mutual information method of feature selection. As Information Gain, the gain of each element in the setting of the target attribute is evaluated. The computation is alluded to as mutual information between two random variables in this slightly modified use. We use the tenfold cross validation approach provided in [21] to verify our conceptual approach. The algorithm seeks to arbitrarily divide the entire dataset into ten equal-sized sections. The model is trained in nine sections, with the last portion being tested. The technique is again repeated 10 times to train and construct an effective classifier capable of detecting intrusions inside the traffic network. Only characteristics that match a predefined condition are employed in model training and validation. The last phase is classification, that gives a category to an instance. As a result, the acquired classifier first from validation and training steps may predict categories of new examples. Build and train an ensemble-based XGBoost system with hyper-parameter optimization for this purpose.

2. **Performances evaluation and results**

Our proposed intrusion detection approach was carefully tested using the BoT-IoT dataset, following the experimental workflow illustrated in Fig. 3. The results strongly support the effectiveness of the model, with all performance indicators showing near-perfect values. As shown in Table 1, the system achieved 99.99% across all standard evaluation metrics, including Accuracy, Precision, Recall, and F1-Score, highlighting its ability to detect both known and unknown attack patterns with high reliability. To provide a deeper insight into the model's performance, Table 1 summarizes the results for both the training and validation stages. During training and testing, the model maintained its strong performance, with an accuracy of 99.99% while requiring only 25 s to complete the process on a GPU. Even more impressive, during the validation phase, the model preserved the same high performance, confirming its ability to generalize well to unseen data, and did so in just 0.30 s. These results suggest that the system is highly efficient and well-suited for real-time intrusion detection, where quick response times are crucial.

The confusion matrices, presented in Fig. 4, further confirm the model's strength in distinguishing malicious and legitimate network behavior. The system achieved an extremely low False Positive Rate (FPR) of 0.00018 and a False Negative Rate (FNR) of 0, meaning the model was able to detect all real threats while minimizing unnecessary alerts. a vital property for any security system deployed in critical environments like healthcare or smart industries. A key part of this success lies in the feature selection strategy, which played a major role in simplifying the model and improving detection performance. As illustrated in Fig. 5, the selected top features were carefully chosen based on their impact on the classification outcome. This allowed the model to focus only on the most meaningful information, reducing noise and improving both training speed and detection accuracy. Feature selection also reduced the computational load, making the model faster and more efficient, especially in large-scale or real-time deployments. Beyond performance, these results highlight the system's practical value in real-world IoT environments, where devices constantly exchange sensitive information and security threats evolve rapidly. The combination of high accuracy, minimal false alarms, and fast processing times makes this approach a promising solution for protecting modern connected systems. Finally,

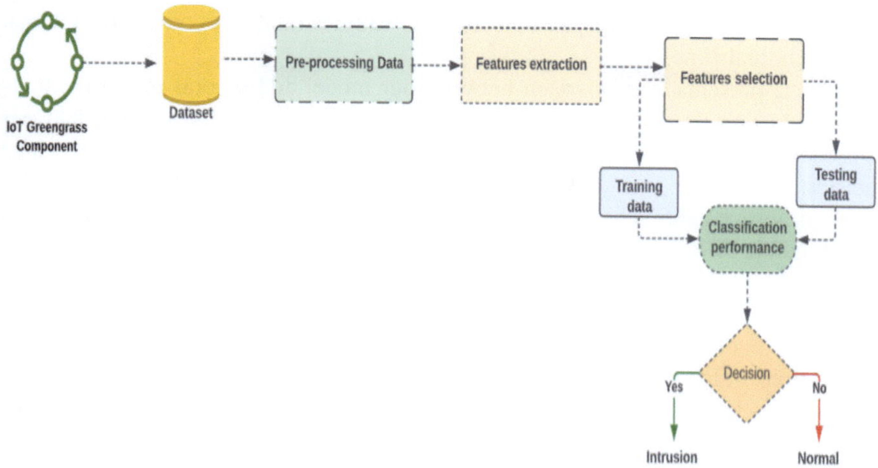

Fig. 3 Implementation of various components of our propose IDS model

Table 1 Performances comparison of our proposed models

	ACC	Precision	Recall	F1-score	Time (s)
Training and test	0.9999	0.9999	0.9999	0.9999	25 s
Validation	0.9999	0.9999	0.9999	0.9999	0.30

Table 2 A comparison of many intrusion detection algorithms on the BoT-IoT dataset

	Algorithm	ACC %	Precision %	Recall %	F-1%	FPR %	Time (s)	Dataset
Shafiq et al. [26]	BayesNet	99.77	100	99	–	–	29.26	BoT-IoT
	C4.5	99.99	100	100	–	–	17.1	
	NaiveBayes	99.79	99	98	–	–	4.03	
	RandomForest	99.99	100	100	–	–	198.83	
	RandomTree	99.99	100	100	–	–	4.32	
Ullah et al. [25]	CNN 1D	99.90	99.75	99.85	99.80	–	–	
Koroniotis et al. [27]	SVM	99.99	99.99	100	86.58	–	6636	
	RNN	97.90	99.99	97.90	20.54	–	6888	
	LSTM	98.05	99.99	98.05	9.85	–	14.073	
Our model	Xgboost	99.99	99.99	100	99.99	0.0058	32	

while the results are promising, ongoing research is still necessary. IoT environments are highly dynamic, and new attack types continue to emerge. Future work will focus on developing adaptive intrusion detection systems that can automatically learn from new data, continuously improve detection capabilities, and respond to novel attack patterns in real-time.

Table 2 presents a comparison between our model and several recent methods evaluated using the BoT-IoT dataset. From this comparison, it is evident that our approach yields superior results in terms of detection accuracy and performance metrics.

The improvements achieved by our model in Fig. 6 can be attributed to the careful use of feature selection techniques. By focusing on the most relevant features within the dataset, our model not only enhanced its ability to detect intrusions but also improved the overall efficiency of the training and testing phases. This resulted in quicker model convergence, making it more suitable for practical, real-time applications in IoT environments.

Furthermore, our model demonstrated robustness when dealing with various attack types and network behaviors found in the BoT-IoT dataset. This suggests

Fig. 4 The matrix confusion of the proposed IDS model

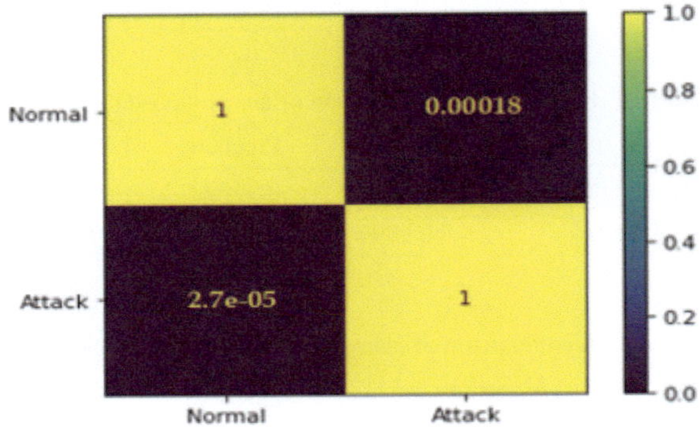

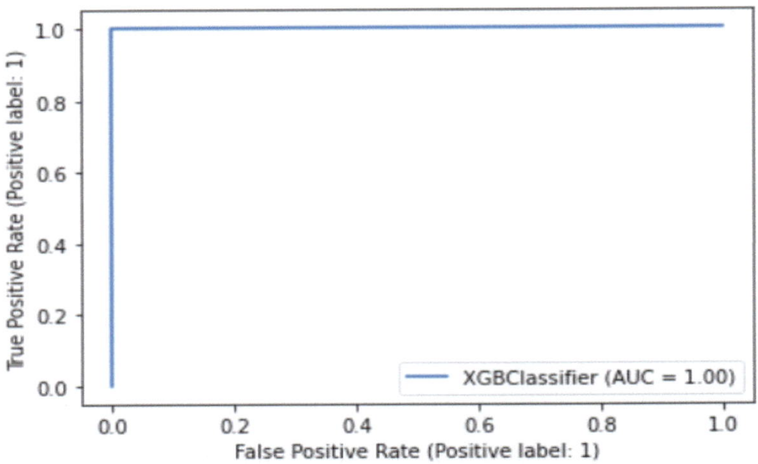

Fig. 5 Learning CURVE of our propose IDS

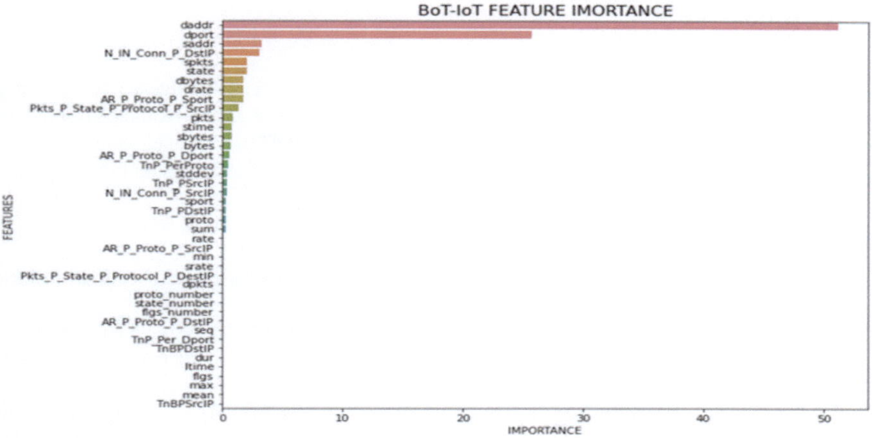

Fig. 6 Importance of features for detection attack on Edge-IIoT dataset

that the method is well-suited to adapt to the evolving nature of cybersecurity threats in diverse IoT ecosystems, where new types of attacks are constantly emerging.

In conclusion, the results underline the effectiveness of our approach in offering a balanced combination of high accuracy, low false positives, and fast processing times, making it a strong candidate for deployment in critical IoT infrastructures, such as those in healthcare or industrial automation.

5 Conclusion and Future Works

This research presents an advanced intrusion detection model tailored to IoT environments, emphasizing both high detection accuracy and efficient real-time performance. By integrating feature selection techniques with machine learning algorithms, the model effectively addresses the security challenges specific to IoT systems. Moreover, the incorporation of blockchain technology enhances the system's security by providing immutable records of detected threats and responses, ensuring transparency and accountability. This makes the model particularly suitable for critical applications in areas like healthcare and industrial IoT systems. Moving forward, as IoT ecosystems continue to grow and face more sophisticated cybersecurity threats, future work will focus on enhancing the model's adaptability through continuous learning methods and integrating deep learning techniques to better handle complex attack scenarios. Additionally, the use of blockchain for secure authentication and data exchange will be further explored to strengthen the model's capability to respond to new and emerging threats in real-time, ensuring a comprehensive and resilient security solution for modern IoT networks.

References

1. Crevier D (1993) AI: Te Tumultuous history of the search for artifcial intelligence. Basic Books
2. Yang L, Cai M, Duan Y, Yang X (2019) Intrusion detection based on approximate information entropy for random forest classification. In: Proceedings of the 2019 4th international conference on big data and computing, ser. ICBDC 2019. New York, NY. Association for Computing Machinery, USA, pp 125–129
3. Alaba FA, Othman M, Hashem IAT, Alotaibi F (2017) Internet of things security: a survey. J Netw Comput Appl 88:10–28
4. Sethi P, Sarangi SR (2017) Internet of things: architectures, protocols, and applications. J Electr Comput Eng 1–25
5. Azrour M, Mabrouki J, Farhaoui Y, Guezzaz A (2021) Security analysis of Nikooghadam et al.'s authentication protocol for cloud-IoT. In: Intelligent systems in big data, semantic web and machine learning, pp 261–269
6. Krishnaveni S, Vigneshwar P, Kishore S, Jothi B, Sivamohan S (2020) Anomaly-based intrusion detection system using support vector machine. In: Dash SS, Lakshmi C, Das S, Panigrahi BK (eds) Artificial intelligence and evolutionary computations in engineering systems. Springer Singapore, Singapore, pp 723–731
7. Liu H, Lang B (2019) Machine learning and deep learning methods for intrusion detection systems: a survey. Appl Sci 9(20):4396
8. Uikey R, Gyanchandani M (2019) Survey on classification techniques applied to intrusion detection system and its comparative analysis. Int Conf Commun Electron Syst 2019:1451–1456
9. Meziane H, Ouerdi N, Kasmi MA, Mazouz S (2021) Classifying security attacks in IoT using CTM method. In: Emerging trends in ICT for sustainable development. Springer, pp 307–315
10. Ibitoye O, Shafq O, Matrawy A (2019) Analyzing adversarial attacks against deep learning for intrusion detection in IoT networks. In: GLOBECOM, pp 1–6
11. Azrour M, Mabrouki J, Guezzaz A, Farhaoui Y (2021) New enhanced authentication protocol for internet of things. Big Data Min Anal 4(1):1–9
12. Von Solms R, Van Niekerk J (2013) From information security to cyber security. Comput Secur 38:97–102
13. Breiman L (1996) Bagging predictors. Mach Learn 24(2):123–140
14. Breiman L (2001) Random forests. Mach Learn 45(1):5–32
15. Freund Y, Schapire RE et al (1996) Experiments with a new boosting algorithm. In: International conference on machine learning, vol 96, pp 148–156
16. Soe YN, Feng Y, Santosa PI, Hartanto R, Sakurai K (2020) Towards a lightweight detection system for cyber attacks in the IOT environment using corresponding features. Electronics 9(1):144
17. Crevier D (1993) AI: Te tumultuous history of the search for artifcial intelligence. Basic Books
18. Butun I, Morgera SD, Sankar R (2014) A survey of intrusion detection systems in wireless sensor networks. IEEE Commun Surv Tutor 16(1):266–282
19. Verma A, Ranga V (2020) Machine learning based intrusion detection systems for IOT applications. Wireless Pers Commun 111(4):2287–2310
20. Aksu D, Üstebay S, Aydin MA, Atmaca T (2018) Intrusion detection with comparative analysis of supervised learning techniques and fisher score feature selection algorithm. In: Czachórski T, Gelenbe E, Grochla K, Lent R (eds) Computer and information sciences. Springer, Cham, pp 141–149
21. Chaudhari R, Patil S (2017) Intrusion detection system: classification techniques and datasets to implement
22. Zainal A, Maarof M, Shamsuddin SM (2009) Ensemble classifiers for network intrusion detection system. J Inf Assur Secur 4:217–225
23. Chanal PM, Kakkasageri MS (2020) Security and privacy in IoT: a survey. Wirel Pers Commun 115:1667–1693

24. Rajasekaran K (2020) Classification and importance of intrusion detection system. Int J Comput Sci Inf Secur 10:44
25. Ullah I, Mahmoud QH (2019) A two-level hybrid model for anomalous activity detection in IOT networks. In: 2019 16th IEEE Annual consumer communications networking conference (CCNC), pp 1–6
26. Shafiq M, Tian Z, Sun Y, Du X, Guizani M (2020) Selection of effective machine learning algorithm and Bot-IoT attacks traffic identification for internet of things in smart city. Futur Gener Comput Syst 107:433–442
27. Koroniotis N, Moustafa N, Sitnikova E (2020) A new network forensic framework based on deep learning for internet of things networks: a particle deep framework. Future Gene Comput Syst 110:91–106

Optimizing Intrusion Detection Systems: A Machine Learning-Based Feature Selection Approach for Enhanced Cybersecurity

Essarghi Hiba Allah and Darouichi Aziz

Abstract The rapid evolution of technologies such as artificial intelligence and big data has led to an increase in sophisticated cyberattacks, making network security an urgent concern. Intrusion Detection Systems (IDS) have been widely studied, with machine learning playing an important role in enhancing their effectiveness. However, developing a high-performing IDS remains challenging due to the complexity of high-dimensional data and limitations in detection accuracy. Effective feature selection is essential for improving IDS performance by reducing dimensionality, enhancing interpretability, and preventing overfitting. In this study, we propose a comprehensive data preprocessing pipeline to optimize feature selection and apply it to the UNSW-NB15 dataset, a benchmark for cybersecurity threat detection. We evaluate the impact of several feature selection techniques on four machine learning models: Logistic Regression, K-Nearest Neighbors (KNN), Decision Tree, and Random Forest. The selection methods include XGBoost-based selection with correlation analysis, Chi-square test, Recursive Feature Elimination (RFE), Fisher's test and CatBoost-based selection. We analyze model robustness across various feature subset sizes, and evaluate performance using Accuracy, Precision, Recall, and F1-score. Our findings demonstrate that feature selection significantly impacts model performance, with Random Forest achieving the highest accuracy (94.96%) using only 25 features selected via XGBoost and correlation. These results highlight the importance of optimal feature selection for designing more effective IDS solutions.

Keywords Intrusion detection systems (ids) · Machine learning · Cybersecurity · Feature selection · UNSW-NB15 dataset

E. Hiba Allah (✉) · D. Aziz
Faculty of Science and Technology, L2IS, Cadi Ayyad University, Marrakech, Morocco
e-mail: h.essarghi5169@uca.ac.ma

D. Aziz
e-mail: a.darouichi@uca.ac.ma

G. H. L. et al. (eds.), *Reliability in Cyber-Physical Systems: The Human Factor Perspective*, Springer Series in Reliability Engineering,
https://doi.org/10.1007/978-3-032-09917-4_10

1 Introduction

Business, governments and individuals are under constant threat of cyberattacks which make cybersecurity a significant concern for them. The existing cybercrimes undermine the security, integrity and confidentiality of the information systems, thereby posing a threat to the finances and security of individuals [1]. In order to handle these emerging dangers, it is important to create effective and preemptive Intrusion Detection Systems (IDS) that can automatically identify and deal with attacks as they happen. By gathering data related to normal and malicious traffic, IDS are able to secure the network and identify the possible threats. Unlike traditional signature-based IDS that depend on predefined patterns, machine learning-matched filter (ML)-based IDS have been developed to tackle zero-day and other sophisticated novel attacks. By making use of historical data, ML based IDS systems can easily detect adaptive threats [2].

Two main classification techniques are rather common in ML-driven IDS: ML-driven IDS uses binary classification to separate benign from malicious traffic and multiclassification to identify particular attack types, including DDoS, malware, or SQL injection [3]. While multiclass classification provides comprehensive threat insights allowing exact countermeasures, binary classification achieves fast threat detection. One still major difficulty is the choice and quality of features applied for model training. Choosing inappropriate features reduces its efficacy against new hazards, increases computational costs and model overfitting. Correct feature selection techniques, according to research, reduce data dimensions while concurrently increasing detection accuracy and processing performance [4].

As cyber-physical systems (CPS) are "cognitive" with the addition of AI and distributed control particularly in Industry environments their interconnectedness and sophistication make them vulnerable to novel security threats at the cyber and physical levels. IDS therefore take on a prominent role in ensuring the integrity and availability of these mission-critical systems [5].

The literature covers a variety of feature selection methods. Tree-based methods such as Random Forest and XGBoost or CatBoost implement importance scores and filter out features with lower scores. Other approaches include tests like the Chi-square or Fisher test which are statistical methods of determining the relationship in a variable and the target value [6].

The objective of this research is to prove a theory that utilizing enhanced multiclass classification feature selection algorithms within an intrusion detection system could increase accuracy while decreasing efficiency cost. To accomplish the objective, this research will measure the evaluation of feature selection for machine learning classifiers in intrusion detection systems and analyze several classification methods in terms of accuracy, precision, recall, and resource consumption. This study focuses on using the most recent developments and extensive empirical research to aid in the security of computer networks and in the area of computer science, particularly in features of artificial intelligence and machine learning, by providing ways of how intrusion detection systems can be improved through feature selection employing machine learning. The rest of the paper is presented in the following order: Sect. 2 presents related works, Sect. 3 presents the dataset and method, Sect. 4 presents the results with future work, and Sect. 5 concludes the paper.

2 Related Work

ML and feature-selection techniques for intrusion detection have received notable attention over domains from healthcare images to Internet of Things security illustrating that if one preprocesses well and utilizes ensemble modeling, the accuracy for detection can largely be enhanced. Halbouni et al. [1] surveyed ML and DL paradigms in cyber security, mentioning autoencoders and the capacity of recurrent networks to embed typical traffic flows, but at the same time suggesting the risks of overfitting without severe feature selection.

Alohali et al. [5] demonstrate how AI-driven IDS can detect both network and control-logic attacks in Cognitive CPS via the combination of multiple modalities (RNN, Bi-LSTM, DBN) and via the selection of discriminative features by an improved Fish Swarm Optimization algorithm; their AIMMF-IDS achieves detection accuracies above 99 % on benchmark ICS datasets, highlighting how bespoke IDS architectures are required to safeguard the tightly coupled cyber and physical processes of modern industrial systems.

Ahmed et al. [7] proposed an optimized ensemble ("Opt-Forest") combining genetic-algorithm feature selection with Random Forest, with more than 95 % accuracy on UNSW-NB15 and demonstrating wrapper-based FS can remove a high percentage of false positives. Khan et al. [8] compared XGBoost, SVM, and neural networks on a multi-step attack dataset for smart city, citing that XGBoost achieved 100 % test accuracy and an F1-score of 88 %, highlighting the effectiveness of gradient boosting for complex IoT tasks. Moudni et al. [9] created a tailored MANET black-hole attack dataset and demonstrated that Random Forest with tailor-made FS achieved high F1-scores. The findings underscore the potential of machine learning in enhancing the security of MANETs by providing an effective means of attack classification, which demonstrates the power of domain-specialized data engineering. Li et al. [2] pitted feature-selection vs. feature-extraction directly against each other for IoT-IDS and found that embedded techniques (tree-based importance) offered the optimum trade-off between detection precision and latency.

Karunamurthy et al. [3] proposed a federated learning-based intrusion detection system (IDS) for IoT edge devices, leveraging deep learning classifiers and the Chimp optimization algorithm for optimal feature selection. Their system was validated on the benchmark MQTT dataset, achieving a maximum detection accuracy of 95.59%, while ensuring data privacy during training. Dash et al. [4] developed an interpretable dimensionality-reduction pipeline for Internet-of-Medical-Things (IoMT), maximizing transparency and performance, This study introduces an interpretable feature selection technique designed to enhance intrusion detection in IoMT by reducing redundant features and improving model efficiency. Random Forest-based explainable AI model provides transparency in attack classification and better decision-making. Logeswari et al. [6] provided an intrusion detection system (IDS) optimized for IoT scenarios, which has a Synergistic Dual-Layer Feature Selection (SDFC) algorithm that combines statistical and model-based methods. By using LightGBM and XGBoost classifiers, the resultant system, as tested against the TON-IoT dataset, significantly improved detection performance, with higher accuracy, precision, recall, and F1-scores compared to traditional methodologies. Saranya and Haldorai [10], which utilized proposed an efficient data preprocessing pipeline integrating SMOTE, normalization, and a deep sparse autoencoder optimized with differential evolution, significantly improving intrusion detection rates and reducing false alarms in general network security and large-scale Internet communications. Mboweni et al. [11] showed that tailored data preprocessing for water treatment system datasets significantly improves the performance of machine learning-based intrusion detection in industrial control systems.

Shiravani et al. [12] cintroduced a novel feature selection approach for intrusion detection systems by modeling the number of selected features as a fuzzy number and integrating it with a correlation-based feature selection heuristic. Evaluated on KDD Cup, NSL-KDD, and CICIDS datasets, their method significantly reduced feature dimensions while achieving higher detection rates, reaching up to 99.9% accuracy, outperforming traditional correlation-based feature selection methods. Ahmad et al. [13] compared machine learning techniques for network intrusion detection and concluded that tree-based ensembles offer an optimal balance between computational expense and accuracy. Metaheuristic FS methods have also continued to advance: Kumari et al. [14] introduced a novel hybrid model of an intrusion detection system with the Flower Pollination Algorithm (FPA), Cheetah Optimization Algorithm (COA), and Artificial Neural Networks (ANN) to improve detection rate, reduce false positives, and increase feature selection and anomaly detection. The Thakkar and Lohiya study [15] focuses on the performance of Intrusion Detection Systems (IDS) based on Deep Neural Network (DNN) by proposing a new feature selection approach that incorporates statistical significance through Standard Deviation and Difference of Mean and Median. The approach prunes features based on rank, acquired by combining statistical significance, for better learning from data and detection efficiency. Gupta et al. [16] proposes a hybrid optimization and deep-learning-centric Intrusion Detection System (IDS) for IoT-enabled smart cities to address challenges such as achieving high detection accuracy and reducing training time, which traditional IDS are inefficient at handling. Du et al. [17] puts forward a deep-learning-based network intrusion detection classification model (NIDS-CNNLSTM) for Industrial Internet of Things (IIoT) wireless sensing environment, combining the learning ability of convolutional neural networks and long short-term memory networks to efficiently classify network traffic data and enhance IIoT security. Hnamte et al. [18] and Hu et al. [19] optimized LSTM-AE and residual-network approaches for adaptable threat patterns. In general, this body of work highlights feature processing, ensemble structure, metaheuristic search, and privacy-protection learning as critical to improving IDS performance in regular, IoT, and smart-city scenarios.

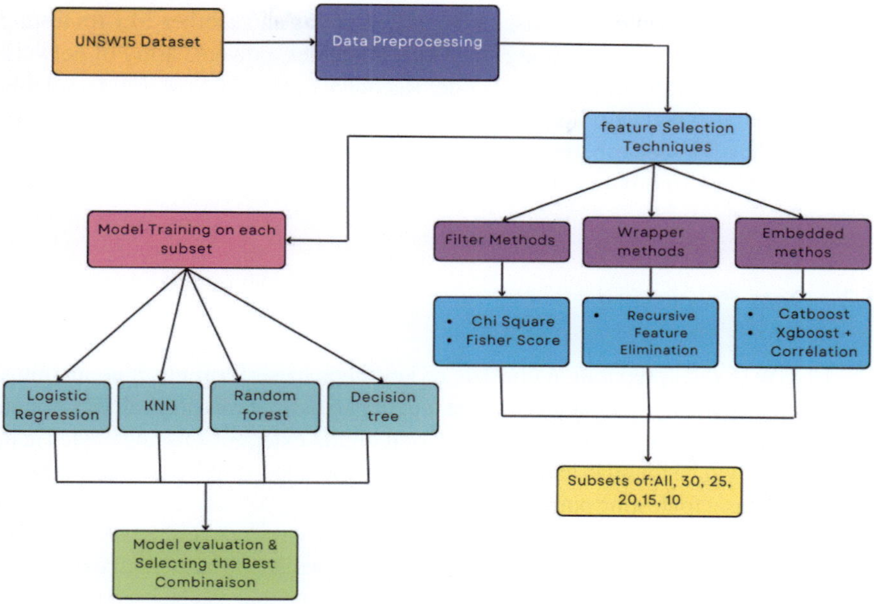

Fig. 1 Feature selection and model training pipeline for UNSW15 dataset

3 Methods

This methodology shown in Fig. 1 uses a formalized sentence-level machine learning pipeline for classification on the UNSW-NB15 dataset. It begins from data loading and preprocessing where missing values are handled, unwanted columns are dropped, features are identified and cleaned based on an accompanying feature description file. It leverages a FeatureSelector class that has a modular design for handling different feature selection algorithms like Fisher Score (f_classif), Chi-Square (chi2), Recursive Feature Elimination (RFE), CatBoost, and XGBoost correlation analysis. All of them select the most relevant features based on statistical significance or model-based importance scores. After features are chosen, these can be used to train models such as Random Forest, Logistic Regression, Decision Trees, and K-Nearest Neighbors for efficient modeling and performance measurement using metrics such as accuracy, precision, recall(Detection Rate), F1-score, and confusion matrix.

3.1 UNSW-NB15 Dataset

Our study leans on the UNSW-NB15 dataset a recognized benchmark when it comes to spotting network intrusions. This dataset carries 42 features after filtering, mixing things like nominal values, integers, binaries, and even floats.Interestingly, UNSW-

NB15 naturally splits into two parts: one for 80% training and another 20% for testing serves as our check for overfitting .The dataset includes a diverse array of network attack types–such as Backdoor, Shellcode, Reconnaissance, Worms, Fuzzers, DoS, Generic, Analysis, and Exploits–accounting. This variety establishes a realistic multi-class classification scenario, making it well-suited for training and evaluating a wide range of machine learning models.

3.2 Data Processing

The UNSW15 data preparation consists of key steps to perform machine learning. The pipeline starts by loading the dataset and then handling missing values through either imputation or deletion, as needed, based on the extent of missing data. Then, types of data will be found in and be converted to a specified kind. We allocate 20% of the entire dataset for testing purposes, while the remaining 80% is used for training the model.Then categorical variables are encoded (LabelEncoder), multiclass variables are assigned a value between 0 and 8. Lastly, we normalise the numerical features to scale them down and convert the attack labels to a format that can be used to classify them. The cleaned data is then finally ready to be used to train the machine learning models.

3.3 Feature Selection

The selection of features is a fundamental component of constructing effective machine learning models, especially when it comes to intrusion detection using the UNSW-NB15 dataset. In feature selection, one evaluates a dataset and identifies the most important features necessary for accurate classification and reduced dimensionality. The key advantages of feature selection include decreased computational complexity, reduced overfitting, improved generalizability, greater interpretability, and decreased data storage. In this work, different feature selection approaches were studied, each with different principles and strategies. Approaches can largely be separated into three types: filter-based, wrapper-based, and embedded methods. Each type provides a different feature importance metric: filter-based uses statistical measures, wrapper-based approaches provide an iterative optimization approach, and embedded methods provide importance based on ranking by models.

3.3.1 Filter-Based Feature Selection Methods

Filter-based feature selection methods evaluate features independently of any machine learning model. They rely on statistical techniques to assess the relevance of features with respect to the target variable. These methods are computationally efficient and robust, as they do not involve model training.

- **Chi-Square Selection**: The chi-square (χ^2) test measures the dependence between feature values and the target class. It is commonly used for categorical data:

$$\chi^2 = \sum_{i=1}^{n} \frac{(O_i - E_i)^2}{E_i} \tag{1}$$

  where O_i is the observed frequency for category i and E_i is the expected frequency for category i under the assumption of independence.

- **Fisher Test Selection**: The Fisher Score is based on the ANOVA F-value between label/feature for classification tasks. It measures the variance between classes relative to the variance within classes. The formula used is:

$$F_i = \frac{\sum_{j=1}^{c} n_j (\bar{x}_{ij} - \bar{x}_i)^2}{\sum_{j=1}^{c} \sum_{k=1}^{n_j} (x_{ijk} - \bar{x}_{ij})^2} \tag{2}$$

  where F_i is the F-score of the ith feature, c is the number of classes, n_j is the number of instances in class j, $\bar{x}_{ij}$ is the mean of the ith feature in class j, $\bar{x}_i$ is the overall mean of the ith feature and x_{ijk} is the kth sample of feature i in class j.

3.3.2 Wrapper-Based Feature Selection Methods

Wrapper methods assess feature subsets by iteratively evaluating their impact on a specific machine learning model. These methods account for feature interactions but are computationally expensive.

- **Recursive Feature Elimination (RFE)**: RFE is an iterative method that starts with all features and progressively removes the least important ones based on their contribution to model performance. The process involves training a model (we used random forest), ranking features by importance, eliminating the least significant feature, and repeating until reaching a desired subset size.

$$\text{Rank } (f_i) = \text{stepwise removal order} \tag{3}$$

The importance score for feature f_i can be approximated by:

$$\text{Importance } (f_i) = \frac{1}{\text{Rank } (f_i)} \tag{4}$$

3.3.3 Embedded Feature Selection Methods

Embedded methods incorporate feature selection directly within the model training process. They combine the advantages of filter and wrapper methods by optimizing feature selection while learning model parameters.

- **XGBoost with Correlation Analysis**: In this method, feature importance is first computed from XGBoost's built-in gain metric:

$$\text{Gain}_f = \sum_{i=1}^{N_f} \Delta\text{Loss}_i \tag{5}$$

  Then, the top features are filtered based on their Pearson correlation with the target variable:

$$r = \frac{\sum_{i=1}^{n}(x_i - \bar{x})(y_i - \bar{y})}{\sqrt{\sum_{i=1}^{n}(x_i - \bar{x})^2}\sqrt{\sum_{i=1}^{n}(y_i - \bar{y})^2}} \tag{6}$$

  where: Gain_f is the total information gain associated with feature f, N_f is the number of splits where feature f was used in the trees, ΔLoss_i is the loss reduction achieved by the i-th split, r is the Pearson correlation coefficient between a feature and the target, n is the number of observations, x_i is the value of the feature for observation i, y_i is the value of the target variable for observation i, $\bar{x}$ is the mean of the feature values across all observations, $\bar{y}$ is the mean of the target values across all observations.
  This method benefits from the high predictive power of gradient boosting (XGBoost), while correlation filtering ensures the selected features are not only important for the model but also statistically associated with the target. It helps reduce redundancy and overfitting by discarding features with low target correlation, even if XGBoost considers them frequent in splits. This hybrid strategy leads to a more interpretable and relevant feature subset.

- **CatBoostClassifier-Based Selection**: CatBoost computes feature importance by measuring the average change in the loss function when the feature is used to split nodes. The importance for a feature f can be estimated as:

$$I_f = \frac{1}{N} \sum_{i=1}^{N} \Delta L_i(f) \tag{7}$$

Where I_f is the importance of feature f, N is the number of splits and $\Delta L_i(f)$ is the reduction in loss due to the ith split on feature f.

3.4 Machine Learning Methods

As cyber-attacks become increasingly sophisticated, intrusion detection relies on advanced machine learning techniques. This paper analyzes widely used models and their effectiveness

3.4.1 Logistic Regression

Logistic regression estimates the probability of an event using the sigmoid function:

$$P(Y = 1|X) = \frac{1}{1 + e^{-(\beta_0 + \sum \beta_i X_i)}} \tag{8}$$

where: $P(Y = 1 \mid X)$ is the probability that the target variable Y equals 1 given the input features X, β_0 is the intercept (bias term), p is the number of features, β_i is the coefficient (weight) associated with the i-th feature, X_i is the value of the i-th feature, e is the base of the natural logarithm (approximately 2.718). It is widely used for its robustness and interpretability [7]. However, it assumes feature independence, which is rarely valid in cybersecurity.

3.4.2 K-Nearest Neighbors (KNN)

KNN classifies a sample based on the majority class among its K nearest neighbors, allowing it to capture complex patterns without strong assumptions about data distribution. However, it is computationally expensive and sensitive to feature scaling [8].

$$\hat{y} = \arg\max_c \sum_{i \in \mathcal{N}_K} ⊮(y_i = c), \tag{9}$$

where $\mathcal{N}_K$ is the collection of K nearest neighbors and $⊮(\cdot)$ is an indicator function. We use neighbors=5 in our code, which is optimal for both binary and multiclass problems. KNN is efficient and simple for detecting complex patterns, but it is sensitive to feature scaling and computationally expensive [8].

3.4.3 Decision Trees

Decision trees partition the feature space by selecting attributes that minimize an impurity measure, such as the Gini index:

$$Gini(S) = 1 - \sum_{i=1}^{C} p_i^2 \tag{10}$$

where: Gini(S) is the Gini impurity of set S, c is the total number of classes, p_i is the proportion of samples belonging to class i in set S. They offer high interpretability but are prone to overfitting, requiring pruning techniques.

3.4.4 Random Forest

Random forest improves upon decision trees by training multiple trees on random subsets of the data and aggregating their predictions:

$$\hat{y} = \arg\max_{c} \sum_{i=1}^{N} \mathbb{1}(h_i(x) = c) \tag{11}$$

where: $\hat{y}$ is the predicted class label for the input sample x, c represents a possible class label, N is the number of decision trees in the forest, $h_i(x)$ is the prediction made by the i-th tree for the input x, $\mathbb{1}(h_i(x) = c)$ is the indicator function, which equals 1 if the prediction of the i-th tree is class c, and 0 otherwise, $\arg\max_{(c)}$ selects the class c that maximizes the sum over all trees. This provides robustness, reduces overfitting, and enhances accuracy [9].

There is no one-size-fits-all model for intrusion detection. The algorithm choice is based on the input accuracy, computational efficiency, and interpretability. The hybrid and ensemble methods produce good results according to recent studies.

4 Results

The results section provides an overview of the performance of different machine learning methods via different feature selection methods. In this section is a collection of comparisons for different evaluation metrics, such as Accuracy, Precision, Recall, and F1 Score that give indication and insights into the impact of feature selection and feature sub-settings to determine and understand its effect on model performance. By comparing evaluation metrics across different models with varying feature sub-settings, we will be able to infer the best feature sub-setting, which indicates the optimal configuration to provide the highest classification performance. Across the comparisons, we explain why one model performs better than another, the pros and cons of each method, and then we can offer more insight to model complexity, accuracy, and computational efficiency.

5 Discussion

Our in-depth analysis (Tables 1, 2 and 3) reveal several key points about feature selection techniques and model performance. Random Forest always performed the best (94.72–94.96%) in all methods, with top performance from XGBoost_Correlation at

Table 1 Performance comparison for Chi2 and Fisher feature selection methods (accuracy, precision, recall, F1 score in %)

		Chi2				Fisher			
Model	Nb Feat.	Acc.	Prec.	Rec.	F1	Acc.	Prec.	Rec.	F1
Logistic Reg.	10	90.11	87.77	90.11	87.59	90.75	88.24	90.75	88.57
	15	90.84	88.22	90.84	88.71	90.77	88.29	90.77	88.61
	20	90.87	88.26	90.87	88.76	90.97	88.32	90.97	88.89
	25	91.18	88.42	91.18	89.14	91.09	88.31	91.09	89.01
	30	91.35	89.57	91.35	89.35	91.39	92.03	91.39	89.58
	All	**91.73**	**90.45**	**91.73**	**90.13**	**91.74**	**90.46**	**91.74**	**90.14**
KNN	10	90.27	89.84	90.27	87.99	91.46	90.44	91.46	90.22
	15	90.72	90.76	90.72	90.22	91.43	90.29	91.43	90.21
	20	90.80	89.83	90.80	90.08	91.06	89.70	91.06	90.12
	25	90.82	89.84	90.82	90.10	90.74	89.69	90.74	89.98
	30	91.06	90.20	91.06	90.43	91.08	90.07	91.08	90.36
	All	**91.38**	**90.75**	**91.38**	**90.94**	**91.38**	**90.75**	**91.38**	**90.94**
Random forest	10	90.23	89.73	90.23	87.99	91.99	91.02	91.99	90.41
	15	92.12	90.90	92.12	90.54	91.94	90.64	91.94	90.41
	20	91.44	90.36	91.44	90.70	91.60	90.57	91.60	90.78
	25	92.81	91.78	92.81	91.64	93.06	92.33	93.06	92.02
	30	93.40	92.62	93.40	92.41	93.69	93.10	93.69	92.92
	All	**94.72**	**94.63**	**94.72**	**94.41**	**94.78**	**94.62**	**94.78**	**94.43**
Decision tree	10	90.23	89.74	90.23	87.99	91.99	90.96	91.99	90.46
	15	92.00	90.80	92.00	90.50	91.92	90.68	91.92	90.42
	20	89.29	89.35	89.29	89.32	89.39	89.39	89.39	89.39
	25	89.87	90.14	89.87	90.00	90.18	90.44	90.18	90.30
	30	91.16	91.38	91.16	91.27	91.62	91.76	91.62	91.69
	All	**93.09**	**93.16**	**93.09**	**93.12**	**93.26**	**93.30**	**93.26**	**93.27**

Table 2 Comparison of model performance—RFE method

Model	Nb features	Accuracy (%)	Precision (%)	Recall (%)	F1 score (%)
Logistic regression	10	90.21	88.36	90.21	87.89
	15	90.26	88.34	90.26	87.98
	20	91.01	89.12	91.01	88.90
	25	91.07	89.23	91.07	88.99
	30	91.08	88.81	91.08	89.05
	All	**91.74**	**90.46**	**91.74**	**90.14**
KNN classifier	10	90.45	89.75	90.45	89.90
	15	91.97	91.69	91.97	91.71
	20	92.40	91.93	92.40	92.06
	25	90.58	89.86	90.58	90.03
	30	90.67	89.93	90.67	90.09
	All	**91.38**	**90.75**	**91.38**	**90.94**
Random forest	10	94.07	93.89	94.07	93.81
	15	94.19	94.02	94.19	93.94
	20	94.49	94.30	94.49	94.22
	25	94.41	94.18	94.41	94.08
	30	94.60	94.45	94.60	94.31
	All	**94.75**	**94.57**	**94.75**	**94.41**
Decision tree	10	92.46	92.68	92.46	92.56
	15	92.45	92.66	92.45	92.54
	20	92.59	92.75	92.59	92.66
	25	92.81	92.90	92.81	92.85
	30	92.86	93.01	92.86	92.93
	All	**93.24**	**93.30**	**93.24**	**93.27**

25 features (94.96%, Table 3). Filter approaches (Chi2/Fisher) experienced increasing performance gains with increasing feature counts, with optimal performance at 42 features (94.72–94.78% accuracy, Table 1). On the other hand, techniques incorporated within gave evidence of greater selectivity of features—XGBoost_Correlation enabled KNN to achieve 93.80% accuracy with 15 features, while CatBoost managed to maintain 94.84% accuracy with Random Forest having only 20 features. RFE demonstrated good performance on subset features (94.75% accuracy at 42 features, Table 2), pointing towards effective ranking of features. Pattern variability by model: Linearity requirements limited Logistic Regression's performance (a maximum of 91.74%), and feature selection sensitivity was witnessed by KNN with its 5.42% method-based difference in accuracy. Decision Trees benefited the most from native methods (93.81% accuracy when using XGBoost_Correlation at 20 features), potentially due to reduced overfitting. These results demonstrate that while filter methods

Table 3 Performance comparison for XGBoost_Correlation and CatBoost feature selection methods (accuracy, precision, recall, F1 score in %)

		XGBoost_Correlation				CatBoost			
Model	Nb Feat.	Acc.	Prec.	Rec.	F1	Acc.	Prec.	Rec.	F1
Logistic Reg.	10	90.62	90.22	90.62	88.37	90.64	88.85	90.64	88.42
	15	90.74	89.06	90.74	88.75	90.81	88.50	90.81	88.63
	20	91.09	89.18	91.09	89.32	90.83	89.21	90.83	88.84
	25	91.39	89.63	91.39	89.77	91.01	89.28	91.01	89.00
	30	91.45	89.88	91.45	89.80	91.11	89.59	91.11	89.28
	All	**91.73**	**90.45**	**91.73**	**90.13**	**91.73**	**90.45**	**91.73**	**90.13**
KNN	10	93.40	93.30	93.40	93.31	**92.77**	**92.47**	**92.77**	**92.55**
	15	**93.80**	**93.63**	**93.80**	**93.65**	92.44	92.05	92.44	92.14
	20	93.64	93.51	93.64	93.53	92.44	92.09	92.44	92.19
	25	93.54	93.37	93.54	93.39	91.72	91.25	91.72	91.35
	30	93.03	92.74	93.03	92.81	91.12	90.34	91.12	90.58
	All	91.38	90.75	91.38	90.94	91.38	90.75	91.38	90.94
Random forest	10	93.99	93.96	93.99	93.96	94.74	94.59	94.74	94.57
	15	94.58	94.48	94.58	94.48	94.61	94.46	94.61	94.36
	20	94.80	94.71	94.80	94.71	**94.84**	**94.70**	**94.84**	**94.58**
	25	**94.96**	**94.83**	**94.96**	**94.81**	94.75	94.62	94.75	94.49
	30	94.88	94.76	94.88	94.67	94.72	94.56	94.72	94.43
	All	94.60	94.48	94.60	94.28	94.67	94.52	94.67	94.32
Decision tree	10	93.35	93.48	93.35	93.41	93.03	93.16	93.03	93.09
	15	93.66	93.80	93.66	93.72	93.04	93.19	93.04	93.11
	20	**93.81**	**93.93**	**93.81**	**93.87**	93.17	93.37	93.17	93.26
	25	93.37	93.54	93.37	93.45	93.10	93.17	93.10	93.13
	30	93.43	93.52	93.43	93.47	93.10	93.15	93.10	93.12
	All	93.02	93.10	93.02	93.06	**93.23**	**93.34**	**93.23**	**93.28**

provide uniform performance, embedded methods provide improved compression of the features without losing accuracy, particularly for deep models such as Random Forest.

Advantages, Limitations, and Future Work

The proposed machine learning feature selection mechanism for intrusion detection systems (IDS) has some critical advantages. Through embedded methods like XGBoost supported by correlation analysis and CatBoost feature ranking, the system yields better classification accuracy using far fewer features, thus minimizing computation, detection times, and model interpretability. Such dimensionality reduction is discriminative in minimizing the risk of overfitting and optimizing resources for successful implementation within resource-constrained domains like IoT and cyber-physical systems.

However, the framework is not flawless. The evaluation was conducted over a static set (UNSW-NB15), which could not possibly map the dynamic attack behavior of live zero-day assaults or evolving intruder tactics in real-world deployments. The algorithms were also benchmarked in sterile settings without addressing potential adversarial attacks or evolving intruders that could potentially exploit model flaws. In addition, while maximizing detection accuracy comes from feature selection, it never automatically ensures future data drift protection or new intrusion vector robustness.

Future research will be aimed at several promising directions. First, hybrid meta-heuristic approaches combining evolutionary algorithms with neural architecture search will be explored for adaptive and dynamic feature selection. Second, multi-objective optimization platforms will be created to balance detection precision, computational cost, and model resilience. Third, incorporating quantum-inspired feature selection techniques could potentially circumvent the issues of extremely high-dimensional cybersecurity data. Finally, live traffic and adaptive adversarial testing in real environments will be conducted to demonstrate and enhance the system's real-world robustness and usability.

Conclusions

This study makes three significant contributions to machine learning-based Intrusion Detection Systems (IDS): (1) demonstrating that Random Forest with XGBoost_Correlation feature selection achieves optimal performance (94.96% accuracy with only 25 features), representing a reduction in computational load compared to using all features; (2) providing a comprehensive comparative analysis of five feature selection methods across four ML models, establishing clear guidelines for method selection based on accuracy-efficiency tradeoffs; and (3) identifying the superior feature compression capability of embedded methods over traditional filter/wrapper approaches. While these results advance IDS development, limitations include static dataset evaluation and adaptability to zero-day threats. Future work should explore hybrid metaheuristics (e.g., Genetic Algorithms with neural architecture search) for dynamic feature selection, multi-objective optimization balancing detection accuracy with resource constraints, and quantum-inspired techniques for high-dimensional feature spaces—directions that could address current limitations while building on this study's robust experimental framework and demonstrated performance improvements.

References

1. Halbouni A, Gunawan TS, Habaebi MH, Halbouni M, Kartiwi M, Ahmad R (2022) Machine learning and deep learning approaches for cybersecurity: a review. IEEE Access 10:19572–19585
2. Li J, Othman MS, Chen H et al (2024) Optimizing iot intrusion detection system: feature selection versus feature extraction in machine learning. J Big Data
3. Karunamurthy A, Vijayan K, Kshirsagar PR, Tan KT (2025) An optimal federated learning-based intrusion detection for iot environment. Sci Rep 15(1):8696
4. Lipsa S, Dash RK, Ivković N (2025) An interpretable dimensional reduction technique with an explainable model for detecting attacks in internet of medical things devices. Sci Rep 15(1):8718
5. Alohali MA, Al-Wesabi FN, Hilal AM, Goel S, Gupta D, Khanna A (2022) Artificial intelligence enabled intrusion detection systems for cognitive cyber-physical systems in industry 4.0 environment. Cognit Neurodyn 16(5):1045–1057
6. Logeswari G, Thangaramya K, Selvi M, Roselind JD (2025) An improved synergistic dual-layer feature selection algorithm with two type classifier for efficient intrusion detection in iot environment. Sci Rep 15(1):8050
7. Ahmed A, Asim M, Ullah I, Ateya AA et al (2024) An optimized ensemble model with advanced feature selection for network intrusion detection. Peer J Comput Sci 10:2472
8. Khan J, Elfakharany R, Saleem H, Pathan M, Shahzad E, Dhou S, Aloul F (2025) Can machine learning enhance intrusion detection to safeguard smart city networks from multi-step cyber-attacks? Smart Cities 8(1):13
9. Moudni H, Er-Rouidi M, Lmkaiti M, Mouncif H (2025) Customized dataset-based machine learning approach for black hole attack detection in mobile ad hoc networks. Int J Electr Comput Eng Syst 15(2)
10. Saranya N (2024) Haldorai A (2024) Efficient intrusion detection system data preprocessing using deep sparse autoencoder with differential evolution. IET Inf Secur 1:9937803
11. Mboweni IV, Ramotsoela DT, Abu-Mahfouz AM (2023) Hydraulic data preprocessing for machine learning-based intrusion detection in cyber-physical systems. Mathematics 11(8):1846
12. Shiravani A, Sadreddini MH, Nahook HN (2023) Network intrusion detection using data dimensions reduction techniques. J Big Data 10(1):27
13. Ahmad Z, Khan AS, Shiang CW, Abdullah J, Ahmad F (2021) Network intrusion detection system: a systematic study of machine learning and deep learning approaches. Trans Emerg Telecommun Technol 32(1):4150
14. Kumari D, Pranav P, Sinha A, Dutta S (2025) A novel approach to intrusion detection system using hybrid flower pollination and cheetah optimization algorithm. Sci Rep 15(1):13071
15. Thakkar A, Lohiya R (2023) Fusion of statistical importance for feature selection in deep neural network-based intrusion detection system. Inform Fus 90:353–363
16. Gupta SK, Tripathi M, Grover J (2022) Hybrid optimization and deep learning based intrusion detection system. Comput Electr Eng 100:107876
17. Du J, Yang K, Hu Y, Jiang L (2023) Nids-cnnlstm: network intrusion detection classification model based on deep learning. IEEE Access 11:24808–24821
18. Hnamte V, Nhung-Nguyen H, Hussain J, Hwa-Kim Y (2023) A novel two-stage deep learning model for network intrusion detection: LSTM-AE. IEEE Access 11:37131–37148
19. Hu X, Meng X, Liu S, Liang L (2024) An improved algorithm for network intrusion detection based on deep residual networks. IEEE Access

Human Factors in Cyber-Physical Systems: Bridging the Gap Between Humans and Technology

Santosh Kumar Birthriya, Priyanka Ahlawat, and Ankit Kumar Jain

Abstract Cyber-physical systems (CPS) are increasingly integral to critical sectors such as healthcare, transportation, and manufacturing. Technological advancements have enhanced the capabilities of CPS, but their design often overlooks human factors, encompassing ergonomic, cognitive, and psychological elements. This oversight can lead to user errors, decreased efficiency, and safety risks. This chapter emphasizes the importance of integrating human factors into the CPS design. It explores the roles of cognitive, psychological, and physical elements in human–machine interaction, identifies challenges in considering human factors, and provides design principles centered on user-centered design, ergonomics, intuitive interfaces, human-in-the-loop systems, adaptive systems, real-time feedback, and multimodal interaction. Furthermore, it discusses future directions, including the integration of AI and machine learning to predict human responses, ethical and societal considerations, regulatory frameworks, and the necessity of interdisciplinary collaboration. Highlighting these aspects, the paper aims to bridge the gap between human capabilities and advanced CPS functionalities, promoting systems that are more effective, trustworthy, and accessible.

Keywords Cyber-physical systems · Human factors · User-centered design · Human–machine interaction · Ergonomics in system design

S. K. Birthriya (✉) · P. Ahlawat · A. K. Jain
National Institute of Technology Kurukshetra, Kurukshetra, India
e-mail: cs.santosh1b@gmail.com

P. Ahlawat
e-mail: priyankaahlawat@nitkkr.ac.in

A. K. Jain
e-mail: ankitjain@nitkkr.ac.in

G. H. L. et al. (eds.), *Reliability in Cyber-Physical Systems: The Human Factor Perspective*, Springer Series in Reliability Engineering,
https://doi.org/10.1007/978-3-032-09917-4_11

1 Introduction

Today's most cutting-edge technologies, Cyber-Physical Systems (CPS), enable the intermixing of computation algorithms and physical processes. These systems consist of interconnected devices, sensors, and actuators that function in real-time to monitor, control, and interact with the physical environment. CPS, centered on advanced information and communication technologies, are present in almost every aspect today, from self-driving cars and smart manufacturing facilities to healthcare professionals' devices, enhancing effectiveness, dependability, and impact of work [1].

All these can be achieved if one can embed the computer inside the physical world and create systems that can reason over the data that they take in. By collecting information about the environment with the help of sensors and exerting physical actions through actuators that respond to computational commands, CPS has the advantage of real-time control with the objective of more efficient rather than excellent system operation. The incorporation of new technologies within a physical place allows users to interact with a computer-simulated model of the selected area to take place seamlessly [2].

The purpose of this chapter is to illustrate the need to integrate human considerations into the processes of designing and developing the CPS. The chapter focuses on the cognitive, psychological, and ergonomic aspects of the human–machine interface and why these are treated so superficially. It also contains guidelines on how to develop a CPS with human factors, which include user-centered design, ergonomics, intuitive design, human-in-the-loop systems, adaptivity, real-time feedback, and multimodal systems.

In addition, the chapter examines the problems of new entries in the field of CPS, including new trends in the significance of human factors considering the growth of artificial intelligence (AI) and machine learning (ML) technologies. The paper is relieving in the sense that it's c overs the ideal and most practical issues, the ethical and societal problems, the legal and political issues, and most importantly the field of necessity of team work across disciplines.

1.1 Components of Cyber-Physical Systems (CPS)

The defining feature of a cyber-physical system (CPS) is that it incorporates computing into the physical systems it manages and controls. These systems connect the real and digital by performing data collection, data processing, and actuating actions in real time. Basically, CPS consists of a sensor, actuator, computation unit, and a communication network. The following are the main components of CPS.

Sensors and Actuators:

- Sensors: These are vital elements for capturing data from the surrounding physical environment in real-time. Sensors have the ability to quantify a multitude of physical phenomenon such as temperature, pressure, light, humidity, motion, vibration, and even a person's health in medical applications. They measure the parameters of the physical world and provide accurate and timely estimates, precisely when needed [2].
- Actuators: These complement sensors since they perform the opposite function. Instead of collecting data, they perform physical actions which were authorized by the computation units. These components can operate mechanical devices, control temperature, fluid flow, or even move robotic arms. Other instances are electric motors, hydraulic pistons, and robotic joints. Control algorithms are the dominant method through which actuators alter physical systems [2].
- Sensor-Actuator Integration: Stable operation of sensors with actuators suggests the correct and adequate response in constantly changing surroundings. In self-driving cars, for example, sensors identify obstacles and actuators perform braking and steering motions [2].

Computation Units:

- Embedded Processors: The computation units of CPS are embedded systems and processors that collect data from sensors and control them using control algorithms. These units function as the head of the CPS where sophisticated computations, modeling and analysis are conducted [3].
- Algorithms and Decision-Making Models: These algorithms have the ability to process information, forecast results, and decide an appropriate action to take. Most of these decision-making processes include anticipatory models, optimization efforts, and artificial intelligence to enhance accuracy and efficiency [3].
- Real-Time Processing: The computation units of CPS should perform data processing and systems' decisions instantaneously. This eliminates waiting periods which lead to system unresponsiveness and unreliability. An example comprises healthcare systems, where continues data processing is crucial for many life-threatening situations, like controlling heart rate and defibrillation signals [3].
- Edge and Cloud Computing Integration: Some modern CPS employ edge technologies around the sensors and actuators for real-time data processing, while the cloud is used for heavy data processing, archiving, and running complex simulations [4].

Communication Networks:

- Data Exchange: Communication networks are in charge of data exchange among the sensors, actuators and computation units. This allows for information and coordination flow among all of their components [4].
- Wired and Wireless Networks: CPS may implement both wired (e.g. Ethernet, industrial control networks) and wireless (e.g. Wi-Fi, Bluetooth, Zigbee, 5G)

networks depending on the particular application or environmental constraints [4].

- Latency and Reliability: The possibility of real time communication poses a huge importance in the environment of CPS, especially on cases where delayed responses may yield failures or catastrophic consequences as with autonomous vehicles or medical equipment [4].
- Internet of Things (IoT) Integration: CPS frequently deals with IoT platforms to foster remote monitoring, data analytics and improved interoperability among systems [4].
- Cybersecurity in Communication: Communication networks ought to guarantee that data integrity, confidentiality, and protection from cyber-attacks are guaranteed. All of these objectives' secure communication protocols, encryption techniques, and authentication mechanisms seek to achieve [4].

Cyber-Physical Systems represent a transformative integration of computation, communication, and physical processes. Through real-time sensing, intelligent computation, and efficient communication, CPS enables advanced automation, precise control, and high reliability in applications spanning healthcare, smart cities, manufacturing, and autonomous transportation. As technologies like IoT, AI, and 5G continue to evolve, CPS is poised to play an increasingly central role in shaping the future of intelligent systems across industries.

1.2 Significance of Human Factors in CPS

Cyber-physical systems are benefits of technological development; human factors turn out to be the weakest link. People interact with cyber-physical systems as operators, end-users, and decision-makers at various levels, and their use will influence the system's performance. The design of these systems must include human factors to ensure they are technically correct, ergonomically correct, operationally appropriate, and safe [5].

Automated systems, or those built-in in loosely coupled systems, are said to be the key shortcomings in CPS, particularly because of security risks. Even though CPS is quite high-tech, people's participation is key in operation supervision, decision-making, and reactions to system alerts. On the other hand, people's perceptions of these systems could be negative, misguided, or protective if the system design neglects the technology's use. Interfaces designed to support user tasks may not function properly due to an overload of orders [6].

Another crucial factor is the existence of trust expectations within a system or among individuals to carry out specific actions and respond appropriately. In this regard, the presumption that the operators will interact positively with the systems and its user interface is bound to fail. When a system breakdown, a lack of feedback, or a lack of clarity erodes trust, operators may reject the system or attempt interactions in situations that don't warrant them, leading to a different outcome than what the automation of the system intended. Therefore, to build trust in such systems and utilize them effectively, CPS designs should encompass the human element in terms of behavioral, mental, and emotional tendencies [7].

1.3 Contributions

This chapter makes the following key contributions:

- The Role of Human Elements in the Design of CPS: This chapter pointed out the importance of human factors, ranging from cognitive processes to physical actions in the functioning of Cyber Physical Systems. Ignoring these elements can result in user mistakes, inefficiency, and even endanger safety.
- Barriers to the Consideration of Human Elements: It attempts to look into reasons why human factors, in particular, are ignored in the design of CPS—concentrating on the technical aspects, belief that automation fully substitutes a human, and the complexity of ergonomic integration within unreasonable timelines.
- Strategies for Human Centered Design of CPS: This chapter suggests user centric or human factor design of CPS systems through implementation of detailed and specific principles debating usability, user interface adoption among non-technical personnel, human decision-making systems for more sensitive decisions, flexible systems, and personalization, feedback for actions taken, and multi-communication systems.
- Providing a Comparative Analysis of CPS Design Aspects: Through a comprehensive table, the paper compares various CPS design aspects in terms of their benefits, challenges, example applications, and design complexity, offering practical insights for designers and engineers.

Addressing these contributions, the chapter seeks to bridge the gap between advanced CPS functionalities and human capabilities, promoting the development of systems that are more effective, trustworthy, and accessible to a diverse range of users.

2 Human Factors in CPS

Human factors, more popularly known as ergonomics, is a scientific field that studies tissues in relation to the mechanism along with its various components. The design of products, systems, and environments must match the most efficient working conditions, safety, and health that people can use. In the case of cyber-physical systems, the human factors emphasize the study of the user's interaction with complex automated systems and the design of such interactions. Ergonomics encompasses both the physical and cognitive aspects of work, aiming to make systems less physically or mentally taxing for users while still achieving desired results [8].

In CPS, the concept of "human factors" automatically adopts the definitions of "interactive" and "physiological" when referring to physical systems. In a specific system, the user population and real-time data are integrated with a physical environment and computerized collaboration, such as in smart manufacturing, healthcare, and autonomous cars. Good design must also allow adequate representation of system data, reliability, and proper interaction with the system so as to reduce errors of actions and decisions both during effective and pivotal conditions [9].

2.1 Role of Cognitive, Psychological, and Physical Elements in Human–Machine Interaction

Cognitive, psychological, and physical elements play a significant role in human–machine interaction in cyberspace, contributing significantly to the assessment of a user's proficiency with the system [10]. The cognitive aspect refers to how users approach the input, how they interpret, and how they understand System of Cyber-Physical Systems screens. These systems perform the majority of their work in a real-time environment, processing vast amounts of data in real-time. As a result, users are required to respond quickly in order to initiate event-forming scenarios. Sometimes, this cognitive aspect exceeds a person's capabilities, resulting in negligent behavior, tardiness in decisions, or difficulties with desk work [9].

The psychological aspect plays a role here as well. In the case of a system that delegated decision-making to the computer, trust plays a crucial role. Users must be confident that the device will always perform correctly and consistently. Users may refuse to use a device that is faulty, erratic, or provides ambiguous information, even if its purpose is to simplify their tasks, thereby negating the benefits of automation. Similarly, psychologically motivated factors such as stress or anxiety tend and are anticipated to augment human ability under stress, like in societal transitions where one has to operate the System of Cyber-Physical Systems in healthcare or defense coliseum [11].

Ergonomic design, while quite conventional, equally plays a significant role in the overall process. The design of the interfaces, control elements, and workspace, including implements, has to address the physical comfort and capabilities of the operator. A badly engineered interface Misuse of the system can sometimes lead to user fatigue, errors in manual data entry, or occupational hazards such as repetitive strain injury. In CPS, which requires end users to respond in real-time, ease of physical interaction contributes to ensuring that operators are capable of sustained performance over long operating periods [12].

2.2 Why Human Factors Are Often Overlooked in CPS Design

The design of the cyberphysical systems significantly considers human factors for task performance. First and foremost, the development of CPS places an excessive emphasis on the performance of technical parameters such as speed, automation, and precision. Therefore, engineers sometimes treat the human aspect as an optional consideration. Engineers, for instance, may be very concerned with the optimization of the computational algorithms or the real-time responses of the physical system without giving deep thought to how the end users, who happen to be the most important of all the included users in any system, will interact with these systems [13].

People often overlook human factors because they believe automation will eliminate the need for human input. While machines can indeed perform many repetitive tasks, humans still manage complex processes like system control, decision-making, and interaction when a system malfunctions or behaves differently, necessitating the development of operational guidance. When the system fails to meet the needs of the people, it can become complicated or burdensome, increasing the risk of errors during critical situations where a response is necessary.

In addition to this, it is common practice that design for human factors is a complex and mutual effort between the engineers, psychologists, and ergonomists, and this is not always possible or prioritized in the speeding timelines of the development of systems. Due to the intricate nature of current system development cycles, most designers create designs with new features and anticipate retraining for the system's end users. However, they often overlook the advantages that come with the system designs, as they are designed to fit the users' needs more naturally in the real world than in the current computer-age setting [14].

2.3 Challenges in Integrating Human Factors in CPS

Integrating human factors with CPS presents several barriers, largely due to the complexity of these systems. One of the key challenges in integrating human factors with CPS is managing the trade-off between the level of automation and human supervision. While some CPS operations can operate independently, there are situations that necessitate human intervention, such as during emergencies or system malfunctions. It is not easy to develop what has been referred to as adequate automation without eliminating human beings from the control point, as this involves evaluating both human and system capacity beforehand [15].

Finally, in terms of the human factors design, in particular in the area of systems integration in a cyber-physical system, the real-time aspect to provide a value may be counterproductive to design. Operators are required to be active decision-makers in the presence of high volumes of information or data to be processed in a timely and accurate manner. Non-intuitive interface design and feedback delay mechanisms overload decision support systems, leading to longer decision times and a rise in decision errors. It is imperative for designers to understand that if at all there is a need to integrate any redundant or extraneous information into the system, it must be very simple and hence minimal operational and cognitive load placed on the operators so that their attention and time can be devoted solely to mission critical operations [9].

3 Designing CPS with Human Factors in Mind

Figure 1 illustrates the central concept of Human-Centered CPS Design, highlighting seven key elements that are essential for effective human–machine interaction in Cyber-Physical Systems. These elements include User-Centered Design, which focuses on aligning systems with user needs; Multimodal Interaction, enabling diverse input methods like voice and touch; Ergonomics and UX, ensuring comfort and efficiency; Intuitive Interfaces, simplifying system navigation; Human-in-the-Loop Systems, allowing human intervention in critical decisions; Adaptive Systems and Personalization, tailoring the system to individual users; and Real-Time Feedback, providing timely, relevant information to support user actions. Together, these elements promote usability, safety, and trust in CPS.

Designing Cyber Physical Systems (CPS) could be interpreted as integrating the human–centred approach to all processes. A CPS is a system with a greater degree of amalgamation between the physical and the informational world, which needs to be simple enough for the intended audience. A good designed CPS optimally integrates processes of human operators, which can be industrial workers, health professionals, or normal customers, and ensures that they can interact with the system as effortlessly as possible [9].

Putting consideration on how users think, act, and move enables product designers to develop devices and systems that minimize the potential of mistakes, boost safety, and improve productivity. The purpose user centered design is to develop the system such that it is efficient from the start by incorporating issues brought up by real people. The system must also be ergonomically designed so that it is easy to interact with. Furthermore, reducing cognitive load means providing information in a manner that is simple to comprehend without overwhelming the user. In the end, by focusing on the human factor of CPS design results more reliable, simpler and effective systems [16].

3.1 Principles of User-Centered Design in CPS Development

It is critical to base the design of systems within the context of the capabilities and the limitations of the users of the system. This information is crucial for ensuring

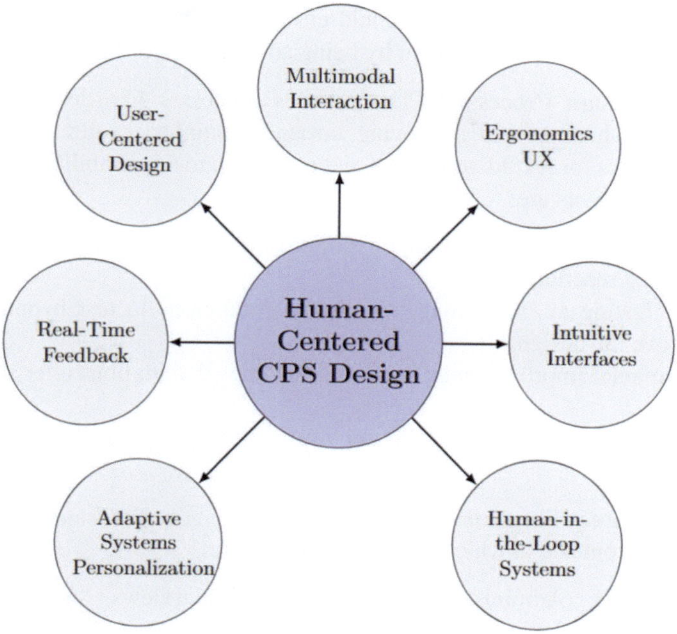

Fig. 1 Designing CPS with human factors in mind

efficiency, safety, and satisfaction. User-centered design (UCD) is a systematic rest approach that requires feedback from the users while understanding the context within which the users interact with the system. This chapter looks at the basic principles, techniques, and the importance of UCD practices in CPS development using specific illustrations and exemplary works sites [17].

3.1.1 Core Principles of User-Centered Design

User-Centered Design operates on several foundational principles that guide the design and development process:

- **Active User Involvement:** Users should be considered an important category of stakeholders that is critical to the success of the design. Their participation is ensured at all levels of the design process. This principle emphasizes:
  - Requirement Gathering: Users should be involved in the defining of the system in as much detail as possible.
  - Feedback Loops: Users should give their input as much as possible, which should be incorporated into the designs and while testing the designs.
  - Collaborative Prototyping: Users should be invited to help in the building of the prototypes and in their evaluation in terms of expectations.

  - User Advocacy: A designer should ensure that the users have always been catered for in the development by being an advocate for the user.

- **Iterative Design Process:** UCD integrates processes like designing, implementing, evaluating, and improving through continued cycles. This iterative process allows for soft identification and resolution to any usability issues. Some of the components are:
  - Rapid Prototyping: Constructing simple and reasonably working prototypes for quick reactions.
  - User Testing at Every Stage: Usability evaluations to test hypotheses and improve the design.
  - Incremental Improvements: Data-informed novel alterations after subsequent cycles.
  - Adaptive Design: Learning through user feedbacks and building on evolving user needs

Focus on User Requirements: The design choices stem from considering user tasks, goals, and the contexts in which they occur. These are:

- User Research: Administering questionnaires, interviews, and ethnographic observations to gain insight into user behavior.
- Task Analysis: Decomposing user activities to determine the essential workflows and pain areas.
- Contextual Inquiry: Capturing users in their working environments and understand their real-world interaction scenarios.
- User Personas: Forming the various types of users as detailed personas that aid in design choices.

Empirical Measurement: Usability is measured through systematic usability evaluation. These are grouped as:

- Usability Testing: Watching users' direct interactions with the system.
- Performance Benchmarking: The counting of the success in the tasks, indices of the errors, and efficiency rating.
- Scenario analysis: Assessing user journeys based on various identified scenarios to detect points of friction.
- Quantitative and qualitative feedback: Returning both countable results and user feedback.

Holistic User Experience: UCD incorporates more than only the design and layout of an application or a website. Hence, it also focuses on:

- Emotional engagement: Creating experiences that once one recall one cannot help but smile.
- Ease of Learning: Designing systems that need little or no training to use the interface.
- System reliability: Providing the system without failures or error messages.

– Contextual awareness: Providing varying experiences depending on other factors like user's device or the environment.

Accessibility and inclusivity: Systems take into account users with different levels of ability, technical knowledge and context. Inclusivity provides equitable access to system features.

3.1.2 Techniques in User-Centered Design for CPS

The application of UCD in developing CPS is done through a set of features that allow users to interact and provide feedback:

- **User Profiles and Personas:** These are used to create user profiles which help the designers define the population and their goals as well as interactions with the system and in case suffering difficulties.
- **Contextual Inquiry**: The users are observed and interviewed while they are working which gives a view into their actual workflows and the constrains and bottlenecks they face.
- **Prototyping:** Users are able to see some parts of the features of the system through low-fidelity prototypes such as paper mockups or high-fidelity prototypes like interactive systems and, as a result, give feedback on important aspects.
- **Usability Testing:** Formal usability testing with representative users helps identify design flaws, navigation issues, and interface bottlenecks.
- **Scenario-Based Design:** Developing realistic scenarios based on user tasks and workflows ensures that the system aligns with actual usage patterns.
- **Design Workshops and Co-design Sessions:** Collaborative workshops foster creativity and allow end-users to contribute design ideas and solutions.

3.1.3 Applications in Cyber-Physical Systems

In regard to user-centered design in cyber-physical systems, people are at the center of the system and therefore UCD contributes towards the effectiveness and safe operation of the system. Some examples are provided below:

- Smart Manufacturing Systems: In industrial CPS, it is critical that operators can easily and effectively control robotic arms and monitor production lines, including their ability to respond to production anomalies. Interfaces that are reliant on multimedia tools are overly simplistic and could cause errors axes, or be overridden by other operators.
- Case Example: Operators of a particular production system were reluctant to turn the system off due to the difficult to navigate interface. Participant feedback helped develop an interface that was much more satisfactory, resulting in quicker reaction times.
- Autonomous Vehicles: User interfaces enabling interaction with self-driving cars must communicate the state of the car, what the autonomous vehicle is planning

to do next, and when the machine is requesting takeover. UCD is beneficial in reducing confusion and increasing trust towards automatic systems.

- Healthcare CPS: Medical devices and monitoring systems are attentive to the fact that their information presentation must be clear and timely for critical information that healthcare professionals need to be alerted to. User testing ensures that the device alarms and displays does not interfere with the workflows of medicine.

3.1.4 Benefits of User-Centered Design in CPS

In the adoption of Cyber-Physical Systems, User-Centered Design (UCD) is important for achieving optimum success. UCD takes the user's needs, preferences, and other relevant aspects of the user's life into consideration throughout the design and development stage with the aim of ensuring that the solutions provided consider any limitations and that they are functional, easy to use, safe, and widely accepted. Below are the key points UCD approach benefits in regards to Cyber-Physical Systems.

- **Usability of the systems is significantly enhanced**: One of the greatest impacts of UCD in the interface design is the use of low-fidelity systems that are user friendly. Through the design participation of the users, developers understand real interactions and difficulties associated with using the system in its intended environment. This not only alleviates the complexity of the systems, but also reduces the effort needed to operate the systems. As a result, the users are able to work with the system effectively in executing different tasks with limited cognitive effort. More so, when people find systems easy to use, the systems become more productive, as people are more encouraged to use the systems. Instead of increasing the errors, greater efficiency can be achieved.

Example: For instance, in industrial CPS environments, equipment monitoring CPS have well designed user interfaces on dashboards that allows operators to identify and respond system anomalies in real time and in a short period of time even in the absence of training.

- **Reduced Safety Risks and Errors**: The safety of a new project in CPS is crucial in areas like healthcare, transportation, and industrial automation. Systems developed with user feedback are often greatly defensive to abuse, misinterpretation, and human errors. UCD aims to be as clear and comprehensive as possible while a user interface is being developed so there is minimal chance for obfuscated instructions causing operational lags.

Example: In medical CPS tending to diabetes patients, like automated insulin pumps and robotic surgery systems, clear visual cues, alerts and controls are crucial. An easy-to-use system will guarantee no inadvertent styling of dosages or blunders during procedures.

- **Improved Adoption Rates and Acceptance**: Systems manufactured with the knowledge of a user's expectations and workflow steps are easier to accept. Users are more likely to adopt a device or application if they are confident that it meets their objectives and eases their daily tasks. UCD eases this gap by allowing users to test and validate skeletal designs, making changes based on real-life user's insights.

Example: In smart city CPS applications like the features of a traffic control system, adoption of citizens is more favorable if traffic alerts, or even available parking spaces can be reported through easy mobile applications.

- **Reduced Development and Maintenance Costs**: Usability issues are more efficiently solved at the earlier stages of system development as compared to attempting to fix them post-deployment and can be significantly more expensive. A user-centered design (UCD) approach focuses on iterative design processes like testing, prototyping, and user feedback, which helps designs to be unfurled before they become expensive issues after deployment.

Example: In the case of energy grid management systems, control panel interface enhancements are implemented at the prototype level in order to avoid complex post-deployment troubleshooting or extensive operator retraining.

- **Increased User Satisfaction, Trust, and Confidence:** Systems that are entirely user-centric in terms of layout and features tend to be free of kinks, which leads to satisfaction of users. Users are more inclined to trust and depend on the solutions offered by CPS systems if they can freely maneuver within the system with minimal challenges. Trust is a vital ingredient for long-term engagement and loyalty.

Example: Autonomous vehicle systems using clear displays on dashboards alongside a more user-friendly interface to operate the vehicle boosts confidence for passengers in the safety and reliability of the vehicle.

Incorporating User-Centered Design on Cyber Physical Systems yields better, reliable and user-friendly solutions. By involving users in all the stages of development, an organization can take an active approach to minimize errors, reduce costs in the long run, and guarantee that their systems are both functional and up to the user's expectations.

3.2 Ergonomics and User Experience (UX) Consideration

The principal aim of Cyber-Physical Systems (CPS) Ergonomics is to enhance user-system relationships for achieving physical and cognitive wellness. CPS domains typically feature interactions with real-world elements like machinery or medical instruments as well as interactions with intangible components like dashboards, control panels and some monitoring systems [12].

3.2.1 Physical Ergonomics

Physical ergonomics addresses the design of hardware interfaces, controls, and workspaces to minimize discomfort, fatigue, and the risk of injury.

- Optimal Placement of Controls: Controls should be positioned within natural reach and movement ranges to prevent repetitive strain and awkward postures. Button sizes, spacing, and tactile feedback should accommodate diverse user needs, including those with limited mobility.
- Minimizing Fatigue: Tools and devices should be lightweight, ergonomically shaped, and require minimal physical effort for operation. Handles should have non-slip grips, and levers or switches should operate smoothly without excessive force.
- Workstation Design: Adjustable chairs, screen angles, and desk heights contribute to long-term usability and reduced physical strain. Workstations should support both sitting and standing postures to reduce prolonged stress on specific muscle groups.
- Environmental Factors: Adequate lighting, noise reduction, and optimal temperature control should be considered to enhance physical comfort and prevent strain-related injuries.

3.2.2 Cognitive Ergonomics

Cognitive ergonomics focuses on reducing mental workload and enhancing user understanding to ensure efficient CPS operation.

- Overload of Information: Applications must filter and present only useful and relevant information to avoid information saturation. Marking systems, alert prioritization, and even visual hierarchy (e.g. color coding, text bolding) serves to help effortless and focus based additional task completion.
- Alerts and Notices: Notifications need to be glaring, but without causing too much distraction. Sounds cues should vary by level of importance, while the visuals should be self explanatory.
- Standardization: Users should spend relatively less time memorizing icons and workflows and more time actually being productive, that is why everything needs to be consistent. Custom designs between systems lead to decreased efficiency and increased training costs.
- Intelligent Agents: Systems are required to have a provision of unsolicited intelligent suggestions to carry out lower level goals to reduce the cognitive effort needed for critical actions.
- Reduction of Error: Interfaces should enable users to automate checks on common mistakes such as confirming when performing critical affairs.

3.2.3 User Experience (UX)

Designing user experiences (UX) focuses on ensuring users' interaction with the CPS interfaces is as efficient and natural as possible.

- Visual Simplification: Dashboards and other user interfaces must be straightforward with no clutter and easy to use. Information must be chunked properly and overarching graphical components should be omitted.
- Factual Clarity: Users must be kept up-to-date regarding the systems status through real-time visual, audible, and tactile reminders. For instance, status indicators could portray system condition, work done, or errors.
- Treatment/Management of Errors: Generating unambiguous informative error alerts and recovery instructions enables the user to fix problems quickly. Visual aids like highlighting the problems or stepwise aiding the user can increase man–machine problems resolving issues.
- Flexibility: Users should be able to modify knobs like font style, font size, color contrast, and outline the types of alarms they wish to receive.
- Adaptability: CPS interfaces also have to meet the criteria of accessibility in the usability of disabled people through screen readers, voice, and other non-standard input devices.

When it comes to UX, it's important to remember that CPS interfaces need to be visually and structurally straightforward and uniform. The goal of UX is to facilitate the user in achieving their objectives by interacting with the system in a swift, efficient, and effort-free manner. Instantaneous relay of feedback, including haptic feedback and errors, keeps the user aware of the system's state and the need to modify their behavior [18].

3.3 Developing Intuitive Interfaces for Non-technical Users

One of the important issues in the design of CPS is addressing the operating interfaces for users who are not technically inclined. As CPS development and implementation progress, a growing number of users may lack a deep technical understanding of the systems in use, particularly in healthcare, smart homes, and agriculture. Therefore, it's crucial to simplify the interface design to eliminate obstacles to efficient operation, as comprehensive training may not be available. Considerable measures that can be taken to address these challenges in the interaction design for non-technical users include [19].

- **Iconography**: The application of well-known symbols and icons has the great advantage of eliminating the burden of language as well as the effort of reading and understanding the materials included to be internalized.

- **Guided Workflows**: Quite often, users may not be familiar with how to operate a feature; in such cases, they may find the necessary instructions employed in the form of user manuals or help guides which may prove to be useful.
- **Error Prevention and Correction**: A good number of users of core interfaces will often make a few mistakes; alleviating this can be done by ensuring that the core interface does not rely fully on the user but is capable of performing some anticipatory actions. A typical example would be an irrigation CPS whose control interface would warn the user against selecting a wrong watering schedule.
- **Customizable Interfaces**: These interfaces, which enable individual users to modify or customize the display, increase the likelihood of effective usage in contemporary interfaces. Absolute beginners would prefer the very basic screens devoid of complexities while low skilled people or those with a bit of experience with the system are allowed to exploit a bit more advanced options.

3.4 Human-In-The-Loop Systems for Critical Decision-Making

Human-in-the-loop (HITL) systems are particularly relevant in situations where a human's intervention becomes necessary to either override or assist the automated decision. In specialized fields such as defense, healthcare, or autonomous driving, one cannot afford to completely trust automation, as there are always unexpected scenarios in which the smartest algorithms might fail. HITL systems enable human operators to intervene and take over control in instances of importance. An effective HITL system design must consider the following factors [20].

- **Clear Role Allocation**: It should be made evident in what situations and in what way humans are expected to assist the system so that the roles of man and machine are not confused.
- **Seamless Handover**: Handover refers to the transfer of control from one occupant to another, which can happen automatically or in response to a command. This ensures that when the system prompts the practitioner to regain control, they are capable of understanding the situation or system and responding promptly. This requires a concise strategy for audio and visual cues that enhance comprehension and context without excessively bombarding the user with content.
- **Trust Building**: An adequate level of trusting requires one to believe that systems meant to automate a majority of the actions would function the way they should, while actors also believe they have an opportunity to act as the situation might demand. You can potentially achieve this by paying close attention to the decisions the system makes. An ideal scenario would involve an electronically operated car displaying the reasons for applying brakes and accelerators, rather than simply carrying out these actions, thereby causing embarrassment to the driver.

3.5 Adaptive Systems and Personalization

One of the increasing areas of interest in the designs of the CPS is the creation of systems that are adaptive to users depending on their behaviors, requirements, and expertise levels. Personalization not only enhances the user's status level but also enhances the efficiency and safety of the environment and system by aligning with the user's intentions. Lastly, adaptive systems use current information to change the presentation, interactivity, and automation level of the system according to the actions performed [21]. For example:

- **Dynamic Interfaces**: A smart grid CPS would have the ability to change the users' interface so that less experienced operators do not get overwhelmed with too much information at a time. A beginner would only see a basic control unit, while an advanced user would be able to change real-time parameters or even have some control over all layers of the structure and operation [21].
- **Behavioral Adaptation**: If the system identifies a user's susceptibility to a particular situation based on their behavior, it can evaluate the user's inadequacies and inadequate assistance in the task, either by halting or modifying the culprits until the activity In the context of healthcare CPS, if a user enters incorrect values, such as high dosages for medication, the system will gently correct these errors through gentle triggers and educational interventions.tle triggers and education [21].

The design of the adaptive systems is an undertaking that requires some degree of caution to avoid irritating the user. The adaptive systems may frustrate the user, potentially leading to overadaptation. This could result in a lack of satisfaction or control, making the user feel powerless and less inclined to utilize the systems [21].

3.6 Real-Time Feedback and Cognitive Load Reduction

In the case of a CPS, real-time feedback is crucial to keep users informed about the system's functioning, potential errors, and necessary steps. However, an excessive amount of information can lead to cognitive overload injuries, a common design issue across various disciplines, especially in scenarios where users must make decisions under time constraints. To effectively manage cognitive load, designers of CPS should take into account the following factors [9]:

- **Minimalist Information Display**: The user should only see the most pertinent information at any given time to aid in information assimilation. Industrial CPS should promptly notify operators of equipment failure, eliminating the need for them to sift through irrelevant information.
- **Feedback Policy Prioritization**: When multiple alerts or notifications are needed, the system must prioritize them according to their severity. For example, a notification concerning a particular error that does not require human attention should

be made in a manner the user should not miss; however, less critical notifications may be addressed subsequently.

- **Visual and auditory cues**: Such cues can serve to eliminate unnecessary selective attention. For instance, when there are color codes (red for danger and green for operations) combined with audio signals such as a horn, people are better able to react than when the warning is just audio.
- **Sequential and Orderly Presentation of Information**: The logical presentation of information is very important in reducing the cognitive load of CPS. Information should be presented in a way that is arranged in accordance with the user's cognitive workflow, from the problem identification to the cause analysis and the remedy formulation. Alerts and notifications should start with concise and straightforward high-level summaries so that users are made aware of the situation, followed by greater details if further probing is necessary. This technique is known as progressive disclosure and helps mitigate excessive detailing of information. Moreover, alerts have to be succinct and devoid of technical terms so that they can be understood immediately. An organized flow of information improves the ability of users to prioritize their actions, and decreases the probability of misinterpretation or errors of omission. Arranging information in an orderly and logical way enables CPS to enhance the cognitive support of users during complex or highly demanding tasks.
- **Incorporating Cognitive Load Management Into the Design of Custom Physical Systems**: The objective is to optimize the design of an ergonomically efficient and user friendly feedback hardware interface in real time." Effective cognitive load management concerning the user of adaptive systems in CPS is an iterative process. In terms of functional ergonomics, adaptive systems need to be further designed to automatically modify information presentation with regard to user workload and system state for efficiency of resource utilization. The AI driven analytics can be used to estimate the probability of a user suffering cognitive overload and automatically adjust the interface as needed in real time to minimize cognitive effort. This is important because, at the design of Custom Physical Systems, these components help operators achieve and maintain situational awareness while reducing errors and improving the system's performance.

Under such conditions, the feedback is presented in an orderly, sequential, informative, and concise manner, resulting in a low psychological demand due to the relevance of the presented information in supporting cognitive processes.

3.7 Multimodal Interaction in CPS

As the design of the CPS expands, users anticipate interacting with it through multiple channels instead of relying on a single one, leading them to demand more dynamic designs. Such systems must provide users with a voice and enhance user friendliness to a greater extent than traditional interfaces. An example can be voice interaction systems that are efficient in operating environments such as industries where the operator's hands are busy. To design a proper multimodal system, one must adhere to the following guidelines [22].

- **Consistent Feedback Across Modalities**: To ensure that a user received a voice command correctly, they should be able to see a response after performing an action.
- **User Choice**: Systems should enable users to choose the modality they prefer, allowing their interaction with CPS to vary based on the task or setting.

Table 1 illustrates crucial CPS design components related to benefit delivery, obstacle resolution, and complexity level. Although user-centered design improves usability, production methods can frequently take advantage of this, which may be time-consuming and expensive.

Ergonomics promotes usability, but this needs careful consideration within the framework of health systems. Maintaining a functional yet user-friendly interface presents a greater challenge, even as navigational systems become more accessible to non-experts, such as in agriculture. Human-in-the-loop system design increases operational complexity and safety, thereby making unmanned vehicles more challenging. Although resource allocation to smart grid systems shapes user experiences, it may also result in the user being incapable of operating the system. Feedback and signaling minimize the strain on the central nervous system during heart surgery, but precise testing is necessary to ascertain the necessary load. Similar to this, multimodal interaction enhances the system's comfort and efficiency, but it requires a consistent operating design or interactivity across all modes of operation, as is the case with medical systems.

4 Future Directions

The integration aspects of cyber-physical systems (CPS) are greatly influenced by emerging advancements in AI, ML, 5G, and edge computing. These innovations increase the level of automation and sophistication in decision-making the systems can achieve, but they also introduce new hurdles regarding human interactions with the system. While a higher number of processes will be automated within the system, human intervention will still be necessary during certain critical junctions where moral, safety, and other intricate judgments are involved. As industrial sectors such as healthcare, transportation, and manufacturing continue to adopt newer technologies in automation, the human to machine control ratio will have to be exercised with caution. This contemporary vision of CPS design captures the need for systems that not just operate effectively, but also produce a greater sense of trust alongside ensuring that the users' human proficiency is not eclipsed by superseding automation. After all, the issue of whether these complex systems are best attended to by humans or machines hinges on the ease with which the two forms of intelligence can be integrated and be articulated within it.

Table 1 Comparative analysis of CPS design aspects

Design aspect	Challenges	Example application	Design complexity
User-centered design (UCD) [17]	Time-consuming and requires constant user feedback and iteration	Manufacturing CPS, where operators monitor and control machinery	High: requires iterative feedback and constant updates
Ergonomics and UX [12, 18]	Requires thorough analysis of human cognitive and physical factors, increasing complexity	Healthcare CPS, where both physical comfort and clear feedback are crucial	Moderate: balancing physical and cognitive elements adds complexity
Intuitive interfaces for non-technical users [19]	Difficult to balance simplicity with functionality for diverse user groups	Agricultural CPS, allowing farmers to manage irrigation systems without technical expertise	Low to moderate: simplicity can be achieved but maintaining functionality is challenging
Human-in-the-loop systems [20]	Ensuring seamless transitions between human and system control can be complex	Autonomous vehicles requiring driver intervention in complex scenarios	High: complex system dynamics require careful coordination between human and machine
Adaptive systems and personalization [21]	Risk of over-adapting, leading to loss of user control or confusing interfaces	Smart grids, adjusting system behavior based on user preferences and needs	High: requires advanced algorithms and machine learning for effective adaptation
Real-time feedback systems [9]	Too much feedback can cause cognitive overload; prioritization is crucial	Industrial CPS providing real-time alerts on machine performance	Moderate: requires careful prioritization of feedback to avoid information overload
Multimodal interaction [22]	Needs to ensure consistency across modalities and avoid confusing interactions	Voice and gesture control in medical CPS, allowing surgeons to control robotic systems	High: integrating multiple interaction methods requires robust design strategies

4.1 *Emerging Trends in CPS and the Evolving Role of Human Factors*

The current directions in Cyber-Physical Systems (CPS) development can clearly be planned out if considered in how they incorporate new advances in artificial intelligence (AI) and machine learning (ML) as well as upcoming communication technologies such as 5G and edge computing. With these developments, tasks such as decision-making, automation, and active cooperation of humans and machines get to a smarter and faster level. However, as more of these systems' functions are unlocked from human control and management the human factors are becoming more intricate. On the contrary, more human cognitive resources are being relied on when operating more intelligent systems [23].

The current and emerging trend is correlating to the general attention of CPS in the areas such as healthcare, autonomous vehicles, and industrial application which are more complex systems. As these systems grow in importance for day to day activities, human users of the systems will still need to have control for the most critical decision points. The forces of automation are being embraced, though, there is a limit to which robots can reach, especially in matters relating to ethics, morals, and safety. Future designs of CPS will be dependent on how effectively skills and wisdom of both machines and humans are harnessed [23].

4.2 AI and Machine Learning to Predict Human Responses

AI and machine learning have had a noteworthy impact in this architecture, since such systems are now able to also model and learn from human behavior and adjust accordingly. Using data obtained from various sensors and interactions with the user, AI Built systems are able to predict when users will make an error, modify the tasks they are undertaking, and request assistance when the users need it. For example, the CPS's that leverage AI in industrial application can, for instance, supervise an operator's workload and restrict it with a timely intervention in case there are signs of cognitive or physical fatigue, hence enhancing productivity and incident management [24].

For instance, in driverless systems, AI can accurately gauge how human drivers will respond and make suggestions or take action in key moments before there is an accident. The same time, it opens up a number of new issues related to users' trust in such predictions and transparency of their implementation. It is necessary to understand and trust these interventions provided by AI if one wishes to take advantage of such capabilities. More comprehension is still needed respecting how AI decisions are made to help users appreciate the purpose of certain system behaviors [24].

4.3 Opportunities for Further Research in CPS-Human Interactions

The human factors in CPS are relatively new areas in research and can therefore be explored further. One such aspect is related to cognitive ergonomics, which seeks answers to considering human cognitive capacities in the design of CPS. For instance, such systems need to be tailored and designed for those human operators who are likely to be in a critical setting, making difficult, swift decisions. By understanding how humans are able to sense and remember information, and how they are able to reason, designers can produce interfaces that do not engage the normal operational effort, that hastens decision making, and that avoids mistakes [25].

The other area for further research is how trust in CPS can formed and developed. In the context of increasing automation of systems, it will be particularly important for users to accept the technologies of CPS. It is important to conduct research in order to understand what factors influence user trust towards automation and what measures can some progressive efforts help create trust over time. Such as giving people the opportunity to override by providing them with feedback, making the system reliable and predictable [25].

Adaptive user interfaces are the last aspect of this section that deserves attention as it is promising research domain. These interfaces change according to the user's activity, level of expertise or knowledge, and even the state of mind. Such adaptive systems could provide an automatic adjustment of feedback offered, or of the interface so that it matches the ability of the user to navigate within the system. While these designs are beneficial, they seem to cause several legal red flags especially given the fact that they are built on perpetual data collection. The solution must involve technological advances that will enable the creation of adaptive interfaces as well as the evaluation of moral issues surrounding their deployment [26].

4.4 Human-Centered CPS: A Vision

As CPS continues to be pervasive throughout daily life and activities, the methodology for constructing these systems must be human-centered. Patient-centered approaches have been limited due to the fact that CPS often reinforce a 'command and control' stance focused on maximizing efficiency. This perspective allows for the creation of systems only capable of performing actions by themselves, after which a user will provide it with feedback integrated from various tasks and decisions from which a user would prefer to participate actively [27].

Barriers to implementations have the following potential features such as bringing the system closer to the operator by designing appropriate and flexible interfaces within the CPS, embedding how-to-use systems into processes in the form of instantaneous and real time guides, and having human op-in capabilities whenever required. For example, in industrial environments, the system could provide information in real time regarding the effect of certain actions, thus guidance is always available, amplifying the chances of project success. In medicine, clinicians could apply the system recommendations to the cases where patient clinical situations are more common while surgeon takes responsibility in critical surgery procedures [28].

4.5 Ethical and Societal Considerations

CPS progress further towards total autonomy, ethical issues concerning decision making and accountability will undoubtedly arise. For example, in the case of self-driving cars, who bears the liability if an AI-powered car malfunctions and causes a car crash? The same applies to the medical domain, if a CPS makes a mistake and gives the wrong diagnosis who can be held liable for that? Such ethical issues must be incorporated in the design of any future CPS to guarantee that human managers are ultimately accountable for what the machines that make decisions do [29].

There is also a societal impact to consider. As CPS become additional integrated into our daily lives the issue of digital divide reverts, where people completely without such advanced technologies get cut off. It will be necessary to ensure that the CPS are consonant to a diverse population of users without regard to levels of expertise of or resources available to the individuals. The wide variety of data collected by the mechanical systems also raises questions of user confidentiality which will be an additional ethical issue to address [30].

4.6 Regulatory and Policy Frameworks

The accelerated evolution of CPS will likewise necessitate the formulation of pertinent regulation and policy frameworks. Policymakers and regulatory agencies should outline the regulations that would enforce the safety, dependability, and protection of the CPS while at the same time encouraging their development. Such frameworks would have to cater both for the technical requirements and the non-technical, e.g. sociological aspects like issues of ethics such as privacy assurance and accountability of CPS. Toward this end, regulations should supports the innovation of algorithmic decision making, and the establishment of human governance over CPS and weaken the threats on the organizational policy regarding the design and utilization of cyber-attacks on CPS.

Meanwhile, other regulatory initiatives and plans should also be aimed at the development of certifying bodies for CPS for markets focused on Critical Industry Safety, such as medicine, self-driving cars and electricity. These certifications would ensure that CPS are exhaustively evaluated prior to being put to active use and that there are preset expectations of the safety and performance of the system.

4.7 The Role of Interdisciplinary Collaboration

Collaboration across disciplines will be even more important in the future of CPS design and development. CPS span across various domains such as engineering, computer science, human factors, psychology, ethics and law. There is a need to integrate the knowledge from respective professions if the systems to be developed are to be sophisticated and people oriented. By bringing together experts in technology and human interfaces, CSA may be enhanced in efficiency while remaining acceptable, safe and ethical [31].

In many views on the future of CPS are favorable for further studies and innovations. Human limitations will continue to be integrated in the design of these systems so that as CPS becomes more complex and autonomous, they will continue to aid the operators, build trust and respond to ethical and social issues.

5 Conclusion

The integration of human factors in performance, safety, and user satisfaction of the user experience in Cyber-Physical Systems (CPS) systems is demanding in nature. With increased advancements and developments in CPS, the human technology interface is becoming more crucial. If the cognitive processes of humans, the ergonomics, or the decision-making process is not considered, there are problems such as inefficiencies, errors, and in some more serious cases even human safety. That being said, incorporation of centimeters within the architecture of human computer interface design maximizes the ease of cooperation and communication between the human and the machine. More importantly, it allows for collaboration with machines which, for the most part, is what would be needed to design. The true nature of how remarkable CPS is comes to the forefront in clinical environments or even in autonomous vehicles. Keeping humans within the loop when the systems are active enables the users to have a chance to intervene when necessary. It adds an extra layer of safety and effectiveness. It not only allows expansion for CPS integration but also allows for prompt human intervention when necessary. Just like all other important factors, human consideration also helps in proving to be beneficial by decreasing user errors or exploitation. Systems of such multi-dimensional functional capability, these engineering designs of CPS are easy to use, but they are hard to work with and when stressed, users are sure to make mistakes or misuse the system's features. This imbalance can be corrected by creating 'smart' systems, employing efficient medicine, and, if necessary, undertaking teaching procedures. In addition, the design of a system that is prone to human error must also incorporate measures of security to avert such unintentional violations and breaches. Finally, future research will include examining methods for better interaction between humans and CS in a changing technological environment. Factors such as the awareness of human cognitive limitations, the enhancement of situational awareness, and the design of mutable systems will be critical in the future, as people's relationships with CPS and its development are on the verge of change. Furthering our understanding of such interactions will allow for improving future systems in usability, safety, and efficiency, among others.

References

1. Baheti R, Gill H (2011) Cyber-physical systems. Impact Control Technol 12(1):161–166
2. Jazdi N (2014) Cyber physical systems in the context of Industry 4.0. In: 2014 IEEE international conference on automation, quality and testing, robotics. IEEE, pp 1–4
3. Moller D (2016) Guide to computing fundamentals in cyber-physical systems. Springer, Heidelberg, pp 307–375
4. Jawhar I, Al-Jaroodi J, Noura H, Mohamed N (2017) Networking and communication in cyber physical systems. In: 2017 IEEE 37th international conference on distributed computing systems workshops (ICDCSW), June 2017. IEEE, pp 75–82
5. Brauner P, Valdez AC, Philipsen R, Ziefle M (2016) On studying human factors in complex cyber-physical systems
6. Geisberger E, Broy M (eds) (2015) Living in a networked world: integrated research agenda cyber-physical systems (agendaCPS). Herbert Utz Verlag
7. Garay-Rondero CL, Thierry-Aguilera R, Schneider AK, Bourguet-Diaz RE, Salinas ML, Zavala G (2021) Designing a cyber-physical-systems and human factors engineering course for Industry 4.0. In: 2021 Machine learning-driven digital technologies for educational innovation workshop. IEEE, Dec 2021 pp 1–8
8. Fantini P, Tavola, G., Taisch M, Barbosa J, Leitão P, Liu Y, Sayed MS, Lohse N (2016) Exploring the integration of the human as a flexibility factor in CPS enabled manufacturing environments: methodology and results. In: IECON 2016—42nd annual conference of the IEEE industrial electronics society. IEEE, pp 5711–5716
9. Wang B, Zheng P, Yin Y, Shih A, Wang L (2022) Toward human-centric smart manufacturing: a human-cyber-physical systems (HCPS) perspective. J Manuf Syst 63:471–490
10. Cross ES, Ramsey R (2021) Mind meets machine: towards a cognitive science of human–machine interactions. Trends Cogn Sci 25(3):200–212
11. Johnson J (2022) The AI commander problem: ethical, political, and psychological dilemmas of human-machine interactions in AI-enabled warfare. J Military Ethics 21(3–4):246–271
12. Cachada A, Barbosa J, Leitao P, Deusdado L, Costa J, Teixeira J, Romero L, Moreira PM (2019) Development of ergonomic user interfaces for the human integration in cyber-physical systems. In: 2019 IEEE 28th international symposium on industrial electronics (ISIE). IEEE, pp 1632–1637
13. Fantini P, Pinzone M, Taisch M (2020) Placing the operator at the centre of industry 4.0 design: modelling and assessing human activities within cyber-physical systems. Comput Ind Eng 139:105058
14. Neumann WP, Winkelhaus S, Grosse EH, Glock CH (2021) Industry 4.0 and the human factor—a systems framework and analysis methodology for successful development. Int J Prod Econ 233:107992
15. Stern H, Becker T (2019) Concept and evaluation of a method for the integration of human factors into human-oriented work design in cyber-physical production systems. Sustainability 11(16):4508
16. Gil M, Albert M, Fons J, Pelechano V (2020) Engineering human-in-the-loop interactions in cyber-physical systems. Inf Softw Technol 126:106349
17. Tekaat JL, Anacker H, Dumitrescu R (2021) The paradigm of design thinking and systems engineering in the design of cyber-physical systems: a systematic literature review. In: 2021 IEEE international symposium on systems engineering (ISSE). IEEE, pp 1–8
18. Pavlovic M (2020) Designing for ambient UX: design framework for managing user experience within cyber-physical systems. L. RAMPINO, I. MARIANI 39
19. Elena Mäkiö-Marusik E, Mäkiö J, Kowal J (2017) Implementation of task-centric holistic agile approach on teaching cyber physical systems engineering. In: AMCIS 2017 proceedings: proceedings of the twenty-third Americas conference on information systems, IS in Education, IS Curriculum, Education and Teaching Cases (SIGED), May 2017, pp 1–10

20. Ma M, Lin W, Pan D, Lin Y, Wang P, Zhou Y, Liang X (2018) Data and decision intelligence for human-in-the-loop cyber-physical systems: reference model, recent progresses and challenges. J Signal Process Syst 90:1167–1178
21. Schlund S, Kostolani D (2022) Towards designing adaptive and personalized work systems in manufacturing. Digit Work Environ Sustain Prod 81
22. Chapman S, Kirks T, Jost J (2019) Study on interaction modalities between humans and CPS in sociotechnical systems. In: Human systems engineering and design: proceedings of the 1st international conference on human systems engineering and design (IHSED2018): future trends and applications, 25–27 Oct 2018, CHU-Université de Reims Champagne-Ardenne, France 1. Springer International Publishing, pp 1044–1050
23. Kadir BA, Broberg O, da Conceição CS (2019) Current research and future perspectives on human factors and ergonomics in Industry 4.0. Comput Ind Eng 137:106004
24. Chopade P, Edwards D, Khan SM, Andrade A, Pu S (2019) CPSX: using AI-machine learning for mapping human-human interaction and measurement of CPS teamwork skills. In: 2019 IEEE international symposium on technologies for homeland security (HST). IEEE, pp 1–6
25. Vanderhaegen F, Jiménez VDB (2023) Opportunities and threats of interactions between humans and cyber–physical systems–integration and inclusion approaches for CPHS. Cyber–Physical–Human Syst: Fundam Appl 71–90
26. Han D, Xing J, Yang Q, Li J, Zhang X, Chen Y (2017) Integrating goal models and problem frames for requirements analysis of self-adaptive CPS. In: 2017 IEEE 41st annual computer software and applications conference (COMPSAC), vol 2. IEEE, pp. 529–535
27. Hadorn B, Courant M, Hirsbrunner B (2016) Towards human-centered cyber-physical systems
28. Zarte M, Pechmann A, Nunes IL (2020) Principles for human-centered system design in industry 4.0–a systematic literature review. In: Advances in human factors and systems interaction: proceedings of the AHFE 2020 virtual conference on human factors and systems interaction, July 16–20, 2020, USA, pp 140–147. Springer International Publishing
29. Henshaw M, van Barneveld J (2016) CPS for community security and safety
30. Khargonekar PP, Sampath M (2020) A framework for ethics in cyber-physical-human systems. IFAC-Pap Online 53(2):17008–17015
31. Sheikh ZA, Singh Y, Singh PK, Ghafoor KZ (2022) Intelligent and secure framework for critical infrastructure (CPS): current trends, challenges, and future scope. Comput Commun 193:302–331

Ensemble-Based Cognitive IDS for IIoT in Cyber-Physical Environments

Lahcen Idougid, Khalid Elfayq, and Said Tkatek

Abstract The industrial internet of things (IIoT) is disrupting industries through greater automation, efficiency and connectivity. But they also make IIoT systems vulnerable to advanced cyber threats like denial-of-service (DoS) attacks and infiltration, necessitating strong security measures to protect these platforms. This paper proposes a cognition computing based intrusion detection system (IDS) for IIoT security. The proposed IDS has the ability to not only detect known threats, but also emerging threats, by incorporating ensemble learning models, together with cognitive principles to achieve computational efficiency. To improve the performance of detection, the system uses feature selection methods such as Recursive Feature Elimination (RFE) and Mutual Information (MI) to select only a refined input data and remove noise. Results: The IDS evaluated on the UNSW-NB15-v2 dataset shows very competitive results compared to the state-of-the-art in terms of accuracy, recall, and ROC-AUC. Ensemble methods including XGBoost and Random Forest surpassed classical methods, with accuracy as high as 98% and a marked decrease in false alarms. We believe that supporting cognitive computing and ensemble learning techniques can provide over IIoT consistent, scalable, and adaptive security solutions. Results illustrate how this method can tackle the distinct security issues of IIoT, leading to more intelligent and robust industrial practices.

Keywords IIoT · IDS · ML · Cybersecurity · Ensemble learning · Feature selection · Anomaly detection · Cognitive computing

L. Idougid (✉) · K. Elfayq · S. Tkatek
Faculty of Sciences, Ibn Tofail University, Kenitra, Morocco
e-mail: lahcen.idouglid@uit.ac.ma

K. Elfayq
e-mail: khalid.elfayq@uit.ac.ma

S. Tkatek
e-mail: said.tkatek@uit.ac.ma

G. H. L. et al. (eds.), *Reliability in Cyber-Physical Systems: The Human Factor Perspective*, Springer Series in Reliability Engineering,
https://doi.org/10.1007/978-3-032-09917-4_12

1 Introduction

The Industrial Internet of Things (IIoT) represent a decentralized paradigm for applying the Internet to advanced manufacturing, which is revolutionizing industries with applications ranging from automation to predictive maintenance and maximizing operational efficiency [1]. Nevertheless, the integration complexity of IIoT systems makes them highly vulnerable to critical cybersecurity attacks, including unauthorized access, data leaks, and advanced persistent threats [2–4]. Intelligent and effective solutions are required to ensure the security and resilience for IIoT infrastructures, due to the increasing sophistication of cyberattacks [5, 6].

Deployment of artificial Intelligence techniques are popular in intrusion detection in cyber security [7]. The diversity of devices, networks, and applications connected through the IIoT ecosystem can effectively serve as multipliers for cyberattacks. Traditional security approaches, including firewalls and antivirus solutions, have proven inadequate to address the specific requirements of IIoT contexts, including real time data processing, scalability, and resource limitation [8, 9]. This limitation necessitates the development of IDS capable of consistently identifying existing and unprecedented threats while preserving computational efficiency [10, 11].

To address these challenges, this paper proposes an Ensemble-Based Cognitive Intrusion Detection System (C-IDS), which integrates machine learning ensemble methods with cognitive computing principles. The main contributions of this work are:

- A hybrid IDS framework utilizing ensemble models (XGBoost, Random Forest, etc.) for high detection accuracy and low false alarms.
- Cognitive design principles that enable adaptability to emerging threats over time without sacrificing computational efficiency.
- Advanced feature selection using RFE, MI, and PCA, optimized for IIoT constraints.
- Comprehensive experimental validation on the UNSW-NB15-v2 dataset, demonstrating superior performance compared to state-of-the-art IDS techniques.
- Performance evaluation in terms of accuracy, ROC-AUC, training time, and prediction time—demonstrating practicality for deployment in IIoT settings.

2 Related Work

IIoT (Industrial Internet of Things) is a game changer for industrial automation. With the cyber attacks becoming more advanced, Intrusion Detection Systems (IDS) are now vital components to secure IIoT networks. We explore existing IDS techniques in this section, demonstrate classical machine learning (ML) methods versus modern ensemble approaches, and identify the areas addressed in this work.

2.1 Existing Related IDS Techniques for IIoT

Intrusion detection can be categorized into two major types: signature-based IDS and anomaly-based IDS. Signature-based systems identify attacks by comparing them to known patterns or signatures. These systems are very effective but are incapable of finding zero-day attacks and frequently need to be updated [12].

In contrast, IDS based on anomaly detection use statistical analysis and unsupervised learning to recognize differences between actual and expected network traffic. Such systems can identify unknown attacks but are challenged by high false positive rates that greatly hamper their practical applicability within IIoT contexts [13].

In recent years, hybrid IDS approaches have attracted researchers by combining signature-based and anomaly-based detection features to achieve both accuracy and adaptability. Especially the use of machine learning as well as deep learning-based techniques for improving the cyber attack detection capabilities in IIoT is becoming more and more popular. Various advanced learning approaches like convolutional neural networks (CNNs) and reinforcement learning identify desirable results in complex IIoT scenarios [14, 15].

2.2 Comparison of Traditional ML and Modern Ensemble Methods

Common techniques used in IDS are Support Vector Machines (SVM), Logistic Regression, K-Nearest Neighbors (KNN) traditional machine learning methods. These models are computationally efficient and moderately interpretable, making them useful for first-pass detection tasks. But, they frequently struggle to generalize effectively with the extensive, imbalanced, and dynamic data found in IIoT networks [11, 16].

The weakness of bagging methods can be overcome by modern ensemble techniques such as Random Forest, XGBoost, or LightGBM (or even more advanced ones) which combine the outputs of several weak learners to generate robust accurate predictions. Ensemble Models work wonders on complex datasets with better accuracy, precision, and grant a lower false positive rate. Many methods based on these models surpassed classical ML models in detection of various types of attacks in IIoT environment [17–19].

2.3 Gaps in Current Approaches Addressed by this Work

Despite the rapid progress from IDS research, some shortcomings still exist in existing methods, preventing their deployment in IIoT systems:

- Scalability: The increasing volume of network traffic in IIoT leads many ML-based IDS to fall short of scalability.
- Feature Engineering: Manual feature selection methods are generally ineffective and often fail to capture salient features.
- Computational Efficiency: IIoT devices are usually resource-constrained and pose challenges to deploy models with high computational overheads.
- Adaptive Learning: Very few of the present-day systems have the ability to learn adaptively to its dynamically changing attack patterns.

To tackle these challenges, this work proposes a novel hybrid IDS framework that employs ensemble learning techniques and sophisticated feature selection techniques. The proposed system integrates scalability, efficiency, and adaptability into a system that can meet the rigorous security demands of IIoT applications.

Despite notable advances, current IDS solutions face key limitations that hinder their practical deployment in IIoT environments. These include poor scalability in high-traffic networks, suboptimal feature engineering methods, high computational overheads unsuitable for resource-constrained devices, and lack of adaptability to new, evolving attack vectors. Our proposed C-IDS addresses these issues by employing lightweight yet powerful ensemble learning algorithms that scale efficiently, incorporating automated and hybrid feature selection to extract meaningful patterns, and embracing cognitive principles to adapt to new threats over time. This unified approach ensures both high detection performance and deployment feasibility in IIoT infrastructures.

3 Methodology

This section describes the methodology of the proposed Intrusion Detection System (IDS) for IIoT during its development and evaluation. This includes dataset preparation, feature engineering, selected models, and evaluation metrics used to assess their performance.

3.1 Dataset Description

UNSW-NB15-v2: This dataset has been used as a benchmark dataset for network intrusion detection research, and it contains data set that can be especially used in IIoT environments. Its generation was done in UNSW Canberra Cyber Range Lab using the IXIA PerfectStorm tool. A combination of real network traffic and simulated modern attack behaviors is represented in this dataset, covering the major IIoTs security threats [13, 20, 21].

3.2 Features and Class Distributions

The dataset includes 45 features grouped into four categories: basic features, content-based features, time-based features, and connection-based features. These features describe different network properties essential to distinguish normal and malicious behaviors. In total, there are 2,390,275 records, with a roughly equal distribution of attacks and normal classes.

Owing to the inclusion of current attack vectors of DoS, backdoor, and infiltration attacks, the dataset is highly relevant to IIoT security in simulating IIoT system threats [22, 23].

3.3 Preprocessing Steps

- Dealing with Missing Values: Rows with missing or incomplete information were dropped.
- Encoding Categorical Data: Some features like protocol type and service were one-hot encoded to get them into numerical representations [24].
- Normalization: Each feature was normalized using Min-Max normalization to maintain consistency and enhance the performance of models [25].

3.4 Feature Engineering

Because every dataset and problem is different, only a combination of feature engineering tailored to a specific problem can achieve the desired level of accuracy while minimizing computation.

Techniques Used

- Recursive Feature Elimination (RFE): Employs model performance to systematically remove the least important ones [26].
- Mutual Information (MI): Defines the dependency of a feature(s) with respect to the target variable, such as the relevance of a feature that has high information gain [27].
- Principal Component Analysis (PCA): A dimensionality reduction technique that transforms features into a set of uncorrelated principal components while maintaining maximum variance [28].

These approaches greatly improve model efficiency by eliminating duplicative information and reducing noise. In addition to this, they also minimize computation complexity making them capable for resource-limited IIoT environments [29, 30].

3.5 Selected Models

Thus, a series of ensemble and classical machine learning models were selected in this study to assure strong performance.

Ensemble Models

- XGboost: Highly optimized gradient boosting algorithm with linear processing and parallel computation, widely used for big data [31].
- LightGBM: It uses a histogram based learning which results in faster training and lower memory usage [18].
- Random Forest: A combination of multiple decision trees that increases accuracy and reduces overfitting [32].
- ExtraTrees: It reduces variance, while improving efficiency using a randomized tree-based ensemble model [33].

Classical Models

- MLP (Multi-Layer Perceptron): A neural network model that can learn complex patterns in data [34].
- Decision Tree: Although being a simple non-parametric model and easily interpretable one, the decision tree splits the data set according to feature thresholds [35].
- KNN (K-Nearest Neighbors): A non-parametric algorithm that labels instances based on their proximity to neighbors [36].
- Logistic Regression: A baseline model that predicts probabilities under linear decision boundaries [37].

3.6 Evaluation Metrics

In order to evaluate the models exhaustively, performance and computational metrics were both used [38].

Performance Metrics

- Accuracy: Proportion of correctly classified instances.
- Precision: Ratio of true positive predictions to all positive predictions.
- Recall (Sensitivity): Ability to correctly identify positive instances.
- F1-Score: Harmonic mean of precision and recall.
- ROC-AUC: Measures the model's ability to distinguish between classes across all thresholds.

Computational Metrics

- Training Time: Time required to train the model.
- Prediction Time: Time required for the model to predict new instances.

After the extensive evaluation, we ensure that the proposed IDS achieves high detection accuracy, while also satisfies the real-time processing demand in IIoT systems.

4 Experiments and Results

4.1 Experimental Setup

Hardware and Software

The experiments were performed in Kaggle's virtual machine which is awesome for data analysis and model training. It includes:

Next, we implemented the experiments based on Kaggle's virtual machine which provides power resources and is most suitable for data analysis and model tests. It consists of a high-performance multi-core CPU (e.g., 16 or more cores from a manufacturer like Intel Xeon, operating at 2.3–3.0 Ghz), facilitating multithreading parallelization. 16 to 32 GB of RAM (virutal machine), needed to load and process big sets of IIOT data. To create, train and assess the models effectively in this computationally intensive environment, popular machine learning libraries and frameworks, including TensorFlow, scikit-learn, and PyTorch, were utilized.

4.2 Dataset Split

The dataset was divided as follows:

- 75/25% Train-Test Split: 75% of the data was used to train the models and remaining 25% was reserved to check the performance of the models on the unseen data.
- Ten-Fold Cross-Validation: In order to confirm reliable results, the dataset was split into 10 parts. The other set was used for training, while the part used for testing. This technique limits the chances of overfitting and confirms that the models execute in similar data.

4.3 Evaluation Metrics

The following metrics were used to evaluate the models:

- Accuracy: Returns the ratio of correct predictions to the total predictions. It provides a general sense of how well the model does.
- Recall: Indicates how many of the real attacks were identified correctly. This is critical, as the cost of failing to spot an attack could be devastating.
- Precision: Determines how many of the attacks predicted were truly correct. This helps prevent false alarms.
- F1-Score: puts Recall and Precision in one score. It helps when the dataset is unbalanced, as in cases when there are less attack cases.
- ROC-AUC (Receiver Operating Characteristic—Area Under the Curve): Shows how the model is able to discriminate between attacks and normal activities across all thresholds." —a higher score indicates better detection by the model.

4.4 Why these Metrics Were Chosen

- Imbalanced data: In IIoT datasets, attacks occur much less frequently than normal behavior. Metrics such as Recall and Precision deal with this imbalance much better than Accuracy.
- True Use: High Recall means the system will catch most attacks, while high Precision means fewer false positives. Finding that balance is crucial for real-world application.
- Thorough: ROC-AUC examination lets properly compare different models, considering their performance across all possible thresholds rather than a single one.

This configuration and the selected metrics allow to make sure that models are thoroughly tested and can yield reliable results used for securing IIoT systems.

4.5 Results Analysis

- **Performance Table**

Table 1 compares results for various ensemble learning models in comparison to classical machine learning models based on relevant evaluation metrics to measure the performance of IDS. These metrics include:

Accuracy, Recall, Precision, F1-Score, and ROC-AUC: These metrics assess the detection quality of the models.

The table of Ensemble learning models demonstrate exceptional performance in the evaluation. XGBoost stands out as the top performer, achieving the highest

Table 1 Model performance comparison: ensemble learning and classical machine learning models

Evaluations metrics	Accuracy	Recall	Precision	F1-score	ROC-AUC
Models	Ensemble learning models				
XGBoost	0.980349	0.978849	0.97747	0.977363	0.997826
Random forest	0.979924	0.978424	0.977	0.976935	0.997356
GBM	0.977494	0.975994	0.974564	0.974506	0.997231
AdaBoost	0.930309	0.928809	0.927292	0.927299	0.979373
ExtraTrees	0.978527	0.977027	0.975714	0.975546	0.99712
	Classical machine learning models				
Logistic regression	0.915977	0.927977	0.913783	0.913102	0.973925
MLP	0.959276	0.971276	0.956369	0.956229	0.992458
KNN	0.965774	0.977774	0.962947	0.962801	0.989099
DecisionTree	0.969175	0.981175	0.966173	0.966174	0.965809

ROC-AUC score (0.997826), along with impressive accuracy (0.980349), recall (0.978849), and precision (0.97747). Random Forest follows closely, with an ROC-AUC of 0.997356 and an accuracy of 0.979924, proving nearly as effective while being computationally efficient. GBM also delivers strong results, with an ROC-AUC of 0.997231 and accuracy of 0.977494, though it lags slightly behind the top two models. ExtraTrees performs admirably, achieving high accuracy (0.978527) and ROC-AUC (0.99712), making it a robust and efficient choice. On the other hand, AdaBoost shows moderate results with an accuracy of 0.930309 and an ROC-AUC of 0.979373, positioning it as less competitive among ensemble methods.

Classical machine learning models show mixed outcomes. Logistic Regression yields the lowest performance with an accuracy of 0.915977 and an ROC-AUC of 0.973925, suggesting limited suitability for complex Intrusion Detection System (IDS) tasks. In contrast, MLP achieves solid results with an accuracy of 0.959276, a recall of 0.971276, and an ROC-AUC of 0.992458, showcasing its potential as a neural network-based option. KNN excels in recall (0.977774) and accuracy (0.965774) but suffers from high prediction time, limiting its real-time applicability. Meanwhile, Decision Tree balances its metrics effectively with an accuracy of 0.969175 and an ROC-AUC of 0.965809, making it suitable for simpler scenarios. These findings highlight the dominance of ensemble models, especially XGBoost and Random Forest, in delivering superior detection performance in IIoT environments, while classical models like KNN and Decision Tree excel in specific, computationally constrained scenarios.

- **Comparison of Ensemble Learning Models Based on Evaluation Metrics**

The histogram graph illustrates the performance of various ensemble learning models based on key evaluation metrics: Accuracy, Recall, Precision, F1-Score, and ROC-AUC. XGBoost and Random Forest show the highest scores across all metrics, with XGBoost leading in Accuracy and ROC-AUC. AdaBoost, on the other hand, exhibits

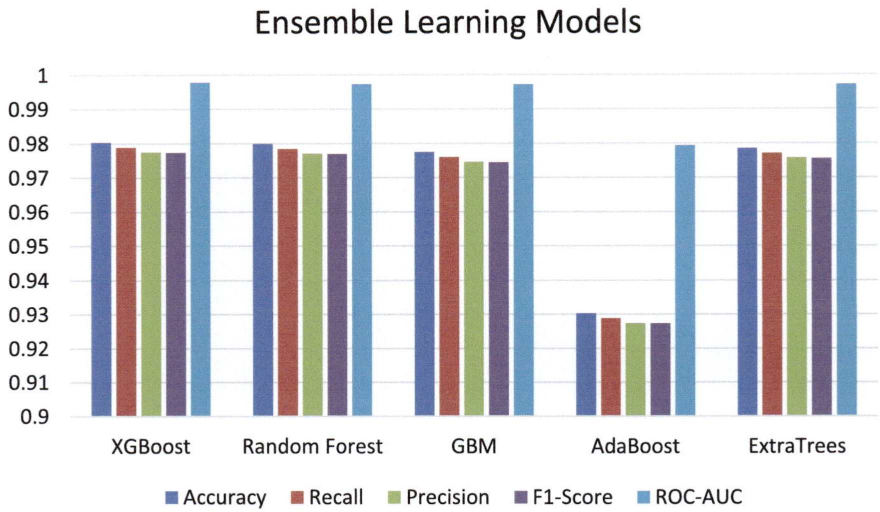

Fig. 1 Comparison of ensemble learning models based on evaluation metrics

the lowest performance, particularly in Accuracy, Recall, and Precision. Overall, the models demonstrate strong results, with XGBoost, Random Forest, and ExtraTrees showing superior performance, particularly in metrics like ROC-AUC, highlighting their effectiveness for classification tasks.

Figure 1 illustrates the accuracy and ROC-AUC superiority of XGBoost compared to other ensemble methods.

- **Comparison of Classical Machine Learning Models Based on Evaluation Metrics**

This histogram graph visually compares the performance of classical machine learning models (Logistic Regression, MLP, KNN, and Decision Tree) across key evaluation metrics: Accuracy, Recall, Precision, F1-Score, and ROC-AUC. MLP stands out as the top performer, especially in Recall and ROC-AUC, followed by KNN and Decision Tree, which demonstrate strong overall performance. Logistic Regression shows lower scores in comparison, particularly in F1-Score and ROC-AUC. Performance comparisons of classical models are visualized Fig. 2, where MLP demonstrates robust recall and ROC-AUC.

- **Training Time and Prediction Time**

The analysis of training and prediction times highlights significant differences between ensemble learning models and classical machine learning models. Among the ensemble methods, ExtraTrees has the shortest training time (0.060 s), showcasing its computational efficiency, while GBM requires the longest time (68.775 s), reflecting its complexity. Similarly, classical models like KNN demonstrate an extremely low training time (0.008 s) but suffer from a substantially higher prediction

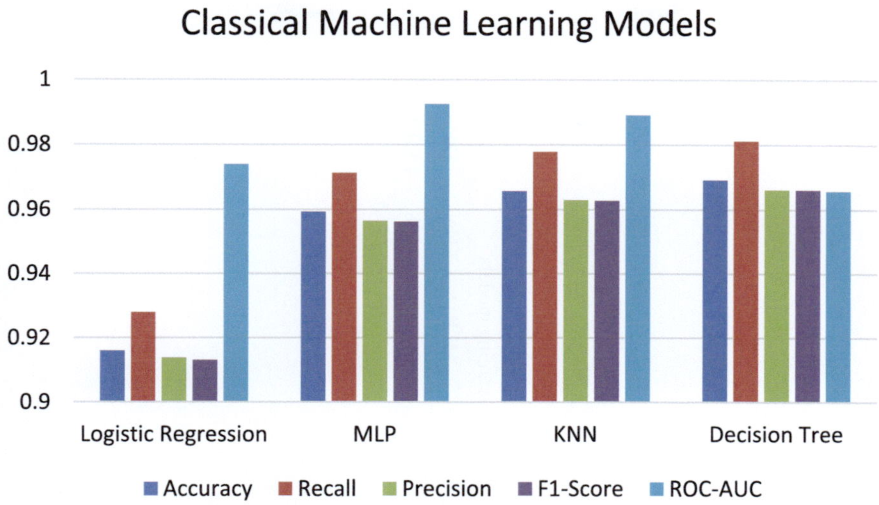

Fig. 2 Performance comparison of classical machine learning models based on evaluation metrics

time (24.837 s), making it less suitable for real-time applications. In contrast, Logistic Regression and Decision Tree exhibit balanced training and prediction times, making them efficient choices for simpler implementations. This comparison underscores the trade-offs between model complexity, training duration, and prediction efficiency.

The training and prediction time differences are summarized Table 2, highlighting ExtraTrees computational efficiency.

This bar chart illustrates the training times (in seconds) for various machine learning models categorized into ensemble and classical types. The ensemble models (in blue) typically have the spending higher training times as they are more complex and require a lot of computation. In contrast, classical models shown in orange have much lower training time on much less resources, making them more fit for environments with resource constraints. In addition to that, this visualization allows us to see how much a model is complex at a certain point in time, which is valuable for applying proper models into appropriate targets (which are IIoT and intrusion detection systems, in this case) detailed in the paper.

Figure 3 shows the training time analysis across models, emphasizing the trade-off between accuracy and computational cost.

Table 2 Training and prediction time analysis of machine learning models

Model	Ensemble learning models					Classical machine learning models			
	XGBoost	Random forest	GBM	AdaBoost	ExtraTrees	Logistic regression	MLP	KNN	DecisionTree
Training time (s)	12.017	10.900	68.775	17.406	0.060	12.272	32.362	0.008	1.070
Prediction time (s)	0.028	0.243	0.057	0.336	0.332	0.002	0.027	24.837	0.004

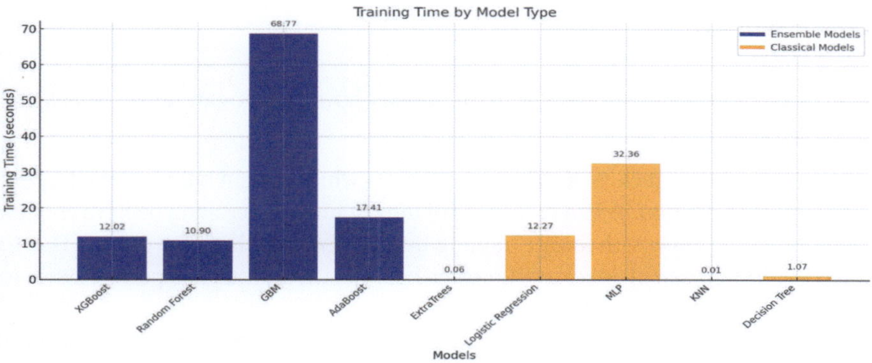

Fig. 3 Training times

5 Discussion

5.1 Key Findings

We can notice some key takeaways from the experiments:

Models like XGBoost and Random Forest showed the best performance as ensemble models. In terms of complexity in IIoT environments, XGBoost delivered the best accuracy (0.980349) and ROC-AUC score (0.997826). Because they merge predictions from many algorithms, these models perform well, providing more accurate results and catching more details.

- The classical models, like Decision Tree and KNN, gave good results but these did not match the performance of the ensemble models, in recall and accuracy at least. The main advantage of these models is their ability to train relatively fast, which may be valuable in environments with scarce resources. For example, KNN showed a strong recall of 0.977774, but its prediction time was a little too slow for real-time applications.
- Models like AdaBoost and Logistic Regression performed worse, especially in accuracy and ROC-AUC. AdaBoost (accuracy: 0.930309) didn't perform as well as the other ensemble models, and Logistic Regression had the lowest accuracy (0.915977), suggesting it might not be the best fit for complex intrusion detection tasks.

5.2 Trade-offs

How much time a model takes to train, versus how well it can detect attacks, is dependent on:

- Intensive models such as XGBoost will take a while to train but yield the highest quality for detection. When security is of utmost importance and you have enough computing to cope with longer training times these models are awesome. Due to their ability to detect attacks with fewer false positives, it makes perfect sense to use them in intricate environments.
- In contrast, more complex models such as Decision Tree and KNN take less time to train. They may not be as accurate as the ensemble models, but they still can perform well in scenarios in which speed is more important than top-most accuracy. For instance, in scenarios where responding to threats quickly matters more than convergence, KNN can be beneficial even if it's less precise.
- Both AdaBoost and Logistic Regression give you an impressive level of CPU efficiency but don't give you the same level of detection power as some of the other models. These models may come in handy when simplicity and efficient use of resources is desired; however, they should be approached with caution in any high-risk setting.

5.3 Comparison with State-of-the-Art

In comparison with current researches on IDS, these results are proof that ensemble models in general and XGboost in particular remain some of the leading performers in the past few years.

- We trained on data until the metric ROC-AUC is taken into account as one of the most important used for IDS models as it describes the model performance given normal/attack traffic. To sum up, both XGBoost and Random Forest achieved very good scores, reflecting the power that can be achieved by using these models on real world, rich and complex problems to be solved in Industrial Internet of Things (IIoT) systems.
- The study also highlights the importance of ensemble learning, combining multiple models to achieve better performance. This is particularly powerful in settings where new varieties of attacks may be introduced, and where the models must adaptively respond to the changes.
- This work also highlights the value of computational efficiency. Although the more complex models, e.g. XGBoost perform very well, they may take longer to train, but they are accurate. Simpler models like Decision Tree and KNN, are probably better suited, where speed and lower resource usage should trump over the absolute best performance.

5.4 Practical Implications

In this regard, the findings of this research can be further implemented in the real-life IIoT systems. This is a great benefit of ensemble models associated to Random forest, that can handle large volumes of data quite efficiently, thus helping secure larger IIOT networks as those scale. Furthermore, high-performing models such as XGboost and Random Forest performed remarkably well with a better ROC-AUC score than Logistic Regression, as a great advantage of these models is their adaptability to new environments, enabling them to detect known as well as new threats in the environment with high reliability. Also, models like KNN are relatively simple and perform well in resource constraints as training time is very low compared to complex models, making it easy to deploy the model in production when faced with limitations in computational power.

5.5 Limitations and Future Work

While the proposed C-IDS demonstrates strong performance and adaptability, certain limitations persist. First, although ensemble models achieve high accuracy, their training time may still be restrictive for some real-time IIoT deployments. Second, the study is based on a single dataset (UNSW-NB15-v2), which may not fully capture all IIoT threat scenarios. Additionally, the current framework lacks real-time streaming capability and integration with hardware sensors in actual industrial setups.

Future work will focus on extending the IDS to support online learning for real-time detection, validating performance across multiple IIoT-specific datasets, and deploying the solution in a simulated IIoT testbed or edge computing environment. Incorporating deep learning and federated learning techniques to further enhance adaptability and security is also planned.

6 Conclusion

In response to the growing cybersecurity challenges in the IIoT, this study proposed a machine learning-based Intrusion Detection System (IDS). The proposed IDS successfully strikes a balance between high detection accuracy and computational efficiency, being particularly applicable in dynamic and resource-limited IIoT scenarios, by orchestrating ensemble learning techniques with cognitive computing principles.

The experimental evaluation was carried out using the UNSW-NB15-v2 dataset and showed that ensemble learning models outperformed classical models in terms of both known and unknown attack detection, with XGBoost and Random Forest consistently ranking as the best performing models. The models outperformed on accuracy, recall, and ROC-AUC metrics, reflecting their robustness in pinpointing various attack patterns while limiting false alarms. Machine learning: With machine learning incorporated into its architecture, the system could learn from previous incidents, improve line vulnerability updates, and better resist potential attacks.

This hones the performance of the system leveraging advanced feature selection processes such as Recursive Feature Elimination (RFE) and Mutual Information (MI). Our research emphasizes the importance of machine learning-based IDS as flexible, adjustable, and strong method of safeguarding IIoT settings at scale.

The future may bring real-time implementation of the results of this better detection system along with its integration in security tools, appear to be in line with current ideas and further improvement could come with deep learning integration. By contributing to the development of intelligent and secure industrial infrastructures, this study will help ensure the ongoing resilience of IIoT systems against an ever-changing cyber threat landscape.

References

1. Umer MA, Junejo KN, Jilani MT, Mathur AP (2022) Machine learning for intrusion detection in industrial control systems: applications, challenges, and recommendations. Int J CritAl Infrastruct Prot 38:100516. https://doi.org/10.1016/j.ijcip.2022.100516
2. M Al-Sharif, A Bushnag (2024) Enhancing cloud security: a study on ensemble learning-based intrusion detection systems. IET Commun cmu2.12801. https://doi.org/10.1049/cmu2.12801
3. Vaiyapuri T, Sbai Z, Alaskar H, Ali N (2021) Deep learning approaches for intrusion detection in IIOT networks—opportunities and future directions. IJACSA 12(4). https://doi.org/10.14569/IJACSA.2021.0120411
4. Brownlee J (2019) How to choose a feature selection method for machine learning. Mach Learn Mastery 10:1–7. https://elearn.daffodilvarsity.edu.bd/pluginfile.php/1225702/mod_label/intro/Feature%20Selection%20with%20numerical%20and%20categorical%20data.pdf
5. Alohali MA, Al-Wesabi FN, Hilal AM, Goel S, Gupta D, Khanna A (2022) Artificial intelligence enabled intrusion detection systems for cognitive cyber-physical systems in industry 4.0 environment. Cogn Neurodyn 16(5):1045–1057. https://doi.org/10.1007/s11571-022-09780-8
6. Laiq F, Al-Obeidat F, Amin A, Moreira F (2023) DDoS attack detection in edge-IIoT using ensemble learning. In: 2023 7th cyber security in networking conference (CSNet). IEEE, Montreal, QC, Canada, pp 204–207. https://doi.org/10.1109/csnet59123.2023.10339784
7. Susanto H, Azim MQ, Fang-Yie L, Susanto AKS, Setiana D, Ibrahim F, Subramanian U (2023) Artificial intelligence, blockchain framework, cyberthreat defenses of resilient digital ecosystems. In: Handbook of research on artificial intelligence and knowledge management in Asia's digital economy. IGI Global, pp 36–63
8. Eid AM, Nassif AB, Soudan B, Injadat MN (2023) IIoT network intrusion detection using machine learning. In: 2023 6th international conference on intelligent robotics and control engineering (IRCE). IEEE, Jilin, China, pp 196–201. https://doi.org/10.1109/IRCE59430.2023.10255088
9. SK Poorazad, C Benzaıd, T Taleb (2023) Blockchain and deep learning-based ids for securing SDN-enabled industrial IOT environments. arXiv: arXiv:2401.00468. Accessed: 22 Jan. 2024. [Online]. Available: http://arxiv.org/abs/2401.00468

10. Arun Kumar A, Krishna Karne R (2022) IIoT-IDS network using inception CNN model. JTCSST 4(3):126–138. https://doi.org/10.36548/jtcsst.2022.3.002
11. L. Idouglid, S. Tkatek, K. Elfayq, A. Guezzaz (2024) A novel anomaly detection model for the industrial internet of things using machine learning techniques. no (1). https://doi.org/10.32620/reks.2024.1.12
12. Yang Z et al (2022) A systematic literature review of methods and datasets for anomaly-based network intrusion detection. Comput Secur 116:102675. https://doi.org/10.1016/j.cose.2022.102675
13. Houichi M, Jaidi F, Bouhoula A (2024) Cyber security within smart cities: a comprehensive study and a novel intrusion detection-based approach. CMC 81(1):393–441. https://doi.org/10.32604/cmc.2024.054007
14. AlHaddad U, Basuhail A, Khemakhem M, Eassa FE, Jambi K (2023) Ensemble model based on hybrid deep learning for intrusion detection in smart grid networks. Sensors 23(17):7464. https://doi.org/10.3390/s23177464
15. Guezzaz A, Azrour M, Benkirane S, Mohy-Eddine M, Attou H, Douiba M (2022) A lightweight hybrid intrusion detection framework using machine learning for edge-based IIoT security. IAJIT 19(5). https://doi.org/10.34028/iajit/19/5/14
16. M Al-Ambusaidi, Z Yinjun, Y Muhammad, A Yahya (2023) ML-IDS: an efficient ML-enabled intrusion detection system for securing IoT networks and applications. Soft Comput. https://doi.org/10.1007/s00500-023-09452-7
17. Al Amien J, Ab Ghani H, Md Saleh NI, Ismanto E, Gunawan R (2023) Intrusion detection system for imbalance ratio class using weighted XGBoost classifier. TELKOMNIKA 21(5):1102. https://doi.org/10.12928/telkomnika.v21i5.24735
18. G Ke et al. (2017) Light GBM: a highly efficient gradient boosting decision tree
19. R Alshamy, M Ghurab, S Othman, F Alshami (2021) Intrusion detection model for imbalanced dataset using SMOTE and random forest algorithm. In Advances in Cyber Security, N Abdullah, S Manickam, M Anbar, Eds., In: Communications in computer and information science vol 1487, Singapore: Springer Singapore, pp 361–378. https://doi.org/10.1007/978-981-16-8059-5_22
20. Thanh HN, Lang TV (2020) Evaluating effectiveness of ensemble classifiers when detecting fuzzers attacks on the unsw-NB15 dataset. JCC 36(2):173–185. https://doi.org/10.15625/1813-9663/36/2/14786
21. Amien JA, YozeRizki, Nasution MAR (2022) Implementasi adasyn untuk imbalance data pada dataset UNSW-NB15 adasyn implementation for data imbalance on UNSW-NB15 dataset. CoSciTech 3(3):242–248. https://doi.org/10.37859/coscitech.v3i3.4339
22. Li S, Wang J, Wang Y, Zhou G, Zhao Y (2023) EIFDAA: evaluation of an IDS with function-discarding adversarial attacks in the IIoT. Heliyon 9(2):e13520. https://doi.org/10.1016/j.heliyon.2023.e13520
23. Ebady Manaa M, Hussain SM, Alasadi SA, Al-Khamees HAA (2024) DDoS attacks detection based on machine learning algorithms in IoT environments. Intel Artif 27(74):152–165. https://doi.org/10.4114/intartif.vol27iss74pp152-165
24. Tien C-W, Huang T-Y, Chen P-C, Wang J-H (2021) Using autoencoders for anomaly detection and transfer learning in IoT. Computers 10(7):88. https://doi.org/10.3390/computers10070088
25. Patro SGK, Sahu KK (2015) Normalization: a Preprocessing stage. Int Adv Res J Sci Eng Technol:20–22. https://doi.org/10.17148/IARJSET.2015.2305
26. J. Brownlee (2019) How to choose a feature selection method for machine learning
27. Battiti R (1994) Using mutual information for selecting features in supervised neural net learning. IEEE Trans Neural Netw 5(4):537–550. https://doi.org/10.1109/72.298224
28. Salo F, Nassif AB, Essex A (2019) Dimensionality reduction with IG-PCA and ensemble classifier for network intrusion detection. Comput Netw 148:164–175. https://doi.org/10.1016/j.comnet.2018.11.010
29. M Bakro (2024) et al. Building a cloud-IDS by hybrid bio-inspired feature selection algorithms along with random forest model. IEEE Access vol 12 pp. 8846–8874. https://doi.org/10.1109/ACCESS.2024.3353055

30. H Liu, H Motoda, RL Cao, D Taniar (2007) Computational methods of feature selection
31. Chen T, Guestrin C (2016) XGBoost: a scalable tree boosting system. In: Proceedings of the 22nd ACM SIGKDD international conference on knowledge discovery and data mining. ACM, San Francisco California USA, pp 785–794. https://doi.org/10.1145/2939672.2939785
32. Cutler A, Cutler DR, Stevens JR (2012) Random forests. In: Zhang C, Ma Y (eds) Ensemble machine learning. Springer New York, New York, NY, pp 157–175. https://doi.org/10.1007/978-1-4419-9326-7_5
33. Geurts P, Ernst D, Wehenkel L (2006) Extremely randomized trees. Mach Learn 63(1):3–42. https://doi.org/10.1007/s10994-006-6226-1
34. Kim KG (2016) Book Review: Deep Learning. Healthc Inform Res 22(4):351. https://doi.org/10.4258/hir.2016.22.4.351
35. Quinlan JR (1986) Induction of decision trees. Mach Learn 1(1):81–106. https://doi.org/10.1007/BF00116251
36. Cover T, Hart P (1967) Nearest neighbor pattern classification. IEEE Trans Inform Theory 13(1):21–27. https://doi.org/10.1109/TIT.1967.1053964
37. Cox DR (1958) The regression analysis of binary sequences. J R Stat Soc Ser B Stat Methodol 20(2):215–232. https://doi.org/10.1111/j.2517-6161.1958.tb00292.x
38. Vujovic ŽĐ (2021) Classification model evaluation metrics. IJACSA 12(6). https://doi.org/10.14569/IJACSA.2021.0120670

Towards Reliable and Secure IoMT: A Deep Learning Perspective on Cyber-Physical Threats

Hafida Assmi, Said Jabbour, and Azidine Guezzaz

Abstract Intrusion detection within Internet of Medical Things (IoMT) environments is complicated by the diversity of communication protocols and the continuous emergence of sophisticated security threats. This research presents a hybrid deep learning framework that harnesses the complementary capabilities of Convolutional Neural Networks (CNN) and Long Short-Term Memory (LSTM) networks to enhance intrusion detection in healthcare IoMT environments. Our CNN-LSTM architecture employs CNN for extracting spatial features and LSTM for modeling temporal dependencies, specifically designed for IoMT data. Evaluated using the CICIoMT2024 dataset, covering Bluetooth, WiFi, and MQTT protocols with 18 attack types grouped into five classes, the model achieved an accuracy of 86.24% in multi-class classification, along with strong precision (0.865), recall (0.863), and F1-score (0.863) metrics.

Keywords Healthcare security · IoMT · Intrusion detection · Deep learning · CNN · LSTM

1 Introduction

The internet of things (IoT) has profoundly impacted multiple sectors by facilitating real-time data collection and communication among smart devices. In the healthcare sector, this advancement has given rise to the Internet of Medical Things (IoMT), which incorporates IoT technologies into medical devices, patient monitoring systems, and clinical applications [1]. This convergence enhances the delivery

H. Assmi · S. Jabbour
CRIL—CNRS, Artois University, Lens, France
e-mail: jabbour@cril.fr

H. Assmi · A. Guezzaz (✉)
SISAR Team, LaRTID Laboratory, Technology Higher School Essaouira, Cadi Ayyad University, Marrakech, Morocco
e-mail: a.guzzaz@gmail.com

G. H. L. et al. (eds.), *Reliability in Cyber-Physical Systems: The Human Factor Perspective*, Springer Series in Reliability Engineering,
https://doi.org/10.1007/978-3-032-09917-4_13

of personalized and efficient medical services, supports remote patient monitoring, and contributes to the development of intelligent health systems, ultimately leading to improved patient outcomes [2].

The sensitivity of healthcare data makes intrusion detection essential in IoMT systems. Compromises in medical data can result in serious outcomes, including privacy violations and critical threats to patient safety. With the growing adoption of IoMT technologies, the associated security risks also expand. As a result, protecting the integrity and security of healthcare systems now hings more than ever on the implementing dependable and proactive intrusion detection mechanisms. To improve intrusion detection capabilities, a hybrid deep learning model integrating CNN and LSTM is proposed. We use the newly released CICIoMT2024 dataset which features the three primary IoMT communication protocols (WiFi, Bluetooth, and MQTT), we conducted comprehensive assessments of 18 attacks organized into five primary classes: DdoS, DoS, Reconnaissance, MQTT, and Spoofing. Three primary contributions were identified:

- Firstly, a hybrid CNN-LSTM deep learning architecture was created to accurately extract spatial and temporel features characteristics from network traffic, boosting intruson detection performance.
- Secondly, advanced preprocessing techniques such as SMOTE oversampling and random undersampling, are employed to correct class imbalance and improve the model's overall generalization performancs.
- Thirdly, the model is developed and tested on an extensive, balanced multi-class dataset demontrating strong capability in accurately classifying multiple types of network attacks with excellent precision and recall.

This document is organized as follows: Sect. 2 offers a comprehensive review of the related studies on IDS and deep learning. Sect. 3 outlines the architecture of the model we proposed. Sect. 4 discusses the experimental results obtained from the model. Finally, this research ends with a summary and suggestions for future work.

2 Related Works

This section presents the fundamental concepts and recent developement in intrusion detection systems (IDS) specifically focused on network security.

In 2025, Berguigua et al. [3] developed HIDS-IoMT, a CNN-LSTM-based hybrid IDS implemented on a Raspberry Pi within a fog computing framework to enhance real-time security in IoMT networks. Evaluated on the IoTID20 and Edge-IIoTset datasets, which feature diverse DoS attack scenarios, the model delivered outstanding performance, achieving 99.92% accuracy along with near-perfect precision, recall, and F1-scores, outperforming contemporary IDS solutions. Assmi et al. [4] Presented a ML-based intrusion detection system (IDS) that employs an ensemble learning strategy, combining KNN, SVM, and RF through Voting Classifier. Evaluated on the NSL-KDD and Edge-IIoT datasets, the model demonstated impressive performance with accuracy ranging from 72 to 99%, precision between 87 and 99%, and recall and F1-score varying from 72 and 99% effectively identifying both normal and malcious activities. Additionally akar et al. [5] introduced a new method for detecting a wide range of intrusion attacks on inernet of Medical Things (IoMT) devices. Building on an improved version of the LSTM deep learning algorithm. They conducted evaluation and comparison of their algorithm utilizing the CICIoMT2024 dataset, which includes a wide range of devices and corresponding attack types. The findings indicate that the novel approach achieved an impressive 98% accuracy across 19 classes, underscoring its strong potential and effectiveness within IoMT environments. Furthermore, Shaikh et al. [6] introduced the Memory Feedback Transformer (MF-Transformer), which embeds Memory Feedback LSTM (MF-LSTM) into every layer of the Transformer to effectively learn and propagate spatial and temporal relationships. The model initially captures spatial feature correlations within individual time steps and then introduces temporal context through MF-LSTM feedback, allowing it to monitor long-term patterns and enhance the detection of both short-lived and persistent anomalies. The MF-Transformer was evaluated on the WUSTL-EHMS-2020, ECU-IoHT, and X-IIoTID datasets, consistently delivering accuracy exceeding 99% in both signature and anomaly detection.

In 2024, Amaouche et al. [7] addressed the security of AVENT networks against threats like DoS attacks by introducing an IDS buit on ensemble learning approach with AdaBoost serving as the primary algorithm for boosting detection performance. Evaluated on the NSL-KDD, UNSW-NB15, and TON-IOT datasets, the model demonstated exceptional performance, surpassing exisitng IDS solutions by achieving almost perfect scores in accuracy, precision, recall, and f1-score. Hazman et al. [8] carried out a study emphasizing the vital role of IDS in smart cities. Their study highlights how crucial these systems are for anomaly detection, data breach prevention, and maintaining the security and dependability of urban digital networks. It also underscores the imperative to consistently advance and implement advanced IDS solutions that utilize cutting-edge ML and DL methods to improve cybersecuriy and ensure the privacy of citizens within smart city environments. Concomitantly, et D et al. [9] designed a hybrid CNN-LSTM approach to monitor and detect any tampering or attacks on data transmitted from wearable IoMT devices to the cloud. The CNN module extracts key features from the data, and the LSTM part analyzes sequential patterns to identif the potential manipulaions. The design features two convolutional layers, with each layer succeeded by a MaxPooling layer. When tested on a large dataset, the model achieved 97.06% accuracy, confirming its proficiency in securing medical data integrity during transfer. Bella et al. [8] developed an IDS-based on Deep Neural Decision Forests called DNDF-IDS, applying four different feature selection strategies to optimize its performance, including SelectKBest, Principal Component Analysis (PCA), Random Forest Feature Importance (RFFI), and LASSO Regression (LR). Evaluations on the NSL-KDD, CICIDS2017, and UNSW-NB15, datasets showed the model attained accuracy rates between 94.09% and 98.84%. Alzubi et al. [11] developed a hybrid model leveraging CNN and LSTM architectures designed to accurately identify intrusions and secure healthcare information. The model specializes in detecting multiple attacks with high accuracy and efficiency during data transmission at the network edge.

In 2023, Faruqui et al. [9] Presented SafetyMed, an advanced IDS leveraging CNN and LSTM for detecting intrusions in IoMT devices across sequential and image data types. The system delivers 97.63% accuracy and strong precision, recall, and F1-scores, successfully balancing detection effectiveness and false positive rates. SafetyMed constitutes a substantial advancement in securing critical sectors, including healthcare, from IoMT-based cyberattacks. Hernandez-jaimes et al. [13] proposed a thorough taxonomy of IoMT intrusion detection schemes, comparing various detection techniques and categorizing datasets to assess their performance. Their study explores specific cybersecurity threats, defines security requirements, and explores the role of Cloud–Fog–Edge architectures. Additionally, it reviews recent AI-driven approaches and discusses the legal and ethical considerations related to IoMT security.

In 2022, Pelekoudas-Oikonomou et al. [10] provide an in-depth review of two primary categories of blockchain-based security approaches to establish a foundation for future research focused on developing reliable and efficiently address the critical security challenges faced by IoMT edge networks. Furthermore, Alamleh et al. [15] designed an MCDM framework to evaluate ML-based IDSs in Federated Learning-enabled IoMT environments, utilizing the Fuzzy Delphi Method to validate 17 evaluation criteria and employing Borda voting to determine model rankings. Analyzing a dataset of over 125,000 entries, the findings highlight CPU time as the most significant evaluation metric, with BayesNet achieving the highest ranking and SVM the lowest. This framework offers a robust, data-driven approach for systematically benchmarking IDS in IoMT settings. In addition, Nandy et al. [11] created an Empirical Intelligent Agent (EIA) employing a unique Swarm-Neural Network (Swarm-NN) approach to detect threats within an IoMT framework focused on edge computing. This methodprioritizes precise attack detection during data transmission while optimizing health data processing at the network edge. The Swarm-NN model underwent testing on the ToN-IoT dataset, wich includes telemetry, operating system, and network data. Their approach outshined traditional classification models, reaching an accuracy of 99.5%.

In 2021, Kumar et al. [17] developed a cyber-attack detection system for IoMT networks by combining ensemble learning methods with a fog-cloud-based infrastructure. The system utilizes the ToN-IoT dataset, a comprehensive and realistic dataset, and applies a layered ensemble strategy, initially using Decision Tree, Naive Bayes, and Random Forest, with XGBoost serving as the final classifier. Deployed as SaaS at the fog layer and IaaS at the cloud layer, the system delivers a 99.98% detection rate, 96.35% accuracy, and a false alarm rate reduced to 5.59%. In addition, Otoum et al. [18] developed an IDS using Federated Transfer Learning with DNNs to safeguard healthcare-connected devices by training edge models locally and aggregation into a global model without exposing sensitive data. When tested with the CICIDS2017 dataset, the system demonstrated high accuracy, improved detection performance, and quicker training, outperforming traditional centralized ML and

DL models by demonstrating enhanced generalization and incremental learning.In 2020, Pustokhina et al. [12] devloped ETS-DNN, a sophisticated deep neural network optimized using a Hybrid Modified Water Wave Optimization technique, aimed at delivering fast and accurate diagnoses in edge-based IoMT systems. Data from IoMT devices is processed on edge nodes to enable quick decision-making before being relayed to the cloud. Extensive experimental results demonstrate that ETS-DNN outperforms existing methods,emphasizing its suitability for real-time healthcare applications. In addition, Rahman et al. [13] studied the concepts and privacy safeguards associated with MEC powered by IoT, highlighting the critical role of information today. They analyzed MEC security mechanisms within IoT systems using autonomous vehicles as a practical example.

Existing intrusion detection studies IoMT networks exhibits several shortcomig: they frequently target a limited variety of attack types, do not effectively combine apatial and temporal data features, and often involve high computational complexity that limit their use on devices with constrained resources. Furthermore, optimization for handling multi-class and imbalanced datasets is seldom adequately addressed.

Our hybrid CNN-LSTM model tackles these challenges by combining apatial and temporal analysis to enhance detection capabilities. Implemeted on a lightweight Fog computing environments. The use of a balanced multi-class dataset improves generalization, and optimization reduce complexity without compromising accuracy, thus providing an efficient solution tailored to the constraints of IoMT network security.

3 Model Analysis

This section offers an in-depth exploration of the architecture of our model, along with the algorithms used in its design and development.

3.1 Designed Architecture

This section describes the design of a hybrid CNN-LSTM model for detecting intrusions in network traffic. The architecture of the model is depicted Fig. 1.

Data pre-processing: The project commenced with a thorough data preprocessing phase to guarantee the integrity and confidentiality of data. This foundational step was crucial for preparing the dataset for effective analysis and modeling.

Modeling: The aim of this phase is to adapt problem for a neural network to effectively learn and identify differents types of network trafic (or attacks) through deep learning methods.

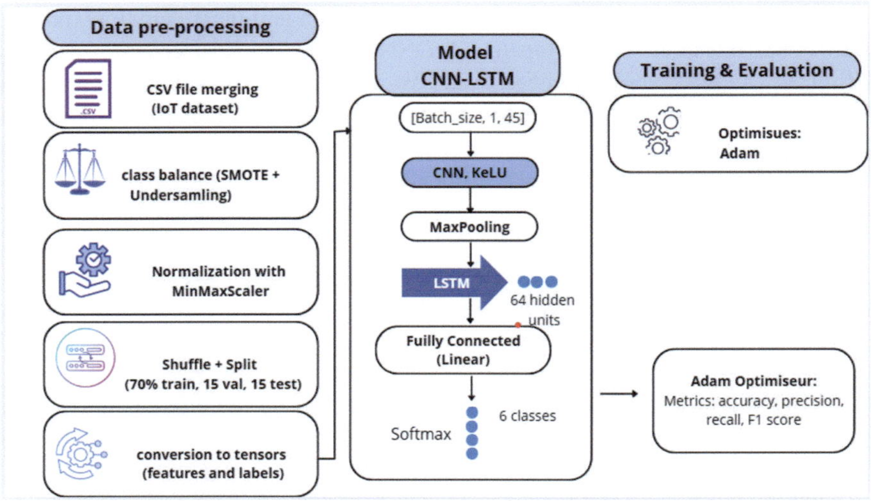

Fig. 1 Proposed architecture based on deep learning on cyber-physical threats

Training & Evaluation: During this phase, the model undergoes training to predict accurately and then evaluated on the new data to prove its effectiveness. It is essential to ensure the model's reliability, robustness, and readiness for real-world applications. Ensuring the model's reliability.

3.2 Methodology

This study describes a hybrid model combining CNN and LSTM networks to extract feature effecively, where a convolutional layer is succeeded by LSTM layer. The CNN specializes in identifying spatial patterns in the data, while LSTM captures temporal dependencies by processing sequences. Margin these two approaches, the CNN-LSTM model leveragestheir complementary strengths to improve feature representation and maintiain information over time.

The CNN architecture begins with an input layer responsible for processing the selected network features, formatted as (N,1) for 1D CNNs, where ***N*** denotes the number of features. In the convolutional layer, neurons uilize shared and biases, called kernels or filters, wich are connected to an ***n*** x ***n*** area of neurons in the previous layer. The response generated by the neurons located at position ***(j,k)*** is computed as follows [14]:

$$yj, k = \sum_{l=0}^{n-1} \sum_{m=0}^{n-1} \omega l, m^x j + l, k + m + b \tag{1}$$

Subsequently, the activation layer applies a nonlinear function, typically the Rectified Linear Unit (ReLU), to add nonlinearity to the model. The function ReLU expressed as:

$$f(x) = \max(x, 0) \tag{2}$$

Eliminates negative values in the features maps by remplace them with zero, allowing positive values to remain unchanged, thereby boosting the model's ability to capture complex features.

Next, the pooling layer subsequently reduces the spacial dimensions of the previous layer's output, usually appliying max-pooling to downsample the output by selecting the highest value within each separate, non-overlapping sub-region. Afterward, the flatten layer transforms the muti-dimensional maps into one-dimensional array, which is then used as input for the LSTM.

LSTM is a recurrent neural network tailored for processing sequential data, capable of learning long-term temporal dependencies. It comprises a memory cell along with input, forget, and output gates, that control the flow of information [15]. After receiving a flattened feature vector from the CNN, the LSTM processes them to extract meaningful temporal patterns.

$$i_t = \sigma\left(W_i.\left[h_{t-1}, X_t\right] + b_i\right) \tag{3}$$

$$f_t = \sigma\left(W_f.\left[h_{t-1}, X_t\right] + b_f\right) \tag{4}$$

$$C_t = \tanh\left(W_c.\left[h_{t-1}, X_t\right] + b_c\right) \tag{5}$$

$$O_t = \sigma\left(W_o.\left[h_{t-1}, X_t\right] + b_o\right) \tag{6}$$

ht is the final hidden state that contains the temporal features and is derived as follow:

$$ht = O_t \tanh(c_t) \tag{7}$$

The CNN-LST hybrid model optimizes key parameters such as model complexity, learning rate, and regularization to enhance both performance and generalization.

4 Model Evaluation

This section delivers an in-depth examination of the data used and summarizes the outcomes achieved by the model we developed.

Table 1 Pre-processed dataset key statistics

Charactiristics	Description
Total number of samples	Approx. 1.2 million (after balancing)
Number of columns (features)	45 numeric colums
Number of classes	6 classes (BENIGN, DdoS, DoS, MQTT, RECON, SPOOFING)
Classification type	Multi-classes
Initial distribution	Very unbalanced (ex: DdoS >4 millions, SPOOFING ~16,000)
Data format	CSV (multiple merged files)
Data separation	Train (70%), validation (15%), test (15%)
Final format	Data converted to PyTroch tensors for learning purposes

4.1 Preprocessing Phase

In the processing phase, we began by loading and combining multiple raw data files containing network features and their associated labels. To address the pronounced imbalance across different attack classes, we implemented data balancing methods:

- SMOTE oversamoling to artificially increase minority classes.
- Sub-sampling to reduce over-represented classes.

To obatin a balanced data set with a similar number of examples by category.

Next, we normalized the numerical features using a Min-Max scaler to standardize the ranges of values, making the model easier to learn. Lastly, the data were randomly shuffled and divided into training, validation, and test sets to facilitate rigorous evauation and promote strong generalization of the model. Table 1 shows the main characteristics of the dataset used.

4.2 Discussion of Result

As illustrate Fig. 2, the dataset is highly imbalanced, with DDoS (4.7 millions samples) and DoS (1.8 millions) attacks dominating the sample distribution, these two types of attacks represent more than 85% of the dataset, indicating their high occurrence in simulated networks environments. In comparison, other categories like MQTT, BENIGN (normal traffic), RECON, and especially SPOOFING with only 16,047 samples are notably underrepresented. To address this issue and enable the model to learn effectively from all classes, balancing methods like SMOTE and undersampling are required.

After applying resampling techniques such as SMOTE and undersampling, the dataset was balanced so that each class, BENIGN (normal traffic), and the five attack types: DdoS, DoS, MQTT, RECON, and SPOOFING containing exactly 200,000 samples, as demonstrated Fig. 3. This uniform class distribution prevents the model

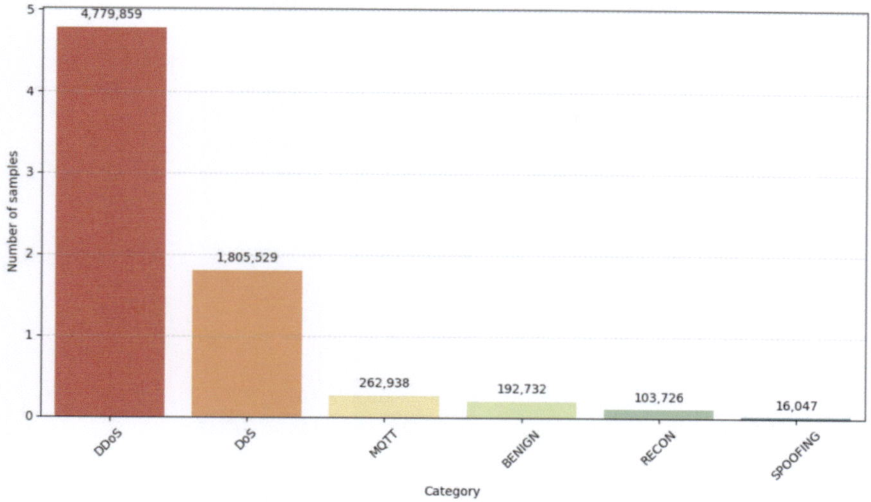

Fig. 2 Breakdown of categories in the CICIoMT2024 dataset, covering Bluetooth, WiFi, and MQTT with 18 attack

from favoring more frequent categories and significantly improves its ability to detect all attacks types, including those that were initially rare.

As shown Fig. 4, the model performs well overall on the test set. The test loss is relatively low at 0.2736, reflecting a moderate average difference between the predict and actual classes. The classification metrics are strong as well: accuracy reaches at

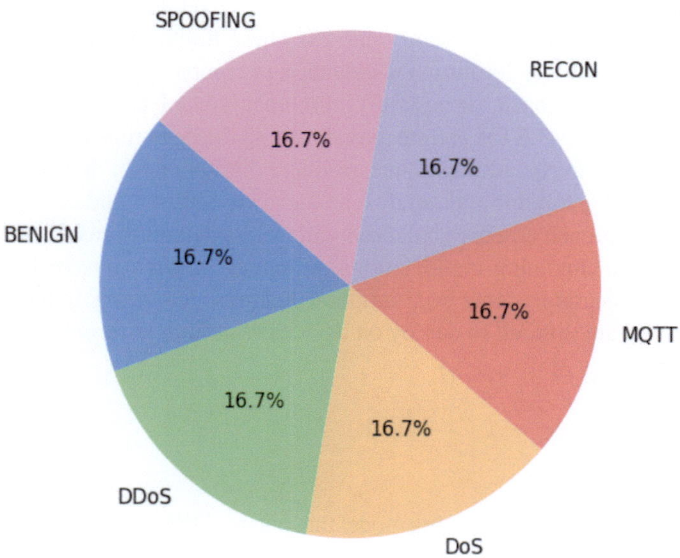

Fig. 3 Balanced distribution of target classes

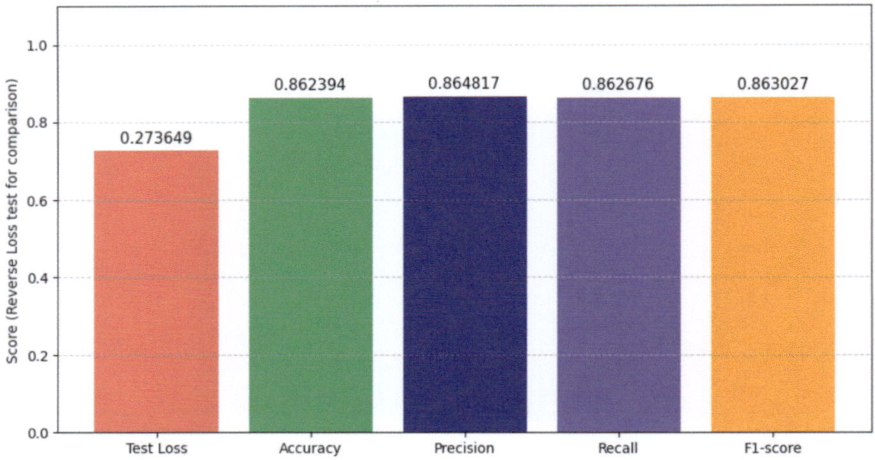

Fig. 4 Model performance on the CICIoMT2024 Test set for our solution

86.2%, indicating that the majority of samples are correctely classified. Precision and recall are similarly at 86.48% and 86.27%, respectively, demonstrating that the model is both precise in its positive predictions and effective in detecting true instances. Finally, the F1-score with 86.31%, which merges precision and recall, emphasizes the model's balanced and stable performance in multi-class classification.

In summary, the CNN-LSTM model presented here effectively handles multi-class network traffic classification of network traffic by margin CNN's spatial feature extraction with the temporal modeling strength of LSTM. Supported by careful preprocessing and class blancing, the model archives over 86% accuracy along with consistent precision, recall, F1-score metrics. These finding highlight the model's robustness and suitability for intrusion detection in complex network environments.

The proposed approach successfully combines spatial and temporal analysis through a hybrid CNN-LSTM architecture, boosting the detection accuracy of subtle intrusions. It effectively addresses the resource limitations of IoMT devices. In contrast to existing solution that tend to be narrowly focused and computationally intensive, our approach utilizes a balanced multi-class dataset to ensure better generalization while maintaining computational efficiency. Current research often falls short in fully integrate spatial and temporal features, managing imbalanced data adequately, and are difficult to deploy on devices with limited resource.

5 Conclusion

We developed a hybrid deep learning model that marges CNN with LSTM networks to improve intrusion detection in network traffic. Our approach leverages CNN's spatial feature extraction and LSTM's temporal learning to effectively classify multiple attack types across six categories. Evaluated on a balanced dataset, the model attained high accuracy, precision, recall, and F1-score demonstrating the effectiveness and reliability of our solution. Future work will focus on investigating federated learning approaches to support cooperative intrusion detection across decentralized MEC nodes, while safeguarding data privacy.

Acknowledgements We are deeply grateful for the support of the CNRST for their backing through the PASS (PhD-Associate Scholarship) program.

References

1. W Villegas-Ch, J García-Ortiz, I. Urbina-Camacho (2023) Framework for a secure and sustainable internet of medical things, requirements, design challenges, and future trends. Appl Sci vol 13 n (11): 11. https://doi.org/10.3390/app13116634
2. P Kumar Das, F Zhu, S Chen, C Luo, P Ranjan, G Xiong (2019) Smart medical healthcare of internet of medical things (IOMT): Application of Non-Contact Sensing. In: 2019 14th IEEE conference on industrial electronics and applications (ICIEA), p 375–380. https://doi.org/10.1109/ICIEA.2019.8833992
3. Berguiga A, Harchay A, Massaoudi A HIDS-IoMT: a deep learning-based intelligent intrusion detection system for the internet of medical things. IEEE Access 13:32863–32882. https://doi.org/10.1109/ACCESS.2025.3543127
4. H Assmi et al. (2025) A robust security detection strategy for next generation IoT networks.|EBSCOhost. Consulté le: 26 juin. [En ligne]. Disponible sur: https://openurl.ebsco.com/contentitem/doi:10.32604%2Fcmc.2024.059047?sid=ebsco:plink:crawler&id=ebsco:doi:10.32604%2Fcmc.2024.059047
5. Akar G, Sahmoud S, Onat M, Cavusoglu Ü, Malondo E (2025) L2D2: a novel lstm model for multi-class intrusion detection systems in the era of IoMT. IEEE Access 13:7002–7013. https://doi.org/10.1109/ACCESS.2025.3526883
6. Al J, Shaikh CW, Saifullah MW, Us Sima M, Arshad M, Rathore WUA (2025) Memory feedback transformer based intrusion detection system for IoMT healthcare networks. Internet Things 32:101597. https://doi.org/10.1016/j.iot.2025.101597
7. S Amaouche, A Guezzaz, S Benkirane, M Azrour (2024) A robust model for predicting abnormal behavior in vehicular networks using AdaBoost and chi-square. Wirel Pers Commun vol 138, n (4): p 2583–2611. https://doi.org/10.1007/s11277-024-11615-0
8. Bella K et al (2024) An efficient intrusion detection system for IoT security using CNN decision forest. Peer J Comput Sci 10:e2290. https://doi.org/10.7717/peerj-cs.2290
9. N Faruqui et al. (2023) SafetyMed: a novel IoMT intrusion detection system using CNN-LSTM hybridization. Electronics vol 12, n (17): p 3541. https://doi.org/10.3390/electronics12173541
10. F Pelekoudas-Oikonomou et al. (2022) Blockchain-based security mechanisms for IoMT edge networks in IoMT-based healthcare monitoring systems. Sensors vol 22, n (7): 7. https://doi.org/10.3390/s22072449
11. S Nandy, M Adhikari, MA Khan, VG Menon, S Verma (2022) An intrusion detection mechanism for secured IoMT framework based on swarm-neural network. IEEE J Biomed Health Inform vol 26, n (5): p 1969–1976. https://doi.org/10.1109/JBHI.2021.3101686
12. Pustokhina IV, Pustokhin DA, Gupta D, Khanna A, Shankar K, Nguyen GN (2020) An effective training scheme for deep neural network in edge computing enabled internet of medical things (IoMT) systems. IEEE Access 8:107112–107123. https://doi.org/10.1109/ACCESS.2020.3000322

13. A Rahman, G Wu, AM Liton (2020) Mobile edge computing for internet of things (IoT): security and privacy issues. Indones J Electr Eng Comput Sci vol 18, n (3): p 1486. https://doi.org/10.11591/ijeecs.v18.i3.pp1486-1493
14. Shaikh JA et al (2025) A deep reinforcement learning-based robust intrusion detection system for securing IoMT healthcare networks. Front Med 12. https://doi.org/10.3389/fmed.2025.1524286
15. EA Adeniyi, RO Ogundokun, JB Awotunde (2021) IoMT-based wearable body sensors network healthcare monitoring system. In: IoT in healthcare and ambient assisted living, G Marques, AK Bhoi, VHC de Albuquerque, H K.S., Éd., Singapore, Springer, p 103–121. https://doi.org/10.1007/978-981-15-9897-5_6
16. C Hazman, A Guezzaz, S Benkirane, M Azrour (2024) A smart model integrating LSTM and XGBoost for improving IoT-enabled smart cities security Clust Comput vol 28, n (1): p 70. https://doi.org/10.1007/s10586-024-04780-1
17. Gopala Krishnan D, Diviya M, Jyoshika Shri K, Subramanian M (2024) Enhanced security for IOMT devices: deep learning techniques for network intrusion detection. In: 2024 5th international conference on communication, computing and industry 6.0 (C2I6), pp 1–6. https://doi.org/10.1109/C2I663243.2024.10895169
18. JA Alzubi, OA Alzubi, I Qiqieh, A Singh (2024) A blended deep learning intrusion detection framework for consumable edge-centric IoMT industry. IEEE Trans Consum Electron vol 70 n (1): p 2049–2057. https://doi.org/10.1109/TCE.2024.3350231
19. Hernandez-Jaimes ML, Martinez-Cruz A, Ramírez-Gutiérrez KA, Feregrino-Uribe C (2023) Artificial intelligence for IoMT security: a review of intrusion detection systems, attacks, datasets and cloud–fog–edge architectures. Internet Things 23:100887. https://doi.org/10.1016/j.iot.2023.100887
20. A Alamleh et al. (2023) Federated learning for IoMT applications: a standardization and benchmarking framework of intrusion detection systems. IEEE J Biomed Health Inform vol 27, n (2): p 878–887. https://doi.org/10.1109/JBHI.2022.3167256
21. Kumar P, Gupta GP, Tripathi R (2021) An ensemble learning and fog-cloud architecture-driven cyber-attack detection framework for IoMT networks. Comput Commun 166:110–124. https://doi.org/10.1016/j.comcom.2020.12.003
22. Otoum Y, Wan Y, Nayak A (2021) Federated transfer learning-based IDS for the internet of medical things (IoMT). In: 2021 IEEE Globecom workshops (GC Wkshps), pp 1–6. https://doi.org/10.1109/GCWkshps52748.2021.9682118

Improved Computational Diffie-Hellman-Based Mechanism for Cyber-Physical Security

Souhayla Dargaoui, Abdeslem Blali, Mourade Azrour, Ahmad El Allaoui, Fatima Amounas, Azidine Guezzaz, and Abdulatif Alabdulatif

Abstract Cyber-physical systems (CPS) generally refer to complex systems with multiple dimensions that combine computation and communication technologies with physical environment interactions. Such systems are widely accepted in smart agriculture, military applications, and other life scenarios. The Internet of drones as a CPS presents various security issues, vulnerabilities, and dangers for customers. Recently, we proposed a PUF-based semi-centralized authentication scheme for Drone-to-Drone communication to secure such networks. Our previous scheme has presented a significant improvement in computation and communication overheads. However, it can encounter difficulties in managing large networks with thousands of drones. In this paper, we provide an improved authentication scheme for peer-to-peer drone communication that mitigates scalability issues. Our protocol significantly improves computation cost, requiring only 0.172 ms while preserving security and

S. Dargaoui · A. Blali · M. Azrour (✉) · A. El Allaoui · F. Amounas
IMIA laboratory, MSIA Team, Faculty of Sciences and Techniques, Moulay Ismail University of Meknes, Errachidia, Morocco
e-mail: mo.azrour@umi.ac.ma

S. Dargaoui
e-mail: s.dargaoui@edu.umi.ac.ma

A. Blali
e-mail: ab.blali@edu.umi.ac.ma

A. El Allaoui
e-mail: a.elallaoui@umi.ac.ma

F. Amounas
e-mail: f.amounas@umi.ac.ma

A. Guezzaz
SISAR Team, LaRTID Laboratory, Higher School of Technology Essaouira, Cadi Ayyad University, Errachidia, Morocco

A. Alabdulatif
Department of Computer Science, College of Computer, Qassim University, Buraydah, Saudi Arabia
e-mail: ab.alabdulatif@qu.edu.sa

G. H. L. et al. (eds.), *Reliability in Cyber-Physical Systems: The Human Factor Perspective*, Springer Series in Reliability Engineering,
https://doi.org/10.1007/978-3-032-09917-4_14

scalability. It also necessitates an acceptable amount of communication overhead compared to similar protocols.

Keywords Cybersecurity · Internet of drones · Authentication

1 Introduction

Cyber-physical systems (CPS) is an emerging term for networked systems in which the computational (cyber) component is closely connected to the physical components [1–4]. In such systems, the computational components process the data collected by physical components and then offer continuous feedback to control the system and influence the environment. Today, the Internet of Drones has become one of the most used Cyber-physical systems. This rise in Unmanned Aerial Vehicle network usage underscores several privacy and security issues [5–12].

Several authentication protocols have been suggested in recent years to mitigate security issues in IoD networks [13–18]. In 2019, Venkatasamy et al. provided a novel architecture for Medical Wireless Sensor networks and presented an ECC-based authentication protocol [19]. Thanks to Burrows–Abadi–Needham logic, Venkatasamy et al. prove the correctness of their protocol. In 2023, the cryptanalysis by Servati et al. [20] proved the traceability issues in Sureshkumar et al.'s protocol [19]. Subsequently, they proposed an improved ECC-based authentication scheme called ECCbAS that surmounts these vulnerabilities. Further, in 2025, Prajapat et al. [21] proposed a novel ECC-based authentication scheme for Brain-Computer Interface systems.

Recently, we proposed an authentication scheme for Drone-to-Drone communication with significant improvement in computation and communication overheads. Nevertheless, it may face scalability issues. This paper proposes an enhanced authentication scheme for peer-to-peer drone communication that minimizes energy consumption while preserving security and scalability. Considering the Dolev-Yao threat model [22], we evaluate the security of our scheme.

The rest of this paper is structured as follows: The Sect. 2 gives details about some related work. In the Sect. 3, we present our proposed scheme. The Sect. 4 provides the performance evaluation results and discussion. Finally, the Sect. 5 concludes the paper.

2 Related Work

The evolution of Internet of Drones authentication protocols over recent years reflects a shift towards more secure, scalable, and privacy-preserving solutions. Early research by Zhang et al. [23] highlighted the need for lightweight cryptographic frameworks tailored for resource-constrained drones, establishing foundational principles for scalable hierarchical authentication. Building on this, Maurya et al. [24] introduced blockchain-based schemes that decentralized authentication processes, enhancing transparency and resistance to single points of failure. Concurrently, Ejiyeh [25] developed energy-efficient mutual authentication protocols suitable for large-scale drone swarms, emphasizing low latency and minimal computational overhead. These advancements significantly improved the robustness of drone networks, addressing key vulnerabilities faced in early implementations.

Recent studies have focused on integrating emerging technologies such as AI and quantum-resistant algorithms. Yang et al. [26] proposed adaptive, machine learning-driven authentication mechanisms capable of detecting impersonation or compromised drones through behavioral analysis. Meanwhile, Abdulkasim et al. [27]developed quantum-resilient protocols based on lattice cryptography, ensuring security against future quantum attacks. In addition, Tanveer et al. [28] emphasized the importance of interoperability standards, advocating for unified frameworks incorporating blockchain, AI, and quantum-resistant techniques. de Jesus Sousa [29] introduced a multi-factor authentication scheme combining cryptographic keys, biometric data, and behavioral analytics to bolster trust levels. Privacy preservation was further addressed by Dwivedi et al. [30], who utilized zero-knowledge proofs to enable confidential authentication. Lastly, Jan [31] explored lightweight, hash-based signatures optimized for real-time drone operations, and Choe and Kang al [32]. discussed deployment challenges and integration strategies for secure authentication protocols in heterogeneous drone ecosystems.

Although these proposed advanced drone authentication, drone still face challenges such as balancing security and efficiency, scalability issues with blockchain, lack of standardization across systems, and so ones methods. Hence, the proposition of new authentication IoD protocol is necessary.

3 Proposed Authentication Scheme

Our network consists of drones and a trusted authority for drone registration. We consider a unique PUF module embedded in each drone, designed to generate response pairs based on given challenges and inherent hardware characteristics. When drones come into communication range with each other, they can initiate mutual authentication. Our scheme consists of three phases;

In the enrollment phase, the trusted authority (TA) selects its private key, an elliptical curve generator point P, and a hash function h (.). Subsequently, TA publishes the generation and reproduction algorithms of the fuzzy extractors Gen (.) and Rep (.), h (.), and P.

3.1 Registration Phase

In this phase, the drone selects the identity IDi, secret key ki, and a random nonce r. Then, it computes its public key $Pubki = ki.\,p$ and pseud-identity $HIDi = h(IDi\|r)$. Then it sends the registration request {HIDi, Pubki} to the Trusted Authority.

Upon receiving the registration request, the TA checks the availability of the drone's pseudo-identity. Then, it computes the challenge $Ci = h(HIDi\|KG)$, stores {HIDi} in its memory, publishes {HIDi, Pubki}, and communicates Ci to the drone.

Upon receiving the challenge Ci, the drone initiates Session CounterSCiand computes $Bi = PUF(Ci)$, $(Ri,\ Pi) = Gen(Bi)$, $Ai = h(HIDi\|Ri\|ki)$, and $Ki = ki \oplus h(HIDi \mid |Ri)$. Finally, the drone stores {HIDi, Ki, Ci, Pi, h(.),SCi Gen(.), Rep(.)} in its memory.

3.2 Authentication and Key Agreement Phase

As illustrated in the figure, the authentication and session key establishment process consists of three steps as shown Fig. 1:

Steps 1 The drone computes bi, Ri*, Ki*, Ai* and verify if Ai* = Ai. Subsequently, if the condition holds, the drone selects a random R1, Ks, generates a timestamp T1, and computes M1, M2, and M3. Then, it sends the authentication request {M1, M2, M3, T1} to the drone Dj.

Step 2 Upon receiving the authentication request, the drone Dj verifies the freshness of the timestamp T1. If the condition holds, the drone Dj computes Bj, Rj*, Kj*, and Aj* and verifies if Aj* = Aj. Then, it retrieves HIDi* and R1 from M3, computes M1*, and verifies if M1* = M1. If the condition holds, the drone Dj generates a timestamp T2 and selects a random R2. Then, it computes the session key SK, M4, and M5. Finally, the drone Dj transmits the authentication response {M4, M5, T2} to the drone Di, increments the session counter SCj, computes Cj^{new}, Bj^{new}, (Rj^{new}, Pj^{new}), Aj^{new}, and Kj^{new}, and updates parameters in its memory.

Step 3 As soon as the drone Di gets the authentication response, it verifies the freshness of T2. If it is fresh, the drone recovers HIDj, R2 from M5and computes the session key SK and M4*. Then, the drone verifies if M4* = M4. If the condition holds, the authentication is accomplished. Finally, it increments the session counter

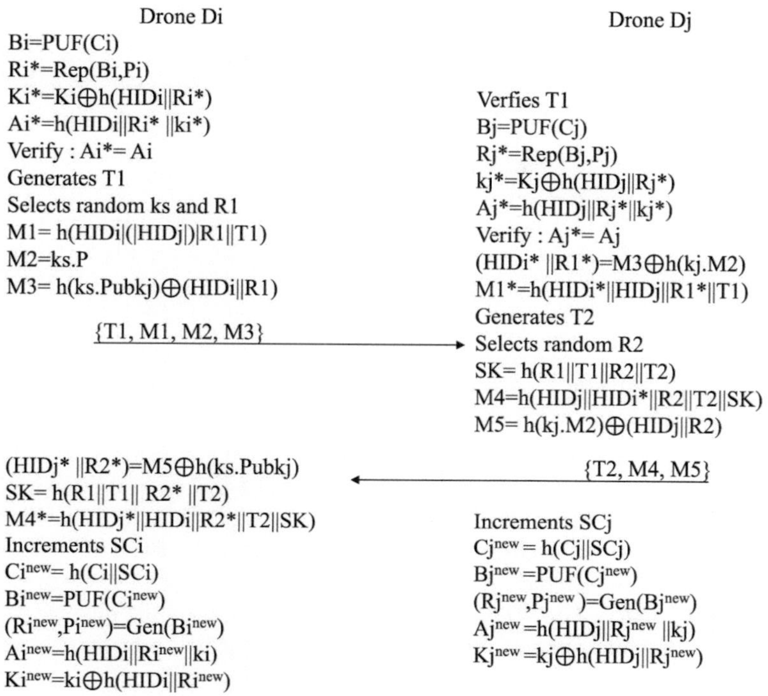

Fig. 1 Authentication and key agreement phase

SCi, computes Ci^{new}, Bi^{new}, (Ri^{new}, Pi^{new}), Ai^{new}, and Ki^{new}, and updates parameters in its memory.

4 Results and Discussions

As illustrated Table 1, all compared schemes ensure the security features. However, the schemes [19, 21] prove their resistance only against impersonation attacks and reply attacks. The scheme [20] also resists DoS attacks and man-in-the-middle attacks while our scheme resists all attacks.

According to [33], Th ≈ 0.0025, Tpuf ≈ 0.0021, Tfe ≈ 0.0098, Tecc ≈ 0.0273, and Tf ≈ 0.442, where they present the timing require1ment of the hash function, PUF, fuzzy extractor, and ECC, respectively.

Figure 2 presents a comparison of the computational cost, measured in milliseconds (ms), associated with executing different protocols, including three specific protocols labeled [19–21], as well as our proposed. In fact, it is clear that our protocol has a significantly lower execution time of just 0.172 ms, compared to the

Table 1 Security features and resistance against attacks

Protocol	F_1	F_2	F_3	F_4	F_5	F_6	A_1	A_2	A_3	A_4	A_5	A_6	A_7
[19]	x	x	x	x	x	x	x	x	–	–	–	–	–
[20]	x	x	x	x	x	x	x	x	–	x	–	–	x
[21]	x	x	x	x	x	x	x	x	–	–	–	–	–
Our protocol	x	x	x	x	x	x	x	x	x	x	x	x	x

F1: mutual authentication, F2: Anonymity, F3: unlinkability, F4: key agreement, F5: key secrecy, F6: perfect forward secrecy, A1: Impersonation attack, A2: reply attack, A3: physical attack, A4: DoS attack, A5: Stolen verifier, A6: Denning-sacco attack, A7: man-in-the-middle.

other protocols, which have execution times of 15,188 ms, 15,63 ms, and 15,671 ms respectively.

Additionally, Fig. 3 illustrates the comparison of communication costs in bits for different protocols, including three specific protocols [19–21], as well as our proposed one. The graph clearly highlights that our proposed protocol has the lowest communication cost, at just 1024 bits, demonstrating a significant reduction compared to protocols [19–21], which have costs of 4416, 3456, and 1632 bits, respectively.

On the other hand, Fig. 4 depicts the relationship between the number of users and the corresponding authentication latency in milliseconds. As the number of users increases from 10^2 (100) to 10^4 (10,000 users), the authentication latency exhibits a dramatic upward trend. Specifically, for 100 users, the latency is approximately 0.1 ms, indicating very quick authentication; however, as the user base expands to 10,000 users, the latency skyrockets to about 50,000 ms (or 50 s).

This exponential increase highlights the scalability challenges faced by the authentication protocol in large networks. In smaller networks, the latency remains practically negligible, making the system suitable for real-time requirements; but as the network size grows, the delay becomes substantial, which could adversely affect user experience and system efficiency. The graph clearly demonstrates the importance of optimizing authentication mechanisms for large-scale deployments, especially in

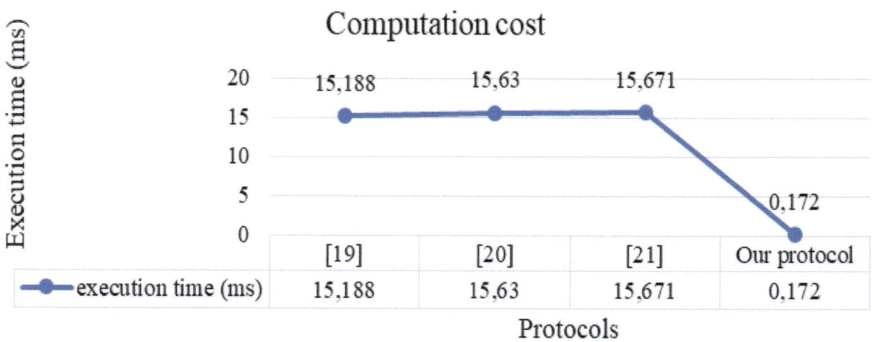

Fig. 2 Computation overheads

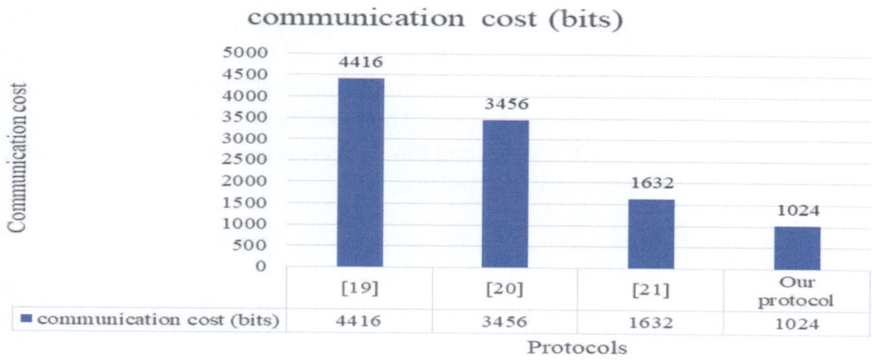

Fig. 3 Communication overheads

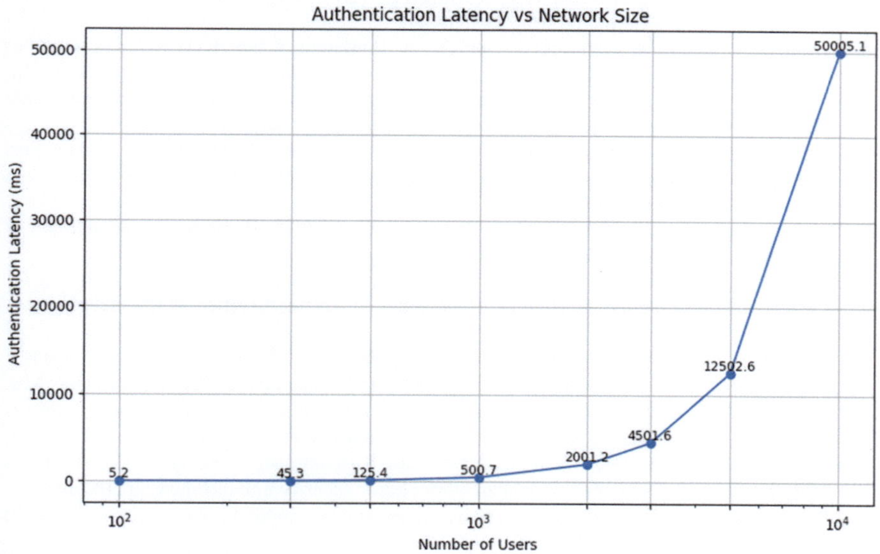

Fig. 4 Authentication latency over network size

environments where high-speed authentication is critical, such as in cloud services, IoT networks, or large enterprise systems.

5 Conclusion

The Internet of Drones plays a crucial role in our daily lives. However, this emerging technology presents new security challenges. Most of the proposed authentication schemes do not secure exchanged data from all raids. Our paper provides a new peer-to-peer authentication scheme for IoD networks, which mitigates scalability issues. The proposed scheme is built on ECC, PUF, and the Computational Diffie-Hellman problem. The evaluation of our paper regarding the state-of-the-art protocols demonstrates its robustness and efficiency. This combination of security features, resiliency against well-known attacks, and low latency makes our scheme a suitable option for real-time IoD network use cases.

References

1. Jamshidi S, Amirnia A, Nikanjam A, Khomh F (2024) Enhancing security and energy efficiency of cyber-physical systems using deep reinforcement learning. Procedia Comput Sci 238:1074–1079
2. Harkat H, Camarinha-Matos LM, Goes J, Ahmed HF (2024) Cyber-physical systems security: a systematic review. Comput Ind Eng 188:109891
3. Azrour M, Dargaoui S, Mabrouki J, Guezzaz A, Benkirane S, Shafik W, Ahmad S (2024) A survey of machine and deep learning applications in the assessment of water quality. In: Mabrouki J, Mourade A (eds) Technical and technological solutions towards a sustainable society and circular economy. Springer Nature Switzerland, Cham, pp 471–483
4. Dargaoui S, Azrour M, El Allaoui A, Guezzaz A, Benkirane S, Alabdulatif A, Amounas F Applications of Blockchain in healthcare:: review study. IoT. Machine learning and data analytics for smart healthcare:1–12
5. Dargaoui S, Azrour M, Mabrouki J, El allaoui A, Guezzaz A, Benkirane S, Alabdulatif A (2023) Security issues in internet of medical things, pp 77–91
6. S Dargaoui, M Azrour, AE Allaoui, A Guezzaz, S Benkirane (2025) Machine learning for security boosting in Internet of Things environments. In: Recent Advances in Internet of Things security CRC Press
7. Dargaoui S, Azrour M, El Allaoui A, Guezzaz A, Benkirane S, Alabdulatif A, Amounas F (2024) IoT-driven smart agriculture: security issues and authentication schemes classification. In: Mejdoub Y, Elamri A (eds) Proceeding of the international conference on connected objects and artificial intelligence (COCIA2024). Springer Nature Switzerland, Cham, pp 61–66
8. S Dargaoui, M Azrour, A El Allaoui, A Guezzaz, S Benkirane, A Alabdulatif, F Amounas (2024) Internet-of-Things-enabled smart agriculture: security enhancement approaches. In: 2024 4th international conference on innovative research in applied science, engineering and technology (IRASET) pp 1–5
9. A Blali, S Dargaoui, M Azrour, A Guezzaz, F Amounas, A Alabdulatif (2025) Analysis of deep learning-based intrusion detection systems in IoT environments. EDPACS 1–35
10. Shah IA (2024) Drone industry security issues and challenges in the context of IoD. Cybersecur Issues ChallS Drone Ind:310–323
11. Shah IA, Jhanjhi NZ, Ray SK (2024) IoT devices in drones: security issues and future challenges. In: Cybersecurity issues and challenges in the drone industry, IGI Global, pp 217–235
12. Dargaoui S, Azrour M, El Allaoui A, Amounas F, Guezzaz A, Attou H, Hazman C, Benkirane S, Bouazza SH (2023) An overview of the security challenges in IOT environment. In: Mabrouki J, Mourade A, Irshad A, Chaudhry SA (eds) Advanced technology for smart environment and energy. Springer International Publishing, Cham, pp 151–160
13. Dargaoui S, Azrour M, El Allaoui A, Guezzaz A, Alabdulatif A, Alnajim A (2024) An exhaustive survey on authentication classes in the IOT environments. Indones J Electr Eng Inform (IJEEI) 12
14. S Dargaoui, M Azrour, A El Allaoui, A Guezzaz, S Benkirane (2023) Authentication in Internet of Things: state of art. In: proceedings of the 6th international conference on networking, intelligent systems and security pp 1–6

15. Dargaoui S, Azrour M, El Allaoui A, Guezzaz A, Alabdulatif A, Ahmad S (2025) ECC-based anonymous and multi-factor authentication scheme for IOT environment. Int J Online Biomed Eng 21
16. Dargaoui S, Azrour M, El allaoui A, Guezzaz A, Alabdulatif A, Alnajim A (2024) Internet of things authentication protocols: comparative study. Comput, Mater and Contin:1–10. https://doi.org/10.32604/cmc.2024.047625
17. Dargaoui S, Azrour M, Allaoui AE, Guezzaz A, Hammoudeh MAA (2025) New efficient puf-based authentication protocol for IOT-driven smart agriculture. JAIT 16:582–593. https://doi.org/10.12720/jait.16.4.582-593
18. H Timouhin, F Amounas, M Badiy Efficient ECC-based RFID authentication for enhanced IoT security. In: recent advances in Internet of Things security CRC Press, pp 39–46
19. Sureshkumar V, Amin R, Vijaykumar VR, Sekar SR (2019) Robust secure communication protocol for smart healthcare system with FPGA implementation. Futur Gener Comput Syst 100:938–951
20. Servati MR, Safkhani M (2023) ECCbAS: an ECC based authentication scheme for healthcare IoT systems. Pervasive Mob Comput 90:101753
21. Prajapat S, Rana A, Kumar P, Das AK, Susilo W (2025) Privacy-preserving authentication protocol for user personal device security in brain–computer interface. Comput Stand Interfaces 94:104009
22. Dolev D, Yao A (1983) On the security of public key protocols. IEEE Trans Inf Theory 29:198–208
23. Zhang Y, He D, Li L, Chen B (2020) A lightweight authentication and key agreement scheme for internet of drones. Comput Commun 154:455–464
24. S Maurya, MMS Rauthan, R Verma, RB Ahmad (2022) Blockchain based secure system for the internet of drones (IoD). In: 2022 6th international conference on electronics, communication and aerospace technology pp 727–735
25. Ejiyeh AM (2024) Secure, robust, and energy-efficient authenticated data sharing in drone to vehicles communications. In: 2024 IEEE european symposium on security and privacy workshops (EuroS&PW), pp 380–389
26. Yang Z, Zhang Y, Jie Z, Yang Y, Yufei J, Song H, Tiejun L, Sun Q, An J (2025) AI-driven safety and security for UAVs: from machine learning to large language models. Drones 9:392
27. Abulkasim H, Goncalves B, Mashatan A, Ghose S (2022) Authenticated secure quantum-based communication scheme in internet-of-drones deployment. IEEE Access 10:94963–94972
28. Tanveer M, Alasmary H, Kumar N, Nayak A (2023) SAAF-IoD: secure and anonymous authentication framework for the internet of drones. IEEE Trans Veh Technol 73:232–244
29. de Jesus SM, Gondim PRL (2025) A multi-factor user authentication protocol for the internet of drones environment. Peer-To-Peer Netw Appl 18:1–22
30. Dwivedi AD, Singh R, Ghosh U, Mukkamala RR, Tolba A, Said O (2022) Privacy preserving authentication system based on non-interactive zero knowledge proof suitable for internet of things. J Ambient Intell HumIzed Comput:1–11
31. Jan SU, Qayum F, Khan HU (2021) Design and analysis of lightweight authentication protocol for securing IoD. IEEE access 9:69287–69306
32. Choe H, Kang D (2025) ECC based authentication protocol for military internet of drone (IoD): a holistic security framework. IEEE Access
33. Nyangaresi VO, Al-Joboury IM, Al-sharhanee KA, Najim AH, Abbas AH, Hariz HM (2024) A biometric and physically unclonable function–based authentication protocol for payload exchanges in internet of drones. E-Prime-Adv Electr Eng, Electron Energy 7:100471

Enhancing PE Malware Detection: A Comparative Study of Feature-Based and Image-Based Representations

Selvaganapathy Shymala Gowri, Vinayakumar Ravi, and S. Sowmya

Abstract Windows Portable Executable (PE) files are a common format for software applications and so they are also a primary target for malware authors. The increasing complexity of malicious software in the form of polymorphic and obfuscated malware, poses a challenge for traditional detection systems that are solely based on static or dynamic analysis. For enhancing the detection accuracy and resilience, this study explores a hybrid malware classification framework. It combines the conventional machine learning models trained on static features with image-based deep learning techniques which transform the metadata into visual representations from the features. Models such as Random Forest, CNN, and MLP are trained on PE file features that are JSON-extracted and a fine-tuned MobileNetV2 model is used for classifying malware and benign files based on their image equivalents. The proposed approach aims to use the strengths of both the structured and spatial representations to for improving the generalization and robustness in detection of malware.

Keywords Malware classification · Windows PE · Image-based malware detection · Static and dynamic malware analysis · Transfer learning

1 Introduction

Windows Portable Executable (PE) files are the fundamental format for executables and dynamic link libraries on Windows, making them prime targets for malware attacks. To automate PE malware detection, cybersecurity has increasingly relied on

S. Shymala Gowri (✉) · S. Sowmya
Department of Information Technology, PSG College of Technology, Coimbatore, India
e-mail: ssg.it@psgtech.ac.in

S. Sowmya
e-mail: 24pb39@psgtech.ac.in

V. Ravi
Center for Artificial Intelligence, Prince Mohammad Bin Fahd University, Khobar, Saudi Arabia
e-mail: vravi@pmu.edu.sa

G. H. L. et al. (eds.), *Reliability in Cyber-Physical Systems: The Human Factor Perspective*, Springer Series in Reliability Engineering,
https://doi.org/10.1007/978-3-032-09917-4_15

machine learning (ML) approaches, which depend heavily on high-quality datasets representing both malicious and benign files. Feature extraction usually has follows two strategies, they are static analysis and dynamic analysis. Static analysis is extracting the structural features like header metadata, imported functions, and entropy without executing the file so that we can have a low-risk detection [1, 2]. However, it does not perform well against obfuscated malware. Dynamic analysis executes files in a sandbox environment to capture runtime behaviors like API calls and network activity so that we can obtain deeper insights but at the cost of higher resource demands and the threat of susceptibility to evasion techniques [3].

To enhance the detection mechanism, hybrid approaches combining static and dynamic features have started gaining popularity. It is implemented by leveraging the efficiency of static analysis with the robustness of the dynamic behavior indicators [4]. Building hybrid-supporting datasets still remains very challenging, especially given the complexity of collecting dynamic data at scale in a consistent manner [5]. Meanwhile, there is a shift toward richer labeling in datasets beyond simple malicious/benign tags, including behavioral traits, targeted platforms, and exploited vulnerabilities. This evolution allows for the development of models that not only detect malware but also analyze its tactics, techniques, and threat levels, reflecting a deeper understanding of adversarial methods.

Large-scale datasets have increased the usage of deep learning models, so that architectures like CNNs and RNNs can learn intricate patterns across millions of labeled samples. However, scaling datasets have introduced challenges like label accuracy, class imbalance, and bias from ground truth that are based on antivirus [6]. There are some specialized datasets that support research into specific areas like metamorphic malware behaviors, structural variations among families, and exploitation of vulnerabilities. These developments help us highlight a growing focus on proactive and nuanced malware detection strategies which go beyond traditional binary classification [7]. The remainder of this article is organized as follows: Sect. 2 provides the literature survey on various malware detection approaches its limitations. Section 3 describes the dataset used, along with exploratory data analysis to understand the underlying distributions and patterns. Section 4 explains the proposed methodology and model architecture followed by Sect. 5, which presents the results and model performance comparison. Finally, Sect. 6 concludes the paper and outlines the directions for future work.

2 Literature Survey

The literature survey reviews recent research that is focused on malware detection and classification techniques for Windows Portable Executable (PE) files. As PE files are usually the primary format for executables and dynamic link libraries on Windows based machines, they are a frequent target for malware attacks by attackers. This survey examines a few studies that use datasets consisting of both benign and malicious PE samples which were collected from malware repositories, open-source platforms, and curated collections and is tabulated in Table 1. It highlights the feature extraction techniques that were used, including the static analysis of structural elements and the dynamic analysis based on runtime behaviors [8].

Each work has been assessed in terms of the characteristics of that particular dataset, the feature engineering strategies, and the advantages and limitations of the proposed methodologies [9]. Approaches that have incorporated dynamic features such as API call traces and hybrid models combining static and dynamic attributes to improve the efficiency of the detection are highlighted [10].

The survey aims to identify trends that are currently prevailing, existing challenges and emerging innovations in PE malware research.

It has an emphasis on the impact of dataset quality and feature design on the performance of machine learning models [11].

In a recent study [12], researchers have focused on classifying malware by observing how the Windows Portable Executable files behave during the execution process. Instead of relying on some traditional static features, the study has considered a dynamic analysis approach by recording API call sequences to capture more authentic behavioral traits of the executable files. The idea is to build the models that could detect malware based not just on what it looks like but also from its runtime behavior. Another work [13], builds a rich dataset which goes beyond simple malware detection. Instead of just labeling files as malicious or not malicious, it has categorized them based on their behaviors, the platforms they target, the vulnerabilities they exploit, and also the packers that are used to hide their code. This broader labeling strategy can help the models recognize not only malware itself but the tactics and technologies behind it in the future.

Another effort [16], meanwhile, returned to static analysis, concentrating on structural features of PE files. By pulling information from file headers, sections, imported libraries, and API calls, this dataset aimed to support detection efforts without the risks associated with executing unknown binaries. Although static methods sometimes miss evolving threats, they still remain valuable, especially when real-time dynamic analysis isn't feasible. Lastly, another benchmark dataset looked at static PE file features without including raw binary files, making it safer to distribute. It grouped its features into categories like header info, byte histograms, and string patterns, providing a strong foundation for models based on feature engineering [18]. Its combination of size and structured feature design has made it a popular choice for many academic and industry projects over the years.

A. Gaps Identified

Malware detection in Windows-based environments is usually been relied on static analysis techniques where the executable files are examined without execution to extract some meaningful features. This technique includes analyzing the metadata from Portable Executable (PE) files like headers, imported DLLs, function names, section sizes, and the entropy measures [5]. These static features are structured, numerical, or categorical, and they are used as input to classical machine learning

Table 1 Windows PE based dataset

S. no.	Title of paper	Publisher and year	Number of samples	Source	Features	Pros	Research gap
1.	Dynamic malware classification and API categorisation of windows PE files using ML [12]	Applied sciences (MDPI), 2024	1020 (582 malware, 438 Benign)	API traces from sandboxed PE files from malware bazaar (malware), GitHub (benign)	API call frequency, categorized into 23 API classes; feature scoring using Chi2 and Gini importance; dynamic analysis using cuckoo sandbox	Focuses on dynamic features like API calls; good for detecting obfuscated malware	Small dataset; dynamic analysis is time- and resource intensive
2.	MalDICT: benchmark datasets on malware behaviors, platforms, exploitation, and packers [13]	CAMLIS conference, 2023	5.5 million (malware only)	Virus total AV reports	Behavior tags (75), platform tags (43), exploit tags (128), Packer tags (79); extracted using ClarAVy from 90 AV engines	Covers multiple under-explored classification types; highly scalable; curated tags from diverse AV products	No benign samples; relies on AV label parsing which may inherit labeling bias
3.	Multi-feature dataset for windows PE malware classification [14]	Independent (GitHub release), 2022	18,551 (malware only)	Malware bazaar	Static features: DLL imports, API functions, 52 PE header fields, 100 PE section fields using PE file	Multiple structured static feature types; supports fine grained static ML analysis	No benign files; limited to 5 malware families
4.	SOREL-20 M: a large scale benchmark dataset for malicious PE detection [15]	Sophos AI reversing Labs, 2020	20 million (10 M malware, 10 M benign)	Virus total (via sophos/reversing labs)	Static features: PE headers, byte histogram, entropy, strings, imported symbols; 11 AV behavior tags; 500 K unlabeled samples	Massive dataset; includes disarmed binaries; useful for deep learning	Behavior tags limited; lacks detailed family classification

(continued)

Table 1 (continued)

S. no.	Title of paper	Publisher and year	Number of samples	Source	Features	Pros	Research gap
5.	A benchmark API call dataset for windows PE malware classification [16]	Independent (GitHub release), 2019	7107 malware (from 20,000+)	GitHub (samples), Virus Total (labels)	API call Sequences (342 unique APIs), frequency analysis; dynamic analysis via cuckoo sandbox	Focus on metamorphic malware; strong dynamic behavioral insights	No benign files; only malware with 10 API calls Included
6.	EMBER: an open dataset for training static PE malware machine learning models [17]	Endgame (elastic security), 2018	2 million (1.1 M malware, 0.9 M benign)	Virus total	Static PE features: Imports, section metadata, byte/ entropy histograms, string statistics	Public, well documented dataset; ready to-use feature vectors	No dynamic features; limited to binary classification (malware vs. benign)

algorithms like Random Forests and Multi-Layer Perceptrons (MLPs), and also to deep learning models like Convolutional Neural Networks (CNNs). These models are trained on the features of the PE files so that they can classify a given file as either benign or malicious by learning patterns present in those labeled datasets. Despite being efficient and safe—since the files are never executed—static analysis is often inadequate against sophisticated malware that uses code obfuscation, encryption, or polymorphic transformations to hide its true intent [6]. As a result, purely static models may suffer from reduced accuracy in real-world scenarios where malware constantly evolves to evade detection.

3 Dataset Description

The PEVuln dataset [19, 20] was created for supporting the research on malware that has exploitable vulnerabilities. It combines data from the Malvuln project, which is a project that catalogs vulnerabilities in malware, and from ExploitDB, which is a database of software vulnerabilities.

The dataset aims to differentiate between the malicious behavior and the software flaws that can be exploited by the attackers. It provides us with a unique perspective on the intersection of malware and software vulnerabilities, an area that is quite underexplored in malware classification.

The dataset contains of 864 vulnerable malware samples across 128 families, 35,241 non-vulnerable malware samples across 520 families, 1425 vulnerable benign samples, and 7905 non-vulnerable benign samples. Malware samples were collected from the Malvuln repository and benign samples are from sources like VirusTotal and GitHub. Vulnerable benign samples were identified in the dataset by cross-referencing ExploitDB. This approach ensures that a diverse range of executables which includes both malicious and legitimate software with exploitable flaws are in the dataset.

The PEVuln dataset focuses on vulnerability labels and classifies samples by the vulnerability status in addition to the malicious or benign nature of the executable. It includes static features like PE header fields, section characteristics, import tables, and string patterns, that are useful for classification models that are based on the file structure. The dataset also can be utilized for the dynamic analysis of runtime behavior, so new opportunities to investigate how vulnerabilities influence malware propagation and exploitation are offered by it [20].

A. Exploratory Data Analysis

The exploratory data analysis of the PEVuln dataset is done by loading and merging metadata from four feature files, vulnerable benign, vulnerable malware, non-vulnerable benign, and non-vulnerable malware. For the analysis, labels are added to each dataset to indicate the vulnerability status (vulnerable or non-vulnerable) and the classification (malware or benign). These labeled datasets are then combined into a single unified data frame so that a comprehensive view of the samples are offered to us.

An important aspect of the EDA involves examining the class distributions in the dataset and they are visualized using count plots. The distribution of malware versus benign samples is compared, and it is followed by an analysis of the vulnerable versus non-vulnerable samples. The data is also assessed for any missing values, and a summary of the dataset shape and the data types was printed to ensure that integrity is maintained.

Further exploration included a correlation matrix of the numerical features that identifies the relationships among the different variables. The heatmap visualization highlights the interdependencies between the different features and it allows the identification of potentially relevant patterns that could inform the feature selection in subsequent modeling. Also the distribution of key features is analyzed through the usage of histograms, and it provides an overview of the data's spread and helps us highlight any skewed distributions or outliers that may require some preprocessing adjustments.

4 Proposed Methodology

For enhancing the accuracy of the detection for improving the resistance against advanced evasion techniques, a hybrid methodology is proposed. This methodology incorporates both structured machine learning and image-based deep learning. Initially, the features are extracted from JSON-formatted metadata of the PE files, so that the static indicators like entropy, section names, and import tables and dynamic indicators like registry, memory and network activity are captured. These extracted features are used to train traditional classifier models like CNNs, Random Forests, and MLPs to see the performance. Building on top of this, the second phase involves converting these structured data samples into image representations using encoding schemes like pixel mapping of the features into a matrix visualization. These generated images are then input into a MobileNetV2 model, where the original classification layer is replaced to suit the binary malware detection task that we are going to perform. The modified MobileNetV2 architecture is fine-tuned on the new image dataset, allowing it to learn the spatial and the structural patterns that are inherent to benign and malicious files.

This dual-stage approach combines the interpretability and efficiency of structured models with the representational power of visual deep learning, so it offers a more robust classification system. Figure 1 represents the flow diagram of the malware classification pipeline using JSON and image-based features.

A. Model Architecture

The proposed system consists of two stages. For the first stage, the PE file metadata like the entropy, section headers, import functions, and other relevant static attributes are extracted in JSON format and preprocessed into tabular structure. The features that were structured are then used to train three ML models: a Random Forest classifier, a Convolutional Neural Network (CNN), and a Multi-Layer Perceptron (MLP). Each model independently learns how to classify samples as benign or malware based on the feature patterns that it learns during the training phase. In the second stage, the same structured feature vectors are converted into grayscale by mapping feature values to pixel intensities. These images are fed as input to a modified version of MobileNetV2 model, which is a lightweight and deep CNN model which is pretrained on the ImageNet dataset. MobileNetV2's last layer has been replaced with a fully connected dense layer that can perform binary classification between malware and benign samples. This model is run for multiple epochs so that it can adapt its learned weights to this domain. This architecture provides interpretability through structured models and a very high expressiveness through feature learning that is based on features extracted from images.

5 Result

A. Classification of Benign and Malware Samples

The task of distinguishing the benign software from the malware involves the extraction and interpretation of a wide range of structural attributes that mostly differentiate the safe files from the harmful ones. In a feature-based classification model, samples are labeled based on the ground truth label that is obtained from trusted antivirus engines or some curated datasets. These samples are then parsed to extract numerical values like entropy, byte size distributions, and imported functions. These features often differ between the clean and infected executables and therefore, can be used to train the classification model. After it is preprocessed and normalized, this data is used to train supervised models, where the benign files display uniform structural patterns and the malware files often shows irregularities like high entropy or some suspicious imports. Random Forests use the ensemble voting method for the decision-making process, whereas models like MLPs and CNNs extract deeper representations through many layers of transformations. During the inference phase, these models output probabilistic scores which indicate whether a sample is likely to be more malicious or benign.

B. Image-Based Classification Using MobileNetV2

In image-based classification, malware detection is approached from the perspective of computer. The idea is to represent the structured features of the PE files as two-dimensional images, where each of the feature value corresponds to a pixel intensity in a grayscale image [21]. These images encapsulate the relationships between the features in a format that deep CNN models can interpret with ease. MobileNetV2, which is a lightweight yet powerful convolutional network, is particularly suited for

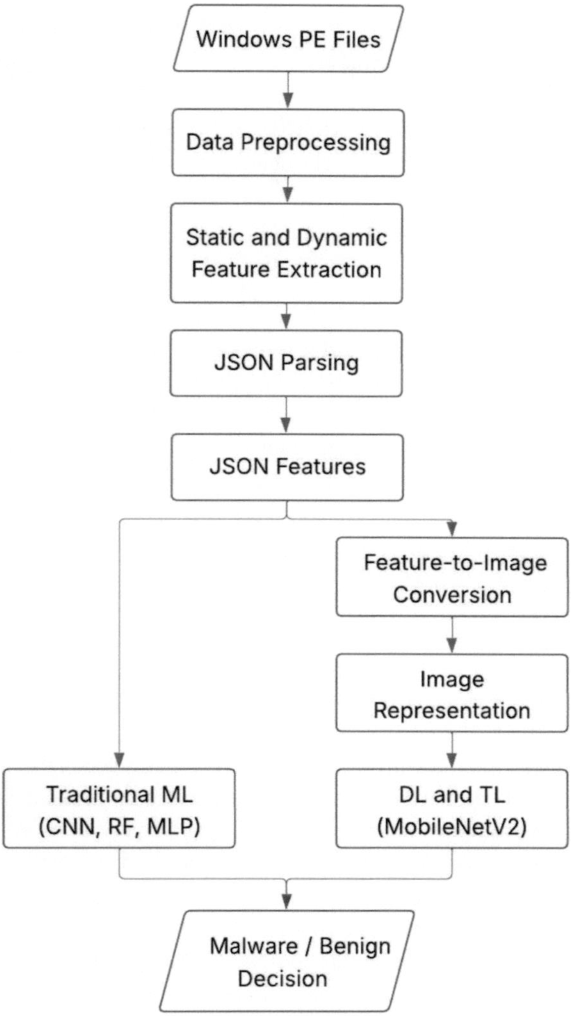

Fig. 1 Flow diagram of the malware classification pipeline using JSON and Image-based features

this task because it has depth wise separable convolutions and it also has an optimized performance on limited data.

The final layer of MobileNetV2 is a classification layer and it is replaced with a dense fully connected layer that matches with the number of output classes we have (malware and benign), and the model is then trained on the image dataset. Through many convolutional and pooling layers, the model learns to recognize the spatial features and the correlations that are often not considered in traditional models.

This technique enables the model to detect the malware patterns that may not be plainly evident from the raw numerical data alone and it has shown promise in improving the detection rates, especially against some obfuscated and relatively new malware variants.

Table 2 Performance metrics of PE malware classification using JSON features and image-based representations

Methodology	Model	Accuracy	Precision	Recall	F1-score
JSON feature-based	CNN	0.8165	0.8892	0.8705	0.8798
	Random forest	0.8195	0.8751	0.8948	0.8849
	MLP	0.8011	0.8679	0.8804	0.8741
Image-based (transfer learning)	MobileNetV2	0.8195	0.8751	0.8948	0.8849

Overall, this method introduces an innovative solution that is also cross-domain which enhances the traditional malware detection methodology. Table 2 represents the performance metrics of PE malware classification using JSON Features and Image-Based Representations.

6 Conclusion and Future Scope

Malware detection in Windows-based environments has usually been relying on static analysis techniques, where the executable files are examined without execution to extract some meaningful features. This includes analyzing the metadata from Portable Executable (PE) files like headers, imported DLLs, function names, section sizes, and the entropy measures. The extracted static features are structured, numerical, or categorical, and they are used as input to standard machine learning algorithms like Random Forests and MultiLayer Perceptrons (MLPs), and also to deep learning models like Convolutional Neural Networks (CNNs).

The models are then trained to classify a given file as being either benign or malicious by learning patterns present in features that were extracted from the labeled datasets. Despite being efficient and safe—since the files are never executed—static analysis is often inadequate against sophisticated malware that uses code obfuscation, encryption, or polymorphic transformations to hide its true intent. As a result, purely static models may suffer from reduced accuracy in real-world scenarios where malware constantly evolves to evade detection.

This work presents a hybrid methodology for the classification of malware and benign files that integrates the structured feature analysis with the image-based deep learning using the MobileNetV2 base model and utilizing transfer learning. The results show us that combining multiple modalities of feature representation will enhance the ability of the models to detect malware, especially against the evasive and obfuscated samples. The model architecture supports adaptability and scalability, so it is more suitable for deployment in real-time and large-scale systems. For future work, certain enhancements aspects can be explored. Incorporating some dynamic analysis features like API call sequences or network behaviors can further help increase the efficiency of the model input. Additionally, experimenting with other deep learning models and the data augmentation techniques can yield further gains in the performance. Developing some explainability tools for both structured and image-based predictions would also help in the investigation of threats and compliance [1]. Ultimately, expanding this hybrid model into a real-world malware detection model holds promise for a more proactive and accurate detection and defense system.

References

1. Azeem M, Khan D, Iftikhar S, Bawazeer S, Alzahrani M (2024) Analyzing and comparing the effectiveness of malware detection: a study of machine learning approaches. Heliyon 10(1). https://doi.org/10.1016/j.heliyon.2023.e23574
2. Song Y, Zhang D, Wang J, Wang Y, Wang Y, Ding P (2025) Application of deep learning in malware detection: a review. J Big Data 12(1). https://doi.org/10.1186/s40537-025-01157-y
3. Shahnawaz M, Gond BP, Mohapatra DP (2025) Dynamic malware classification of windows PE files using CNNs and greyscale images derived from runtime API call argument conversion, May 2025 [Online]. Available http://arxiv.org/abs/2505.24231
4. Kale G, Bostancı GE, Çelebi FV (2024) Evolutionary feature selection for machine learning based malware classification. Eng Sci Technol Int J 56. https://doi.org/10.1016/j.jestch.2024.101762
5. Maniriho P, Mahmood AN, Chowdhury MJM (2024) A systematic literature review on Windows malware detection: techniques, research issues, and future directions. J Syst Softw 209. https://doi.org/10.1016/j.jss.2023.111921
6. Brown A, Gupta M, Abdelsalam M (2024) Automated machine learning for deep learning based malware detection. Comput Secur 137. https://doi.org/10.1016/j.cose.2023.103582
7. Ferdous J, Islam R, Mahboubi A, Islam MZ (2025) A survey on ML techniques for multi-platform malware detection: securing PC, mobile devices, IoT, and cloud environments, 01 Feb 2025, Multidisciplinary Digital Publishing Institute (MDPI). https://doi.org/10.3390/s25041153
8. Mustafa Majid AA, Alshaibi AJ, Kostyuchenko E, Shelupanov A (2023) A review of artificial intelligence based malware detection using deep learning. Mater Today Proc 80:2678–2683. https://doi.org/10.1016/J.MATPR.2021.07.012
9. Hasan R, et al. (2025) Enhancing malware detection with feature selection and scaling techniques using machine learning models. Sci Rep 15(1). https://doi.org/10.1038/s41598-025-93447-x
10. Singh S, Krishnan D, Vazirani V, Ravi V, Alsuhibany SA (2024) Deep hybrid approach with sequential feature extraction and classification for robust malware detection. Egypt Inf J 27:100539. https://doi.org/10.1016/J.EIJ.2024.100539
11. Gopinath M, Sethuraman SC (2023) A comprehensive survey on deep learning based malware detection techniques. Comput Sci Rev 47:100529. https://doi.org/10.1016/J.COSREV.2022.100529
12. Syeda DZ, Asghar MN (2024) Dynamic malware classification and API categorisation of Windows portable executable files using machine learning. Appl Sci (Switz) 14(3). https://doi.org/10.3390/app14031015
13. Joyce RJ, Raff E, Nicholas C, Holt J (2023) MalDICT: benchmark datasets on malware behaviors, platforms, exploitation, and packers, Oct 2023, [Online]. Available http://arxiv.org/abs/2310.11706
14. Yousuf MI, Anwer I, Shakir T, Siddiqui M, Shahid M (2022) Multi-feature dataset for windows PE malware classification, Oct 2022, [Online]. Available http://arxiv.org/abs/2210.16285
15. Harang R, Rudd EM (2020) SOREL-20M: a large scale benchmark dataset for malicious PE detection, Dec 2020 [Online]. Available http://arxiv.org/abs/2012.07634

16. Catak FO, Yazı AF (2019) A benchmark API call dataset for windows PE malware classification, May 2019 [Online]. Available http://arxiv.org/abs/1905.01999
17. Anderson HS, Roth P (2018) EMBER: an open dataset for training static PE malware machine learning models, Apr 2018 [Online]. Available http://arxiv.org/abs/1804.04637
18. Noever DA, Miller Noever SE. VIRUS-MNIST: a benchmark malware dataset
19. nross12/PEVuln: PEVuln: a benchmark dataset for using machine learning to detect vulnerabilities in PE malware. Accessed 30 June 2025 [Online]. Available https://github.com/nross12/PEVuln/
20. Ross N, Olukoya O, Carlin J (2024) PEVuln: a benchmark dataset for using machine learning to detect vulnerabilities in PE malware, 2024 [Online]. Available https://pure.qub.ac.uk/en/persons/domhnall-carlin
21. Bensaoud A, Kalita J (2024) Deep multi-task learning for malware image classification. https://doi.org/10.1016/j.jisa.2021.103057

A Lightweight Attention-Enhanced Deep Learning Framework for Malware Detection in IoT: A Comparative Study of Structured and Image-Based Data Representations

Selvaganapathy Shymala Gowri, Vinayakumar Ravi, and A. Reveetha

Abstract The runaway growth in IoT and IIoT devices has revolutionized sectors like healthcare and manufacturing while increasing the attack surface for cyber-attacks. The connectivity explosion highlights the need for effective, IoT-focused intrusion detection and mitigation systems (IDMS). At their core is having access to realistic, heterogeneous datasets that reflect actual network behaviour and attack patterns. Recent data sets cater to these requirements by providing increased protocol support, device heterogeneity, and diverse attack settings, which enhances both federated and centralized IDMS research. In order to overcome existing scalability and interpretability issues, a new lightweight deep learning framework is presented that relies on the use of autoencoders for compressing features and attention mechanisms to bring emphasis to essential patterns in IoT traffic. A compact multi-layer perceptron is used to classify with less computational expense suitable for deployment at the edge. The model is evaluated on both structured and image data formats, showing high accuracy in malware detection with performance sufficient for real-time IoT intrusion detection.

Keywords Internet of Things (IoT) · Malware classification · Autoencoder · Attention mechanism · Lightweight neural network

S. Shymala Gowri (✉) · A. Reveetha
Department of Information Technology, PSG College of Technology, Coimbatore, India
e-mail: ssg.it@psgtech.ac.in

A. Reveetha
e-mail: 24pb37@psgtecg.ac.in

V. Ravi
Center for Artificial Intelligence, Prince Mohammad Bin Fahd University, Khobar, Saudi Arabia
e-mail: vravi@pmu.edu.sa

G. H. L. et al. (eds.), *Reliability in Cyber-Physical Systems: The Human Factor Perspective*, Springer Series in Reliability Engineering,
https://doi.org/10.1007/978-3-032-09917-4_16

1 Introduction

The inclusion of smart sensing, autonomous control and cloud connectivity that defines Internet of Things (IoT) and Industrial Internet of Things (IIoT) systems has changed the way modern systems interact, monitor and respond to their environments. As intelligent infrastructures become embedded into essential areas of our lives, performing well and securely has become a research focus. The evolving threat landscape continues to necessitate developing adaptive, context-aware IDMS that meet the needs of IoT environments. One of the challenges to develop effective IDMS solutions is the absence of appropriately complete and representative datasets that simulate IoT behaviour and types of attacks.

To tackle this issue, several benchmark datasets have been developed. The MU-IoT dataset [1] includes over 34.8 million records from a smart lab testbed, covering network and application-layer traffic, 121 features, and 16 types of attacks, such as DDoS, botnets, and injection attempts. It also features machine learning baselines using models like LSTM, CNN, and Random Forest for performance comparison. For behaviour-based detection, IoTGeM [2] uses the UNSW-NB15 dataset to train generalizable models that perform well in different IoT attack scenarios, showing the value of cross-context learning. In consumer IoT [3] it is also examined adversarial attacks, this time focusing on AI-based malware classifiers for consumer IoT, as well as the need for explainable and robust models. Moreover, Edge-IIoTset [4] supports both centralized and federated learning methods and contains telemetry, flow data, and 14 attack scenarios from an industrial testbed. It allows evaluation of distributed detection models that protect device-level privacy while maintaining accuracy.

The ECU-IoFT dataset [5] is an example of a UAV-selection approach, with Wi-Fi based cyberattacks against drones in a controlled academic environment. Its focus on telemetry data from devices like DJI Tello helps develop lightweight, federated intrusion models aimed at threats such as deauthentication and command injection. In the wireless domain, AWID3 [6] provides a large-scale dataset containing features for 802.11 and non-802.11 based approaches to application-layer attack detection that can enhance intrusion detection systems (IDS) for smart home and enterprise systems. In the case of SDN-based IoT systems, InSDN [7] captures attacks at the path level and controller and switch interactions, using flow-based monitoring records; this dataset has been useful in training federated models for dynamic network environments, such as denial-of-service, probing, identification services, and brute-force attacks.

Thus, these datasets and approaches allow for intelligent, scalable, and context-aware IDS to be built for the heterogeneous and ever-changing context of the IoT and IIoT datasets for these heterogeneous and ever-changing IoT and IIoT network contexts. The rest of the article is organized as follows, Sect. 2 discusses the literature survey of recent malware research work, and Sect. 3 provides a detailed description of the dataset with the exploratory data analysis, followed by Sect. 4 which presents the proposed methodology in a detailed manner. Section 5 contains the results and analysis of the results. Finally, Sect. 6 contains the conclusion of the paper and offers directions for future work.

2 Literature Survey

Recent studies on IoT/IIoT intrusion detection employ various machine learning approaches across domains such as edge computing, federated learning, UAV security, Wi-Fi IDS, and SDN networks. These works emphasize effective feature extraction, tailored attack modelling, and a mix of traditional, deep, and federated models. Transfer learning methods have been studied for image-based malware classification in IoT, where structured network data is represented as grey-scale images and analyzed using pre-trained deep CNNs to enhance detection accuracy and generalization across threat types [8]. Table 1 provides a summary of the datasets utilized, techniques implemented, and findings of the papers model in between Scouting the Current Landscape and What are the Gaps. The overview presented in the tab offers context for the conversation about scalability, explainability, adaptability, presents shortcomings, and illustrates opportunities for proposing new special purpose IDMS frameworks, particularly in the context of IoT.

Several recent studies have introduced specialized datasets and detection methods to improve cybersecurity in IoT and IIoT environments. The MU-IoT dataset focused on identifying threats at both the network and application layers using LSTM, CNN, and Random Forest models. It addressed attacks such as SQL injection and data theft while showing the need for edge-based, privacy-protecting intrusion detection systems [1]. The IoT GeM framework focused on model generalizability by using behaviour-based modelling with datasets like UNSW-NB15. It showed effective detection in various environments using deep learning techniques [2]. Edge-IIoTset combined telemetry, logs, and protocol-level data to support both centralized and distributed learning for identifying MITM, injection, and scanning attacks in industrial IoT settings [4]. For UAV-specific situations, the ECU-IoFT dataset captured telemetry and Wi-Fi-based attacks on drones like the DJI Tello, allowing the creation of lightweight detection models such as LSTM and SVM suitable for distributed environments [5]. In the wireless domain, AWID2 used 802.11 MAC and physical-layer features to detect impersonation and flooding attacks in limited IoT devices [6]. Meanwhile, InSDN provided flow-level and controller-switch interaction data to enable detection of DDoS, probing, and brute force attacks in SDN-based IoT networks through distributed learning [7].

Malware classification in these environments has improved greatly using deep learning techniques. They effectively recognize complex patterns in high-dimensional traffic data. Deep learning reviews [9] confirm that structures like CNNs, LSTMs, and Autoencoders are leading the field by enabling hierarchical feature extraction from large datasets. Although traditional models like Decision Trees and

Table 1 Literature survey on recent IoT/IIoT intrusion detection systems and datasets

S. No.	Title (research work)	Feature type/ Focus area	Dataset + Sample size + Analysis type	Tools/Models used/Platforms used	Research findings	Future Scope/gaps
1	MU-IoT: A new IoT intrusion dataset for network and application layer attacks analysis [1]	IoT Network & Application Layer IDS	MU-IoT (34.8 M + flows, 15.8 GB CSVs)—Supervised—Flow-based/ Protocol-specific	LSTM, CNN, RF/ CICFlowMeter, Custom IoT Testbed	Detects network and app-layer IoT attacks (e.g., DDoS, botnets) with high accuracy	Enhance detection of new IoT attack types; improve edge privacy-preserving models
2	IoT GeM: Generalizable Models for Behaviour-Based IoT Attack Detection [2]	Behaviour-Based IDS for IoT	UNSW-NB15 (2.54 M samples)—Supervised—Flow/ Time/Content-based	LSTM, CNN, RF/ IXIA, Argus, Bro	Behaviour-based generalized models improve real-world IoT threat detection	Need dynamic model updates and real-time generalization to emerging threats
3	Edge-IIoTset: A Realistic Dataset for Centralized and Federated Learning [4]	IIoT Security, Federated IDS	Edge-IIoTset (600 k + samples)—Federated—Flow, Logs, Telemetry	CNN, RF, DT/ Wireshark, MITMproxy, Python	Enables intrusion detection and anomaly classification in IIoT via federated learning	Improve FL algorithms, cover more attack types, integrate newer IoT protocols
4	ECU-IoFT: Dataset for Cyber-Attacks on Internet of Flying Things [5]	UAV Cyberattack Detection	ECU-IoFT (54,566 UAV data points)—Anomaly Detection—Telemetry + PCAP	LSTM, CNN, SVM/DJI SDK, Wireshark, Scapy	Detects UAV-targeted attacks like Deauth, Tello API exploit with federated models	Expand to include UAV-specific zero-day attacks; enhance real-time detection

(continued)

Table 1 (continued)

S. No.	Title (research work)	Feature type/ Focus area	Dataset + Sample size + Analysis type	Tools/Models used/Platforms used	Research findings	Future Scope/gaps
5	Detecting Application Layer Attacks via 802.11 Features [6]	Wireless IDS, 802.11 Analysis	AWID2 (Millions of frames)—Supervised—MAC + RadioTap Features	SVM, RF, CNN/ Wireshark, Tshark, Aircrack-ng	Wireless IDS using MAC layer and custom 802.11 features; detects injection & flooding	Add dynamic attack types; enhance 802.11 feature extraction and model tuning
6	InSDN: A Novel SDN Intrusion Dataset [7]	SDN-based Federated IDS	InSDN (343,939 flows)—FL-based—SDN-specific Flows	DNN, RF, KNN, LSTM/Mininet	Achieves strong SDN security using flow-based FL; high attack detection accuracy	Adapt to dynamic SDN environments

Random Forests have been assessed using datasets like TON_IoT, their effectiveness often hinges on addressing class imbalance using strategies like random undersampling or SMOTE [10]. CNNs, in particular, are very effective at identifying malware variants in network traffic by extracting spatial flow-based features. In fact, CNNs outperform traditional classifiers with regard to precision and recall metrics [11]. Transfer learning has emerged as a good technique to transform structured traffic data into grayscale image formats, which can then be processed utilizing pre-trained CNNs like ResNet18, and VGG16. Transfer learning has significantly improved the detection task for many types of malware, and even zero-day threats [8, 12].

Classification accuracy can also be improved when utilizing hybrid frameworks. For example, multi-scale temporal feature learning involving Bi-LSTM and SPP-Net has improved the ability to detect anomalies within smart environments [13]. MalBoT-DRL is based on techniques from the world of reinforcement learning, and dynamically identifies botnet behavior, and continuously adapts to the actions and strategies of attackers by acting agent-environment interactions [14]. Lightweight models like HoleMal cut down on computational demands by efficiently extracting host-level features without losing classification performance [15]. Distributed learning methods have also become popular because they allow decentralized training across IoT nodes. This approach addresses privacy issues and improves security, as evidenced by datasets like Edge-IIoTset and InSDN [4, 7]. Additionally, Threat Trekker uses Random Forest classifiers. It combines proactive threat hunting with under-sampling and over-sampling techniques to achieve high detection accuracy for multiple classes and identify attacks in their early stages [16]. Ensemble models that integrate Random Forest and XGBoost, along with statistical features and synthetic oversampling, have been effective in detecting rare types of malware [17]. Ongoing surveys assess these advancements. They outline strategies and discuss the challenges of using deep learning models for real-world IoT malware defense [18, 19].

A. **Gaps Identified**

Despite major advances, current methods for IoT malware classification have several weaknesses. Many CNN-based solutions, while effective, require significant computation and are not suitable for use on limited edge devices found in IoT ecosystems [20]. Transfer learning approaches that contain image representations such as deep learning, may enhance accuracy but can decrease interpretability and require models that are very large size, which can affect real-time use [8, 12]. MalBoT-DRL Could utilize reinforcement learning as it also has its limitations when attempting to learn with training stability, converging slowly, and needing large volumes of interaction data that may be difficult and possibly unsafe to gather, especially in IoT contexts [14].

Some hybrid models, such as Bi-LSTM with SPP-Net, have a great ability to capture temporal and spatial connections, however can also bring additional computational cost and complexity. More computational costs and complexity may depict if the real-time detection will be delayed [13]. Lightweight models like HoleMal, although optimized, may struggle to detect more complex polymorphic malware since they depend on host-level features alone [15]. Ensemble models can improve detection rates but may overfit smaller classes if not fine-tuned properly, and they can consume a lot of memory [17]. The nature of conventional models converting categorical variables to numerical values can result in complete loss of information that affects classification when dealing with imbalanced datasets such as TON_IoT [10].

Federated Learning models have to deal with the non-IID issues, synchronization, and poison attacks from compromised nodes [4, 7]. Threat Trekker, while performing well with static datasets, may not be adaptable to an evolving threat environment since the model relies on fixed features [16]. Explainability AI methods are being developed in the cybersecurity models, and many are black box models which make it hard to validate predictions under critical situations [3, 18]. Overall, the literature continues to emphasize clear, scalable, and effective solutions that balance performance and efficiency in the real world IoT cases [9, 19].

3 Dataset Description

The ToN_IoT [21] datasets created by UNSW Canberra at ADFA seek to further cybersecurity research in IoT and IIoT environments by facilitating AI-based solutions, such as intrusion detection systems (IDS) and fraud detection. Our datasets were collected from a large testbed that simulated real-world IoT and IIoT networks. The dataset comprises data collected from virtual and physical devices across the IoT, Edge/Fog and Cloud. The devices used various operating systems and sensors. It comprises five main parts: raw datasets (including telemetry, network traffic, and audit logs), processed datasets (which contain cleaned and labelled data), train-test datasets (for evaluating models), descriptive statistics (metadata and explanations of features), and ground truth security event logs (which provide precise attack labels and timestamps). This makes the ToN_IoT datasets a valuable resource for creating and testing AI-based cybersecurity models for modern connected environments.

A. **Exploratory Data Analysis**

Figure 1 shows the exploratory data analysis of the network traffic dataset, which includes 667,300 samples and 49 features. This dataset is useful for network intrusion detection, capturing details like IP addresses, protocols, and flow sizes. The correlation matrix in Fig. 1 points out multicollinearity and redundancy among numerical features. This information helps in optimizing feature selection and simplifying the dataset. Although there are complications from parsing column headers incorrectly, which can cause names to show up as data values or labels, the dataset is structurally sound and can be used for machine learning after preprocessing. Preprocessing involves renaming columns, removing unnecessary fields, and encoding categorical data. Once the dataset is pre-processed, it should be possible to build effective predictive models with it.

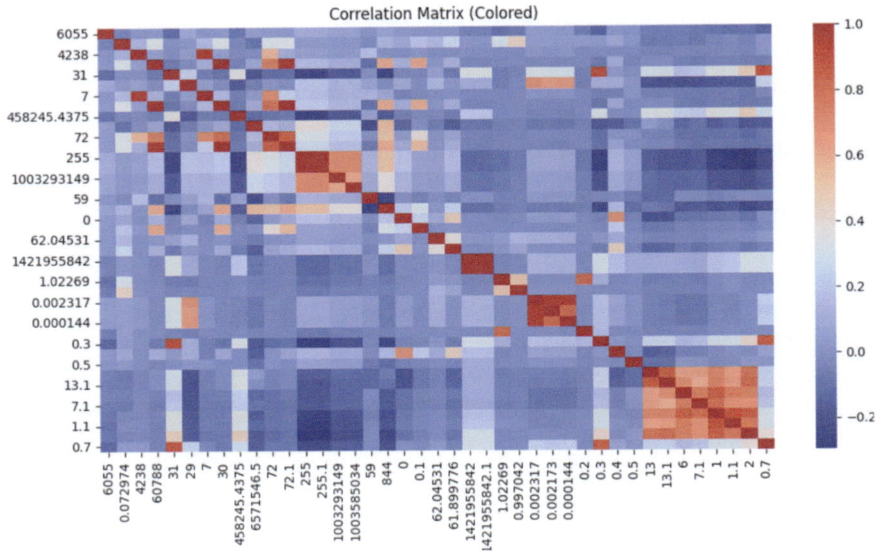

Fig. 1 Correlation matrix heatmap of dataset features

4 Proposed Methodology

To reduce limitations on current malware classification, we recommend a lightweight model that includes attention mechanisms and autoencoders. This approach enables efficient feature learning and cuts down dimensionality while keeping important patterns in IoT traffic and decreasing the computational load. Attention mechanisms improve classification accuracy, especially for rare attacks. They also make the model easier to understand. We use a small multi-layer perceptron (MLP) rather than heavier models like convolutional neural networks (CNNs) or ensembles. This choice makes it better suited for an IoT environment. SMOTE addresses class imbalance and enhance performance without losing generalization. This method provides a clear and scalable solution for both binary and multi-class malware detection.

Figure 2 depicts the comprehensive structure of the proposed AEL-MalNet model for malware classification in the IoT environment. The starting point is the UNSW-NB15 dataset, which contains labelled network traffic in the form of raw CSV. After some preliminary preprocessing, which consists of label encoding and feature extraction, we split the data into training and testing sets or our models. Next, to improve some of the models' predictive abilities, especially for rare attack instances, we used the Synthetic Minority Oversampling Technique (SMOTE) to increase the number of samples of the minority class: the attacks. Then, we processed the data through an autoencoder and removed some feature dimensions while retaining the key characteristics of the traffic. Attention mechanisms are applied next to highlight important features before the data is fed into a lightweight Multi-Layer Perceptron (MLP) classifier. This efficient approach allows the model to make reliable predictions with

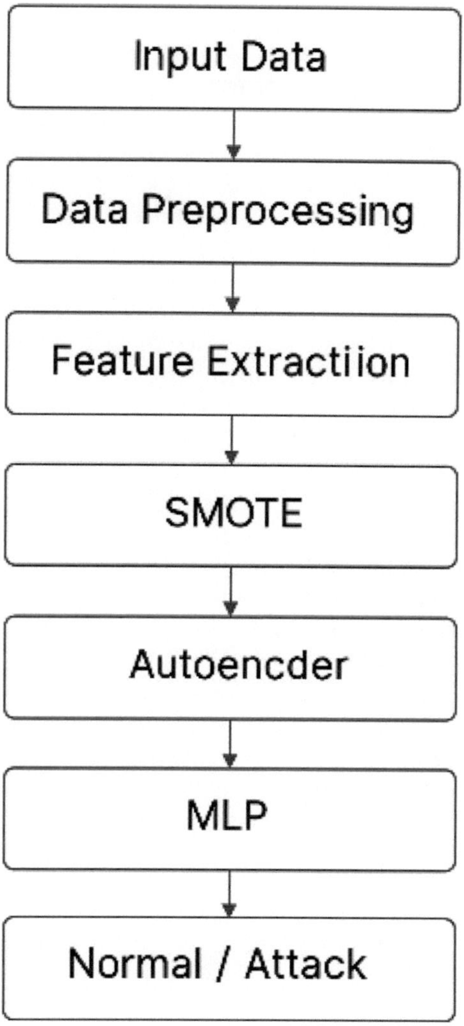

Fig. 2 Flow diagram of the proposed architecture for malware classification

low computational cost. The system ultimately provides a binary classification that shows whether a traffic instance is benign (Normal) or malicious (Attack), aiding in real-time intrusion detection in resource-limited IoT environments.

5 Result Analysis

To evaluate how data formats affected the classification of malware, we investigated two methods: raw CSV input classified by an Attention-Enhanced Lightweight Classifier, and a conversion of the original dataset to grayscale images that classified the images to the ResNet18 architecture. Performance was evaluated, using "precision", "recall", "F1 score", and "accuracy" for each of the 'Normal' and 'Attack' classes.

Table 2 shows the results from the UNSW-NB15 dataset. The Attention-Enhanced Lightweight Malware Classifier produced an overall accuracy of 91%. Its precision and recall scores for classifying normal instances correctly were 0.93 and 0.90, respectively. For attack instances, the scores were 0.90 (precision) and 0.92 (recall). ResNet18 achieved a higher overall accuracy of 93%, with precision and recall scores of 0.99 for normal instances, and 0.95 for attack instances. ResNet18 also scored higher than the AEL-MalNet in the class-wise and aggregate macro average and weighted average scores, with 0.97 and 0.99, respectively while AEL-MalNet maintained a overall score of 0.91, and its lowest aggregate score at also 0.91, (primarily in normal instance detection).

Figure 3 depicts the performance of Attention-Enhanced Lightweight Malware Classifier using raw CSV inputs. In summary, the model correctly classified 116,505 normal instances and 120,136 attack instances but falsely classified 12,932 normal instances as attacks (false-positives) and falsely classified 9,328 attack instances as normal (false-negatives). This shows that the model has a balanced detection capability, with a precision of 0.93 for normal traffic, 0.90 for attack traffic, and an overall accuracy of 91%. The matrix shows that while the model performs well, it slightly tends toward over-predicting attacks, as indicated by the higher number of false positives.

Figure 4 shows the confusion matrix for the ResNet18 model. This model processes IoT traffic by converting structured CSV data into grayscale images. It identified 128,855 normal instances and 10,000 attack instances. It also recorded only 523 false positives and 495 false negatives. These results demonstrate ResNet18's strong classification performance. It achieved precision and recall scores of 0.99 for

Table 2 Performance comparison of malware classification methods using raw CSV and CSV as image representations

Model	Class	Precision	Recall	F1-score	Support
Data format	Raw CSV				
Attention-enhanced lightweight malware classifier	Normal	0.93	0.90	0.91	129,437
	Attack	0.90	0.93	0.92	129,464
	Accuracy	–	–	0.91	258,901
	Macro average	0.91	0.91	0.91	258,901
	Weighted average	0.91	0.91	0.91	258,901
Data format	CSV as Image				
ResNet18	Normal	0.99	0.99	0.99	129,350
	Attack	0.95	0.95	0.95	10,523
	Accuracy	–	–	0.93	139,873
	Macro average	0.97	0.97	0.97	139,873
	Weighted average	0.99	0.99	0.99	139,873

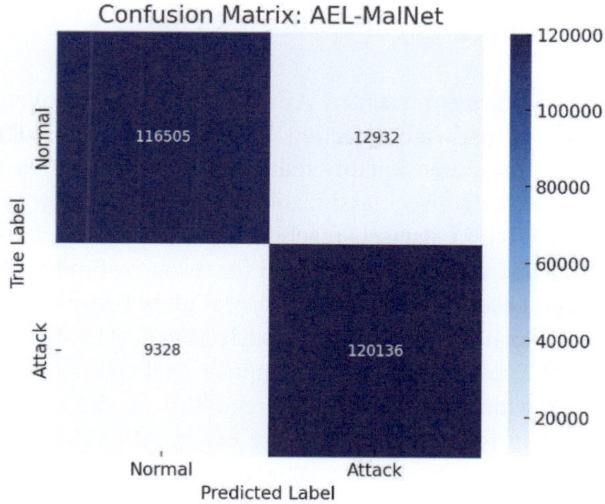

Fig. 3 Confusion matrix for attention-enhanced lightweight malware classifier model

normal traffic and 0.95 for attack traffic, leading to an overall accuracy of 93%. The matrix confirms the model's strong generalization and low misclassification rate, particularly in distinguishing benign activity.

These results also illustrate that although ResNet18 is computing vastly superior as raw accuracy and precision, AEL-MalNet can still be a very viable consideration based on its cheaper model and computing cost.

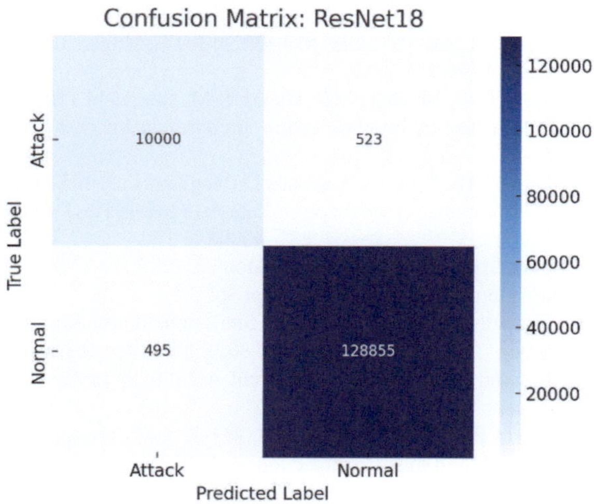

Fig. 4 Confusion matrix for ResNet18 model

6 Conclusion and Future Scope

This work has presented AEL-MalNet as a lightweight, simple and interpretable model of malware detection designed within an IoT context. It combines autoencoders for dimensionality reduction with an attention-enhanced multi-layer perceptron for effective classification. The model showed strong performance on the UNSW-NB15 dataset, reaching 91% accuracy while keeping low computational demands, which makes it ideal for resource-limited devices. Despite the ResNet18 image-based model having a very slightly better accuracy of 93%, the computation cost was higher. But as previously stated, AEL-MalNet can strike a good balance between performance and computational cost, as well as a good attention mechanism, which helps with interpretability, as you can see what salient features were found in the input. In future research, AEL-MalNet will be extended to incorporate online learning for adaptation to evolving attack patterns. It shall be evaluated on other IoT datasets such as TON_IoT and Edge-IIoTset to ensure its generalizability. Future enhancements shall involve utilizing explainable AI methods in order to make decision-making more understandable and employing a federated learning framework for preserving privacy in malware detection across various IoT nodes.

References

1. Clinton UB, Hoque N (2024) MU-IoT: a new IoT intrusion dataset for network and application layer attacks analysis. IEEE Access 166068–166092
2. Kostas K, Just M, Lones MA (2023) IoTGeM: generalizable models for behaviour-based IoT attack detection. arXiv
3. Yang C-H, Mwangi MB, Cheng S-M, Lee H-M (2025) An adversarial attack on artificial intelligence malware detection in consumer internet of things. IEEE Consum Electron Mag 53–66
4. Ferrag MA, Friha O, Hamouda D, Maglaras L, Janicke H (2022) Edge-IIoTset: a new comprehensive realistic cyber security dataset of IoT and IIoT applications for centralized and federated learning. IEEE Access 40281–40306
5. Ahmed M, Cox D, Simpson B, Aloufi A (2022) ECU-IoFT: a dataset for analysing cyber-attacks on internet of flying things. Appl Sci
6. Chatzoglou E, Kambourakis G, Smiliotopoulos C, Kolias C (2022) Detecting application layer attacks through 802.11 and non-802.11 features. Sensors
7. Elsayed MS, Le-Khac NA, Jurcut AD (2020) InSDN: a novel SDN intrusion dataset. IEEE Access 165263–165284
8. Panda P, CU OK, Marappan S, Ma S, S M, Veesani Nandi D (2023) Transfer learning for image-based malware detection for IoT. Sensors (Basel)
9. Song Y, Zhang D, Wang J, Wang Y, Wang Y, Ding P (2025) Application of deep learning in malware detection: a review. J Big Data
10. Kaddour H, Das S, Bajgai R, Sanchez A (2024) Evaluating the performance of machine learning-based classification models for IoT intrusion detection. In: IEEE opportunity research scholars symposium (ORSS), Atlanta, GA, USA
11. Li Q, Mi J, Li W, Wang J, Cheng M (2021) CNN-based malware variants detection method for internet of things. IEEE Internet Things J 8(23):16946–16962

12. Bendiab G, Shiaeles S, Alruban A, Kolokotronis N (2020) IoT malware network traffic classification using visual representation and deep learning. In: 6th IEEE Conference on network softwarization (NetSoft)
13. Jeon J, Jeong B, Baek S, Jeong YS (2022) Hybrid malware detection based on Bi-LSTM and SPP-Net for smart IoT. IEEE Trans Ind Inform 4830–4837
14. Al-Fawa'reh M, Abu-Khalaf J, Szewczyk P, Kang JJ (2023) MalBoT-DRL: malware botnet detection using deep reinforcement learning in IoT networks. IEEE Internet Things J 11(6):9610–9629
15. Chen Z, Xia W, Li Z, Xiong G, Gou G, Zhang H, Li H, Xiao J (2025) HoleMal: a lightweight IoT malware detection framework based on efficient host-level traffic processing. Comput Secur
16. Casanova Bienzobas Á, Sánchez-Macián A (2023) Threat trekker: an approach to cyber threat hunting, p 11
17. Yang Z, Liu X, Li T, Wu D, Wang J, Zhao Y, Han H (2022) A systematic literature review of methods and datasets for anomaly-based network intrusion detection. Comput Secur 116:102675
18. Darwish R, Abdelsalam M, Khorsandroo S (2024) Deep learning based XIoT malware analysis: a comprehensive survey, taxonomy, and research challenges. Cryptography and Security
19. Chen Y, Guo Y, Gao Y, Liu B (2025)A novel lightweight deep learning framework using enhanced pelican optimization for efficient cyberattack detection in the Internet of Things environments. J Eng Appl Sci
20. Jeon J, Park JH, Jeong YS (2020) Dynamic analysis for IoT malware detection with convolution neural network model. IEEE Access
21. U. C. a. ADFA The TON_IoT Datasets, 2021 [Online]. Available: https://research.unsw.edu.au/projects/toniot-datasets

Machine Learning for Cyber-Physical Systems: A Short Survey

Moez Krichen

Abstract This chapter focuses on the application of Machine Learning (ML) for cyber-physical systems. These systems consist of both software and hardware components that interact with their physical surroundings. Using these systems can present a number of security, reliability, and performance concerns. In this context, the application of ML could be a viable approach for ensuring these critical features. This chapter focuses specifically on the application of ML to ensure the appropriate operation of CPSs in unexpected contexts and scenarios. We begin by offering an overview of both CPSs and ML. Next, we outline the benefits of employing ML for CPSs and provide two particular examples. The first example is related to a road state scanning application while the second one is about the detection of electricity theft in a smart grid. We then outline the key issues and constraints in this scenario. Finally, we provide some recommendations and guidance for ML users in CPSs, as well as future directions in this subject.

Keywords Cyber-physical systems (CPS) · Machine learning (ML) · Advantages · Limitations · Future directions

1 Introduction

Cyber-physical systems (CPS) are widely used in modern daily life, including healthcare, transportation, industry, and agriculture [34]. They contain software and hardware components that interact with their physical environments via sensors, actuators, and particular control algorithms [25]. For example, in the transportation industry, CPS provides driver assistance technologies that help lower the chance of accidents

---

M. Krichen (✉)
Al-Baha University, Al-Baha, Saudi Arabia

ReDCAD Laboratory, National School of Engineers of Sfax, University of Sfax, Sfax, Tunisia
e-mail: moez.krichen@redcad.org
URL: https://www.redcad.org/members/mkrichen/

G. H. L. et al. (eds.), *Reliability in Cyber-Physical Systems: The Human Factor Perspective*, Springer Series in Reliability Engineering,
https://doi.org/10.1007/978-3-032-09917-4_17

while increasing vehicle efficiency. CPSs are also useful in the healthcare area, allowing for continuous monitoring of patient states and prompt intervention in the event of medical emergency.

The complexity of these systems has repercussions for cyber security [14]. CPSs can pose substantial security, reliability, and performance problems. Several catastrophic incidents can occur while using CPSs, ranging from the loss of crucial data for a large number of users to the possibility of damaging calamities and human fatalities. As a result, it is critical to constantly monitor the appropriate operation of these systems in order to avoid (or at least mitigate) this type of harm. At this level, the usage of Machine Learning (ML) may provide a sufficient and effective answer.

The purpose of this chapter is to explore the various uses of ML for CPSs while highlighting their advantages, potentials, and limitations [17, 18, 24, 26]. We study the different techniques and methodologies used in this field while identifying the challenges encountered and difficulties to overcome [1, 2, 15, 27, 31]. We also propose some recommendations for the use of ML for CPSs and provide an overview of future directions in this field [6, 7, 11, 15, 32]. In addition, we present two specific examples of the use of ML for CPS. The first example is related to road state scanning application [29] which is based on the use of mobile phones sensors and machine learning techniques. The second example is about electricity theft detection application [40] based on the use of smart meters and ML. The remainder of the chapter is organized as follows. Section 2 provides an overview of CPSs. Section 3 provides an overview of ML techniques. Section 4 focuses on the use of ML for CPSs providing two specific examples. In Sect. 5, we identify the main challenges and main future directions. Finally, Sect. 6 concludes the chapter. The structure of the chapter is illustrated in Fig. 1.

Fig. 1 Structure of the chapter

2 Overview of Cyber-Physical Systems

CPSs are now commonly employed in many aspects of our daily life. They are made up of hardware and software components that enable real-time interaction with the physical world [16]. Control components in these systems can be physical or computer-based [37]. Sensors, actuators, and control algorithms enable CPSs to operate autonomously or semi-autonomously in many dynamic situations [36]. CPSs employ specialized communication networks to communicate data and messages among system components [9]. Figure 2 illustrates the general construction of a CPS.

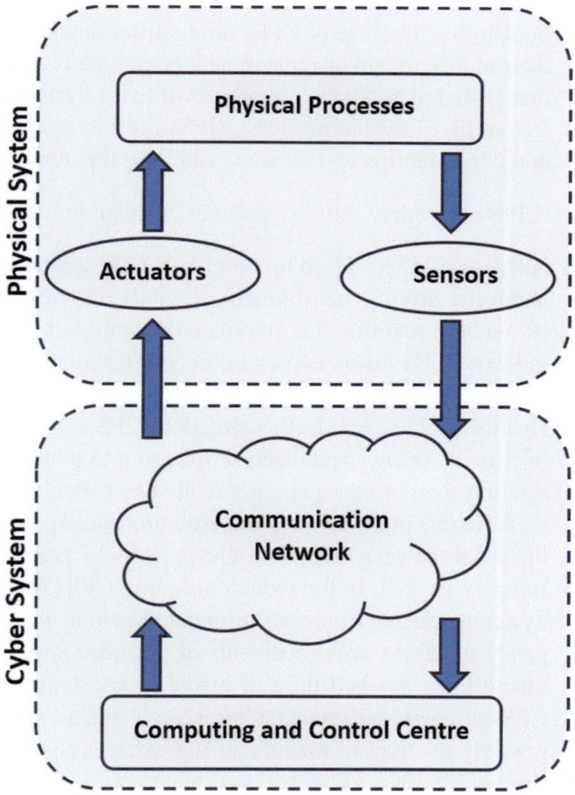

Fig. 2 CPS general structure

The main characteristics of CPSs are:

- Real-time: To guarantee safe and effective functioning, they must be able to react appropriately and within reasonable timeframes to changes in their surroundings.
- Heterogeneity: They are typically based on the application of various technologies, which may have diverse origins and be produced by different firms, and which rely on the type of sensors, devices, and computers employed.
- Interconnection: To guarantee the sharing and interchange of information and messages as well as the exchange of control orders, CPSs can make use of a number of communication networks.
- Autonomy: Another crucial feature of CPSs is their independence, which enables them to function constantly, securely, and in real-time without the assistance of outside human operators.
- Scalability: The size of CPSs might differ depending on the context, which limits their ability to add or remove new components. It also affects the volume of data that is shared and how it changes while the system is running.
- Reliability: Last but not least, CPSs need to ensure a minimal degree of security and dependability so that users can have the utmost faith in the systems they use.

CPSs are widely utilized in many different fields, such as:

- Automotive [23, 35]: In this situation, CPS sensors are used to collect information about the driving environment of smart vehicles so that suitable judgments can be made in real time for automated or semi-automatic driving. In the automotive industry, CPS errors can cause deadly accidents, resulting in both financial loss and human lives.
- Healthcare [33, 38]: In this situation, CPS systems are used to monitor the state of patients using specialized sensors and to intervene quickly when necessary by sending alerts to the patient's relatives or medical institutions that are watching them. Errors in the CPSs in this situation could possibly result in the loss of human life or, at the very least, complicate patients' health conditions.
- Industry [8, 13]: In the industrial sector, CPSs improve manufacturing processes by automating task monitoring and intervening when necessary by providing emergency orders to avoid accidents or optimize specific procedures, lowering manufacturing costs and time. If errors or accidents occur in the industrial sector's CPSs, tremendous financial harm can result, as well as endangering the safety and possibly the lives of manufacturing workers and technicians.
- Energy [28, 39]: CPSs are used in smart grids to monitor power consumption by different nodes of the grid, as well as to ensure that electricity is distributed optimally and evenly. CPSs can also help to avert catastrophic electrical accidents by sending out emergency notifications or even making emergency shutdown decisions when needed. If the CPSs do not perform well in this setting, there is a possibility of blackout or deterioration of particular equipment sensitive to electrical disturbances, resulting in significant financial losses.
- Agriculture [10, 12]: Today, agriculture has undergone a significant revolution in terms of the use of modern technologies, and CPSs have enabled better monitoring

of farmers' activities in order to ensure optimal conditions for plants and animals, resulting in cost savings, reduced damage, and increased harvests and income. In agriculture, CPS faults can damage crops or contaminate the soil, posing financial or environmental problems, among other things.

- Home automation [3, 30]: The use of CPSs in smart homes can assist the automation of some repetitive chores, hence enhancing inhabitants' quality of life, such as temperature monitoring and automated activation of the air conditioning or heating system, or the automatic opening and closing of window shutters, among others. Another example is fire detection, alarm system activation, and emergency fire extinguishment. However, these benefits can be lost if the CPSs malfunction or exhibit irregularities.

As listed (Table 1), numerous risks threaten the use of CPSs that can cause major damage in terms of financial loss or risk to human life, etc. This requires significant efforts to ensure the effective operation of CPSs, including the application of ML.

3 Overview of Machine Learning

The difference between classical programming and machine learning is mainly based on the fact that classical manual programming relies on the programmer's talents to write the source code of the programs developed while for machine learning and as its name suggests the codes and models are developed automatically using learning

Table 1 Applications of cyber-physical systems (CPSs) in various fields

Field	Description
Automotive	CPS sensors gather details about the driving environment for real-time decisions in automated driving. Errors can lead to fatal accidents and financial losses
Healthcare	CPS systems monitor patient conditions with sensors and send alerts for urgent interventions. Errors can result in loss of life or health complications
Industry	CPSs automate monitoring in manufacturing, issuing emergency orders to avoid accidents and optimize processes. Errors can lead to financial damage and risk to worker safety
Energy	CPSs monitor power consumption in smart grids and ensure balanced electricity distribution. Malfunctions can cause blackouts and financial losses
Agriculture	CPSs enhance monitoring of farming activities, optimizing conditions for plants and animals. Malfunctions can damage crops and lead to financial or environmental issues
Home Automation	CPSs automate tasks in smart homes, improving quality of life. Malfunctions can result in serious issues, such as fire hazards

techniques from progressively and iteratively collected datasets [5, 19, 20, 22]. Each technique has its own advantages and disadvantages and can be chosen according to selection criteria depending on the needs of the users and the available means. The difference between the two notions is illustrated in Figure 3.

As illustrated in Fig. 4, the machine learning lifecycle consists of the following steps:

1. Data gathering: Corresponds to the collection of data from different sources.
2. Data preparation: Data at this level is cleaned by removing unnecessary data.
3. Data wrangling: At this stage, the data is appropriately formatted to facilitate and automate its use.
4. Data analysis: The data is analyzed to maximize the value derived from its content.
5. Model training: At this level, the learning model is built from the collected and analyzed data.

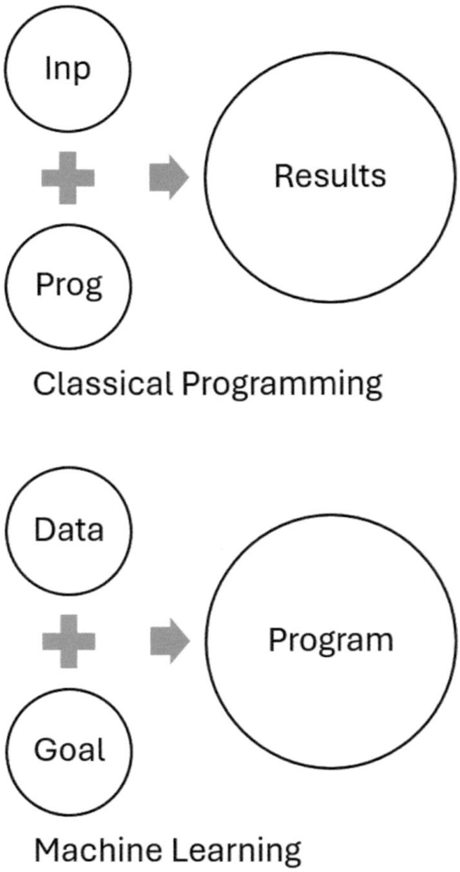

Fig. 3 Classical programming versus machine learning

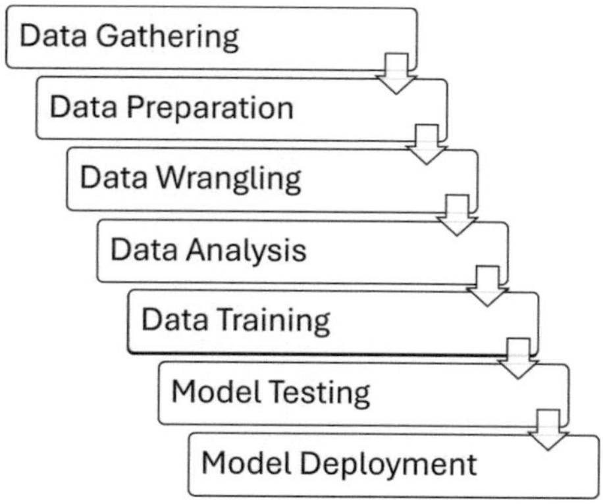

Fig. 4 Machine learning lifecycle

6. Model testing: The validity and quality of the developed model are evaluated at this level.
7. Model deployment: At this stage, the validated model is deployed in the execution environment and will be effectively used in a real-world context.

4 Integration of ML in CPS

The use of machine learning techniques for CPSs can bring considerable benefits, allowing both to improve the performance of the systems considered as well as to minimize the risks of errors and accidents. This is mainly due to the potential offered by ML techniques based on the collection of appropriate data and the production of intelligent models allowing CPSs to make optimal decisions automatically or semi-automatically. In this way, costs are minimized given the reduction of human intervention, generally synonymous with high cost and high probability of errors and imperfections. For example, in the industrial domain, ML techniques can help adapt the work process according to current conditions and create a market balance in terms of supply and demand by offering future forecasts for upcoming events and consumer behavior. Similarly, other domains can benefit from the advantages offered by ML techniques. In the following we consider two specific examples that we have developed in previous work: a first example related to the detection of anomalies in road networks and the second example related to the detedction of theft of electricity in smart grids.

4.1 Road State Scanning

In modern cities, monitoring the quality of road networks is a major concern for officials aiming to maximize citizen comfort and reduce the risk of accidents and financial and human damage. Conventional methods to achieve this goal are generally

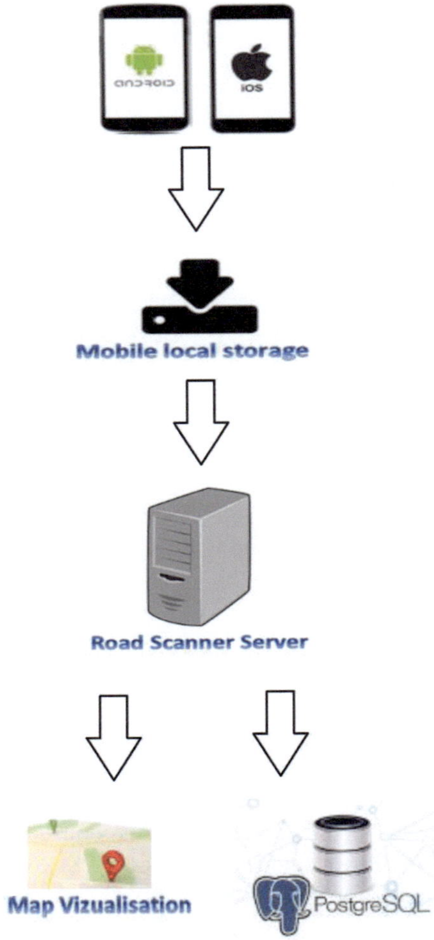

Fig. 5 Solution architecture

expensive and slow, as they require heavy equipment and a complicated process for data collection, analysis, and decision-making. To overcome these limitations, we considered using a simpler and less expensive approach based on the use of sensors integrated into mobile phones. These sensors allow us to measure the irregularities of the roads traveled by vehicles to which mobile phones are attached. Once the data is collected, ML techniques were used to analyze this data and build appropriate models to predict the quality of road networks as well as the presence of breaks or other anomalies. The results obtained were quite satisfactory for the different models adopted. Figure 5 presents the architecture of our solution. Figure 6 illustrates the work steps from data collection to model deployment. Finally, Fig. 7 presents the results obtained for the different ML models adopted. More details about this case study can be found in [21].

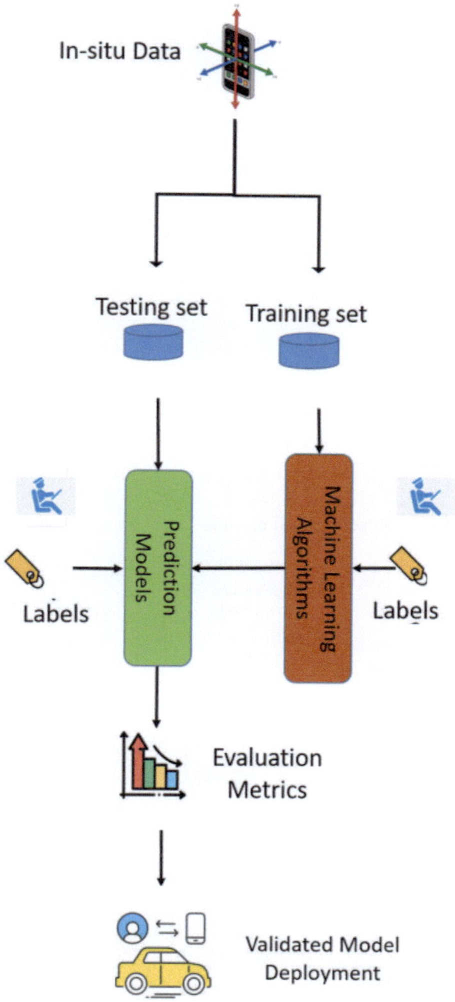

Fig. 6 Steps from data collection to model deployment

4.2 Electricity Theft Detection

Another problem encountered in modern cities is the theft of electricity in smart grids. Detecting these thefts is generally quite difficult and often requires sophisticated means and well-trained experts, hence the high cost, which could sometimes exceed the cost of the stolen electricity itself. For this, we thought of developing a new low-cost method based on machine learning techniques while allowing us to obtain acceptable and satisfactory results. More precisely, we developed a Lightweight AI-based Intrusion Detection System (IDS) designed for real-time security in smart meters. Our system was tested on Raspberry Pi, demonstrating its effectiveness

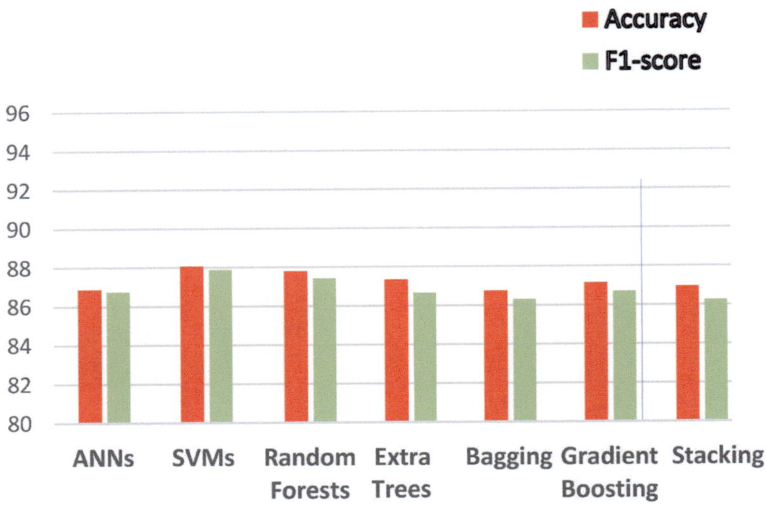

Fig. 7 Results of all models

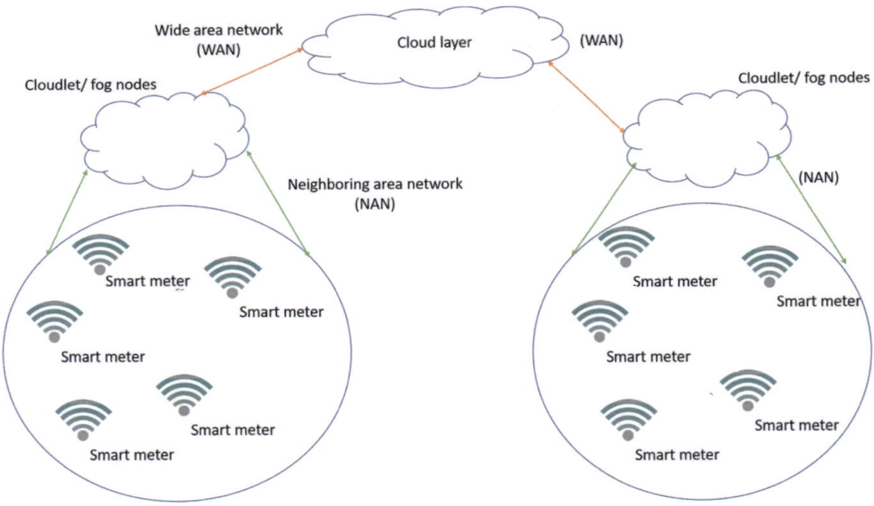

Fig. 8 Advanced metering infrastructure

even in environments and architectures with reduced storage, computing, and battery capacity. Figure 8 illustrates the metering infrastructure and Fig. 9 illustrates the block diagram. In addition, Fig. 10 represents the flow of the TensorFlow lite conversion.The results obtained were quite satisfactory both in terms of anomaly detection percentage and in terms of resource consumption and computing time. More details about this example can be found in [4].

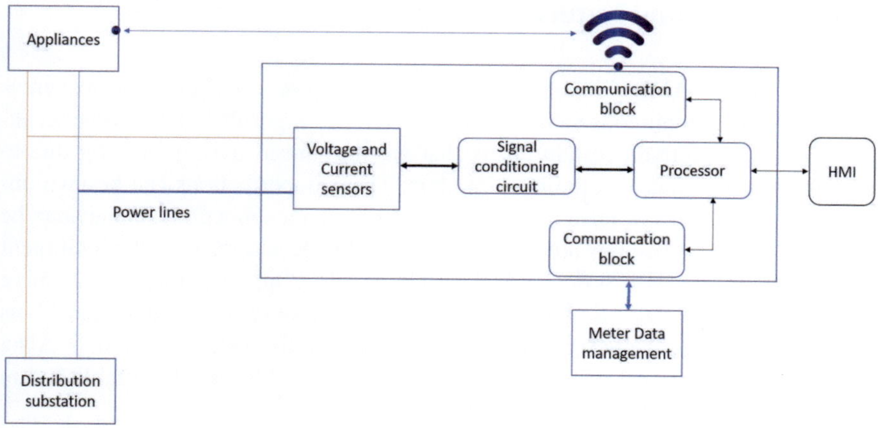

Fig. 9 Block-diagram

Fig. 10 Flow of the TensorFlow lite conversion

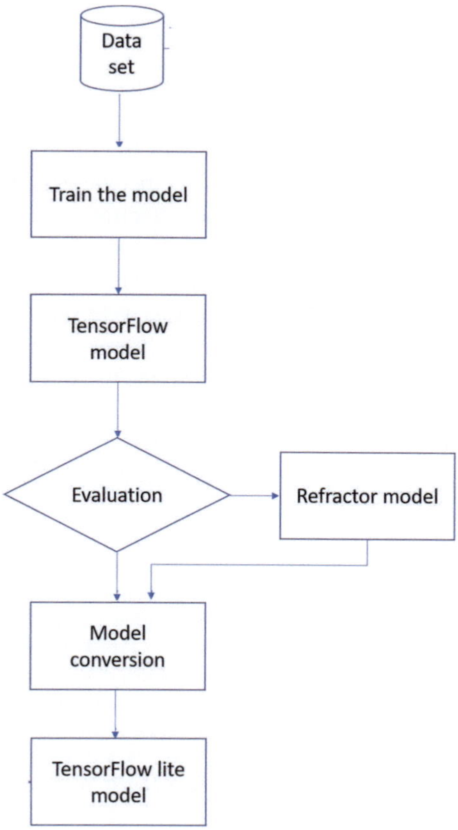

5 Challenges and Future Directions

The integration of AI for CPS presents several challenges, especially in critical areas such as health, autonomous vehicles, etc. Indeed, ML algorithms have some weaknesses and may contain certain errors that are introduced during learning due to imperfections contained in the learning data. These imperfections can be involuntary due to anomalies or false manipulations during data collection. As they can be voluntary and introduced by adversaries to distort the development of the intelligent model. Intelligence also poses ethical problems when applying them to sensitive areas. To overcome these limitations, it is necessary to develop learning algorithms and provide more guarantees related to the correction of the developed models. Also, well-studied regulations organizing the use of ML for CPS must be established in order to avoid any possible ethical overstepping.

6 Conclusion

In this chapter, we have provided an overview of the possibilities of using ML techniques for CBS. The benefits of this use are multiple, namely the automation of several human tasks, the increase in production, the improvement of product quality, etc. We have given two specific examples of the use of ML for CPSs. The first example is related to the detection of anomalies in road networks, combining the use of mobile phone sensors and AI techniques. The second example corresponds to the detection of electricity theft in smart grids using lightweight AI techniques. For both examples, the results were satisfactory. As future directions, it is necessary to circumvent existing limitations such as non-robust models, the risk of errors, and also ethical issues related to the use of AI in general.

References

1. Adil M, Farouk A, Abulkasim H, Ali A, Song H, Jin Z (2024) NG-ICPS: Next generation industrial-CPS, security threats in the era of artificial intelligence, open challenges with future research directions. IEEE Internet Things J
2. Albasir A, Naik K, Manzano R (2023) Toward improving the security of IoT and CPS devices: an AI approach. Dig Threats Res Pract 4(2):1–30
3. Amin S, Salahuddin T, Bouras A (2020) Cyber physical systems and smart homes in healthcare: current state and challenges. In: 2020 IEEE International conference on informatics, IoT, and enabling technologies (ICIoT). IEEE, USA, pp 302–309
4. Ben Dhaou I, Mihoub A, Krichen M (2026) Lightweight ai-driven real-time intruder detection system for IoT-enabled smart meters. In: International conference on service-oriented computing. Springer, Singapore, pp 251–262
5. Ben Fredj O, Mihoub A, Krichen M, Cheikhrouhou O, Derhab A (2020) Cybersecurity attack prediction: a deep learning approach. In: 13th International conference on security of information and networks, pp 1–6

6. Chae J, Lee S, Jang J, Hong S, Park KJ (2023) A survey and perspective on industrial cyber-physical systems (ICPS): From ICPS to AI-augmented ICPS. IEEE Trans Ind Cyber-Phys Syst 1:257–272
7. Chopade P, Edwards D, Khan SM, Andrade A, Pu S (2019) CPSX: using AI-machine learning for mapping human-human interaction and measurement of CPS teamwork skills. In: 2019 IEEE International symposium on technologies for homeland security (HST). IEEE, USA, pp 1–6
8. Dafflon B, Moalla N, Ouzrout Y (2021) The challenges, approaches, and used techniques of CPA for manufacturing in industry 4.0: a literature review. Int J Adv Manuf Technol 113:2395–2412
9. Duo W, Zhou M, Abusorrah A (2022) A survey of cyber attacks on cyber physical systems: recent advances and challenges. IEEE/CAA J Automatica Sinica 9(5):784–800
10. Dusadeerungsikul PO, Nof SY, Bechar A, Tao Y (2019) Collaborative control protocol for agricultural cyber-physical system. Procedia Manuf 39:235–242
11. Fulton N, Platzer A (2018) Safe AI for CPS. In: 2018 IEEE International test conference (ITC). IEEE, USA, pp 1–7
12. Guo P, Dusadeerungsikul PO, Nof SY (2018) Agricultural cyber physical system collaboration for greenhouse stress management. Comput Electron Agric 150:439–454
13. Hoffmann MW, Malakuti S, Grüner S, Finster S, Gebhardt J, Tan R, Schindler T, Gamer T (2021) Developing industrial CPS: a multi-disciplinary challenge. Sensors 21(6):1991
14. Kayan H, Nunes M, Rana O, Burnap P, Perera C (2022) Cybersecurity of industrial cyber-physical systems: a review. ACM Comput Surv (CSUR) 54(11s):1–35
15. Kim H, Ben-Othman J (2020) Toward integrated virtual emotion system with AI applicability for secure CPS-enabled smart cities: AI-based research challenges and security issues. IEEE Netw 34(3):30–36
16. Kim S, Park KJ, Lu C (2022) A survey on network security for cyber-physical systems: From threats to resilient design. IEEE Commun Surv Tutor 24(3):1534–1573
17. Klumpp M (2018) Innovation potentials and pathways merging AI, CPS, and IoT. Appl Syst Innov 1(1):5
18. Lee S, Chae J, Jeon H, Kim T, Hong YG, Um DS, Kim T, Park KJ (2025) Cyber-physical AI: systematic research domain for integrating AI and cyber-physical systems. ACM Trans Cyber-Phys Syst 9(2):1–33
19. Mian Qaisar S, Alyamani N, Waqar A, Krichen M (2022) Machine learning with adaptive rate processing for power quality disturbances identification. SN Comput Sci 3(1):1–6
20. Mihoub A, Fredj OB, Cheikhrouhou O, Derhab A, Krichen M (2022) Denial of service attack detection and mitigation for internet of things using looking-back-enabled machine learning techniques. Comput Electr Eng 98:107716
21. Mihoub, A., Krichen, M., Alswailim, M., Mahfoudhi, S., Bel Hadj Salah, R.: Road scanner: A road state scanning approach based on machine learning techniques. Appl Sci 13(2):683
22. Mihoub A, Snoun H, Krichen M, Salah RBH, Kahia M (2020) Predicting covid-19 spread level using socio-economic indicators and machine learning techniques. In: 2020 First international conference of smart systems and emerging technologies (SMARTTECH). IEEE, USA, pp 128–133
23. Munir A, Koushanfar F (2018) Design and analysis of secure and dependable automotive cps: A steer-by-wire case study. IEEE Trans Dependable Secure Comput 17(4):813–827
24. Odyurt U, Sapra D, Pimentel AD (2021) The choice of AI matters: alternative machine learning approaches for CPS anomalies. In: International conference on industrial, engineering and other applications of applied intelligent systems. Springer, Berlin, pp 474–484
25. Oks SJ, Jalowski M, Lechner M, Mirschberger S, Merklein M, Vogel-Heuser B, Möslein KM (2024) Cyber-physical systems in the context of industry 4.0: a review, categorization and Outlook. Inf Syst Front 26(5):1731–1772
26. Radanliev P, De Roure D, Van Kleek M, Santos O, Ani U (2021) Artificial intelligence in cyber physical systems. AI Soc 36(3):783–796

27. Rahman Z, Yi X, Khalil I (2022) Blockchain-based AI-enabled industry 4.0 CPS protection against advanced persistent threat. IEEE Internet Things J 10(8):6769–6778
28. Ramesh MV, Devidas AR, Rangan V et al (2017) Using CPS enabled microgrid system for optimal power utilization and supply strategy. Energy Build 145:32–43
29. Sghaier S, Krichen M, Ben Dhaou I, Elmannai H, Alkanhel R (2023) Identification, 3d-reconstruction, and classification of dangerous road cracks. Sensors 23(7):3578
30. Song J, Guo Y, Yu W (2025) Edge intelligence in smart home CPS. In: Edge intelligence in cyber-physical systems. Elsevier, pp 247–264
31. Song J, Lyu D, Zhang Z, Wang Z, Zhang T, Ma L (2022) When cyber-physical systems meet ai: a benchmark, an evaluation, and a way forward. In: Proceedings of the 44th International conference on software engineering: software engineering in practice, pp 343–352
32. Soon RJ, Sang DV, Chng CB, Chui CK (2023) Explainable AI for CPS-based manufacturing workcell. In: 2023 International conference on system science and engineering (ICSSE). IEEE, USA, pp 332–337
33. Tiwari S, Agarwal S (2022) Data stream management for cps-based healthcare: a contemporary review. IETE Tech Rev 39(5):987–1010
34. Tyagi AK, Sreenath N (2021) Cyber physical systems: analyses, challenges and possible solutions. Internet Things Cyber-Phys Syst 1:22–33
35. Weyer S, Meyer T, Ohmer M, Gorecky D, Zühlke D (2016) Future modeling and simulation of cps-based factories: an example from the automotive industry. Ifac-Papersonline 49(31):97–102
36. Yu Z, Gao H, Cong X, Wu N, Song HH (2023) A survey on cyber-physical systems security. IEEE Internet Things J 10(24):21670–21686
37. Zhang K, Shi Y, Karnouskos S, Sauter T, Fang H, Colombo AW (2022) Advancements in industrial cyber-physical systems: an overview and perspectives. IEEE Trans Industr Inf 19(1):716–729
38. Zhang Y, Qiu M, Tsai CW, Hassan MM, Alamri A (2015) Health-CPS: healthcare cyber-physical system assisted by cloud and big data. IEEE Syst J 11(1):88–95
39. Zhong S, Deng S, Zhao J, Wang Y, Su P (2019) A case study of operation optimization on a renewable energy building by E-CPS method: from both sides of supply and demand. Energy Procedia 158:6145–6151
40. Zidi S, Mihoub A, Qaisar SM, Krichen M, Al-Haija QA (2023) Theft detection dataset for benchmarking and machine learning based classification in a smart grid environment. J King Saud Univ Comput Inf Sci 35(1):13–25

Formal Methods for Cyber-Physical Systems

Moez Krichen

Abstract In this chapter, we focus on the use of formal methods for cyber-physical systems (CPS). These systems are composed of both software and hardware components that interact with their physical environment. Thus, several challenges in terms of security, reliability, and performance can arise when using these systems. In this regard, the use of formal methods could present an effective solution to guarantee these crucial properties. Formal methods are indeed rigorous mathematical methods that allow the rigorous verification and validation of different types of these systems. In particular, this chapter focuses on the use of formal methods to guarantee the proper functioning of CPSs in unforeseen environments and conditions. We begin by providing an overview of CPSs and an overview of formal methods. Next, we highlight the benefits of using formal methods for CPSs providing two specific examples: a first example about Controlling Drone Access to Critical Zones amd a second example about Optimizing Traveler Behavior Between Two Cities. We then identify the main challenges and limitations encountered in this context. Finally, we present some recommendations and advice for users of formal methods for CPSs as well as future directions in this field.

Keywords Cyber-physical systems (CPS) · Formal methods (FMS) · Advantages · Limitations · Future directions

1 Introduction

Cyber-physical systems (CPS) are ubiquitous in various areas of our modern daily life such as healthcare, transportation, industry, agriculture, etc. [32]. They include software and hardware components that interact with their physical environments

M. Krichen (✉)
Al-Baha University, Al-Baha, Saudi Arabia

ReDCAD Laboratory, National School of Engineers of Sfax, University of Sfax, Sfax, Tunisia
e-mail: moez.krichen@redcad.org
URL: https://www.redcad.org/members/mkrichen/

G. H. L. et al. (eds.), *Reliability in Cyber-Physical Systems: The Human Factor Perspective*, Springer Series in Reliability Engineering,
https://doi.org/10.1007/978-3-032-09917-4_18

using sensors, actuators, and specific control algorithms [26]. For example, in the transportation world, CPS offer driver assistance solutions that help reduce the risk of accidents and increase vehicle efficiency. Also in the healthcare field, CPSs allow continuous monitoring of patient conditions to intervene urgently in case of medical emergencies.

The high complexity of these systems is not without consequences [17] . Indeed, CPSs can present significant risks in terms of security, reliability, and performance. Several serious accidents can occur during the use of CPSs, ranging from the loss of critical data for a large number of users to the risk of destructive disasters and the loss of human lives. Thus, it is necessary to constantly ensure the proper functioning of these systems in order to avoid (or at least minimize) this type of damage. At this level, the use of formal methods (FMs) can present an adequate and effective solution.

Formal methods are rigorous reasoning techniques based on the use of precise and well-founded mathematical tools. These methods are primarily based on the modeling of various types of safety, performance, and proper functioning properties using specific mathematical formalisms. These properties are then analyzed using dedicated tools to ensure the correctness and performance of the systems under consideration. Thus, formal methods, thanks to their precision and mathematical rigor, increase confidence in CPSs and minimize the risk of accidents of all types [7, 16, 29].

However, the application of formal methods can also present certain challenges given their complexity and relatively high cost in terms of computational resources and execution time. This problem becomes even more critical for large systems, which consequently require much more significant resources. In addition, formal methods generally require in-depth mathematical knowledge, making them accessible only to experts in this field. Finally, CPSs are most of the time composed of several heterogeneous elements communicating often via complex protocols, which makes their modeling and analysis quite cumbersome and difficult. Thus, it is necessary to find effective solutions to overcome these difficulties when using formal methods for CPSs [3, 37].

The purpose of this chapter is to explore the various uses of formal methods for CPSs while highlighting their advantages, potentials, and limitations [14, 35]. We provide two examples: a first one about Controlling Drone Access to Critical Zones and a second one about Optimizing Traveler Behavior Between Two Cities. We study the different techniques and methodologies used in this field while identifying the challenges encountered and difficulties to overcome [2, 23]. We also propose some recommendations for the use of formal methods for CPSs and provide an overview of future directions in this field. The remainder of the chapter is organized as follows. Section 2 provides an overview of CPSs. Section 3 provides an overview of formal methods. Section 4 focuses on the use of formal methods for CPSs providing two specific examples. In Sect. 5, we identify the main challenges and limitations encountered. Section 6 is devoted to future work. Finally, Sect. 7 concludes the chapter. The structure of the chapter is illustrated in Fig. 1.

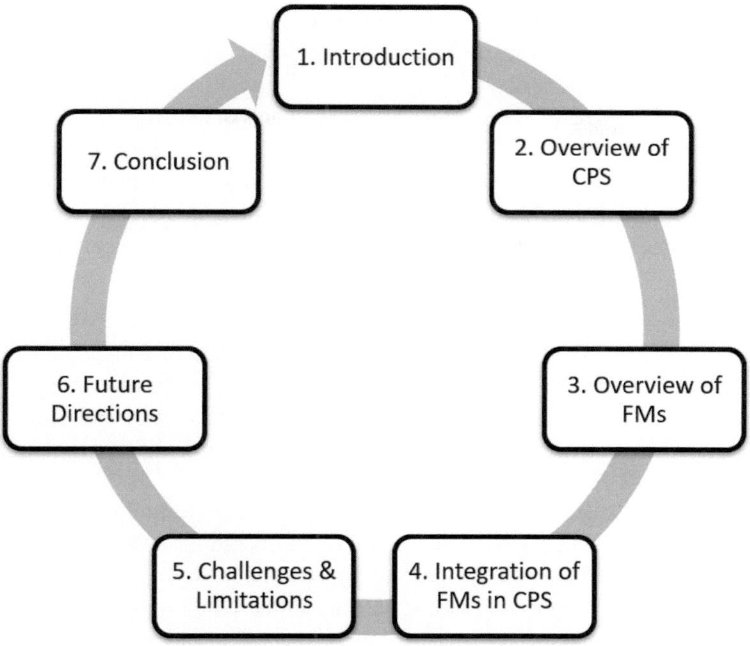

Fig. 1 Structure of the chapter

2 Overview of Cyber-Physical Systems

CPSs are now widely used in various areas of our daily lives. They consist of hardware and software components that allow them to interact in real time with their physical environment [18]. The control components present in these systems can be either physical or computer-based [38]. Thanks to sensors, actuators, and control algorithms, CPSs can be autonomous or semi-autonomous in dynamic environments of various types [36]. CPSs use dedicated communication networks for the exchange of data and messages between the different components of the system [8].

CPSs are primarily characterized by:

- Real-time: They must be fast enough to interact properly and within acceptable timeframes with changes in their environment to ensure safe and efficient operation.
- Heterogeneity: They are generally based on the use of different technologies that depend on the nature of the sensors, devices, and computers used, and which may have different origins and be developed by different companies.
- Interconnection: Multiple communication networks can be used by CPSs to ensure the sharing and exchange of data and messages, as well as the exchange of control commands.

- Autonomy: Another important characteristic of CPSs is their autonomy, allowing them to operate in real-time, continuously, and securely without the intervention of external human operators.
- Scalability: The size of CPSs can vary from one context to another, imposing constraints on them in terms of adding or removing new components, as well as the volume of data exchanged and its variation during system execution.
- Reliability: Finally, CPSs must guarantee a minimum level of security and reliability that allows users to have maximum confidence in the systems used.

CPSs are used quite frequently in various fields, including:

- Automotive [25, 34]: In this case, CPS sensors are used to gather the necessary details about the driving environment of smart vehicles in order to make appropriate decisions in real time for automated or semi-automated driving. In the automotive industry, errors at the CPS level can lead to fatal accidents resulting in both financial damage and loss of human life.
- Healthcare [31, 39]: In this case, CPS systems are used to monitor the condition of patients using specialized sensors and to intervene urgently when necessary by sending alerts to the relatives of the patient under surveillance or to medical institutions monitoring these patients. Errors in the CPSs in this case could also lead to the loss of human lives or at least to the complication of the health condition of patients.
- Industry [6, 15]: In the industrial sector, CPSs improve manufacturing processes by automating the monitoring of certain tasks and intervening when necessary by issuing emergency orders to avoid accidents or optimize certain steps, thereby reducing manufacturing costs and time. If errors or accidents occur in the CPSs in the industrial sector, enormous financial damage can occur as well as putting the safety and even the lives of workers and technicians in the factory at risk.
- Energy [28, 40]: CPSs are used in smart grids to monitor power consumption by different nodes of the grid as well as the optimal and balanced distribution of electricity between these different nodes. Also, CPSs are useful in preventing serious electrical accidents by sending alerts in case of emergency or even making emergency shutdown decisions when necessary. If the CPSs do not function properly in this context, there could be risks of blackout or deterioration of certain equipment sensitive to electrical disturbances, hence the possible considerable financial losses.
- Agriculture [9, 13]: Today, the field of agriculture has undergone a great revolution in terms of the use of modern technologies and CPSs have enabled better monitoring of farmers' activities in order to guarantee optimal conditions for plants and animals, thus leading to the optimization of costs, the reduction of damage and the increase of harvests and income. In agriculture, malfunctions of CPSs can damage crops or poison the soil, leading to risks of financial losses or environmental pollution, etc.
- Home automation [1, 30]: In smart homes, the use of CPSs can facilitate the automation of certain repetitive tasks, improving the quality of life of residents,

such as temperature monitoring and automatic activation of the air conditioning or heating system, or the automatic opening and closing of window shutters, etc. Another example is fire detection, triggering of the alarm system, and emergency fire extinguishing system. However, these advantages can turn into misfortune in the event of anomalies or malfunction of the CPSs.

As mentioned in the examples cited above, several risks threaten the use of CPSs that can cause significant damage in terms of financial loss or the risk of human lives, etc. For this, it is very important to make considerable efforts to guarantee the proper functioning of CPSs, among other things through the use of formal methods.

3 Formal Methods: An Overview

Formal methods are rigorous techniques based on the use of specific mathematical formalisms. These techniques allow the behavior of the systems studied to be modeled in order to verify their correctness and that they do not contain errors or anomalies from the preliminary design phases. In particular, with the remarkable evolution of software and hardware systems that modern life has experienced, the use of formal methods is increasingly necessary, especially for critical systems whose operating errors can cause enormous damage.

As illustrated in Fig. 2, formal methods and simplified way allow to verify that the model of the considered system described using a particular mathematical language satisfies a set of properties which are in turn described using the same mathematical language or another more adequate language. This operation is carried out using dedicated tools which will give as verdict either that the considered system is compliant or that it contains errors. In this second case, details about the detected anomaly can be provided in order to repair the system model and launch a second verification iteration. As shown in Fig. 3, there are several types of formal methods [11, 33], including: Semantic Static Analysis [5], Abstract Interpretation [12], Model Checking [24], Proof Assistants [10], Design by Refinement [4], Deductive Verification [27], and Model-Based Testing [19, 20], etc. Formal methods have several advantages, including precision, reliability, and rigor, which increase confidence in the verified systems. In addition, they allow errors and anomalies to be discovered in early stages, which reduces the costs of possible accidents after implementation and implementation of the system. However, these methods have certain limitations, such as the need for significant resources for computing and storing data necessary for model verification. Also, formal methods require in-depth knowledge and expertise to use them. In addition, formal methods suffer from the problem of scaling, especially for systems of considerable size.

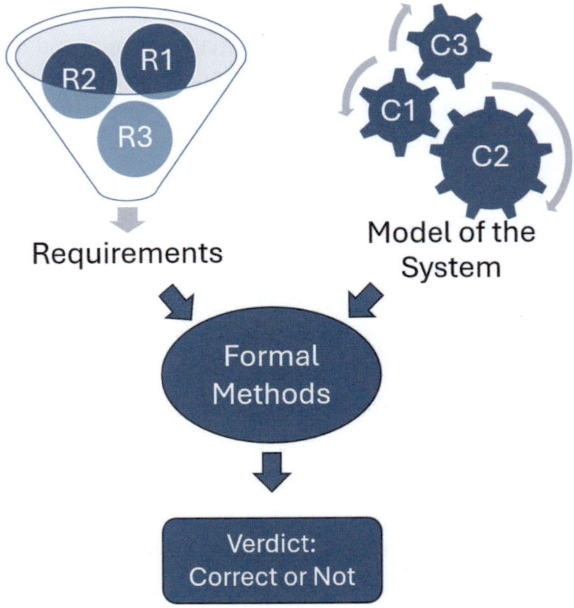

Fig. 2 Illustration of the principle of formal methods

4 Integration of Formal Methods in CPS

Formal methods, thanks to their precision and high accuracy, can be used to ensure the proper functioning of CPSs. This could be done by modeling the behavior of these systems and their environments using appropriate mathematical languages that can detect existing differences in these systems during the preliminary development phases, which could lead to a considerable reduction in costs due to repairs after development and deployment. In the following, we consider two examples of the use of formal methods for CPSs. The first is related to the use of formal methods for controlling drone access to critical areas. While the second example is related to the use of these methods for optimizing a traveler's journey between two cities with the possibility of using different means of transport (train, bus, car).

4.1 Controlling Drone Access to Critical Zones

In this first work, we used modeling and verification techniques based on the use of the UPPAAL model for drone access control in critical areas. The UPPAAL tool is based on the use of the timed automata model, which are directed graphs enriched with clocks that allow us to measure the time delays between different events. We modeled this CPS as a product of timed automata: a first automaton describing the behavior of the drone (Fig. 4), a second automaton modeling the behavior of the

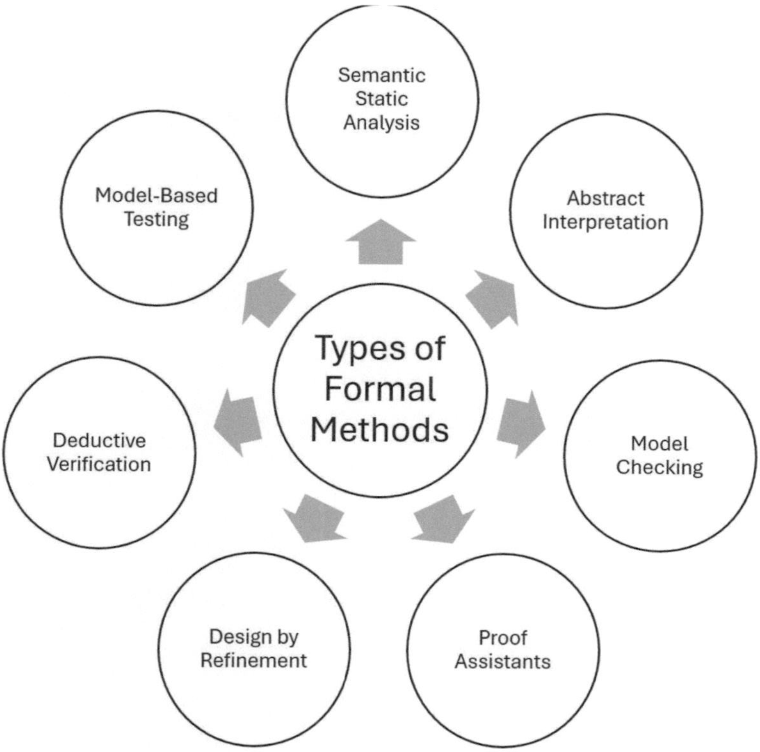

Fig. 3 Different types of formal methods

controller (Fig. 5), and a third modeling the drone queue. The goal was to ensure that the controller imposes on the drone network the presence of a single drone in a critical area. The results were satisfactory and the adopted method was successful (Fig. 6) despite some limitations related to the state explosion problem and high computation time, especially when the number of drones becomes high. More details about this example can be found in [21].

4.2 Optimizing Traveler Behavior Between Two Cities

In this second work, we also used the UPPAAL tool for modeling and optimizing the behavior of a traveler between two cities who has the choice to use different means of transport which are the train, the bus or his personal car (Fig. 7). The model takes into account the state of the road network: if it is loaded or not. Also, the model uses statistical distributions to model road traffic. The results obtained were also satisfactory and the adopted technique proved adequate for this CPS. Figure 8 shows an example of a concrete simulation of the model of the CPS and Figure 9 is a

Fig. 4 Timed automaton modeling drone behviour

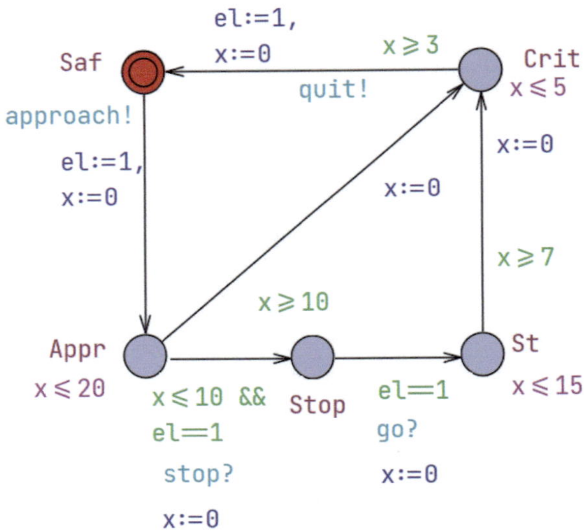

Fig. 5 TA modeling the behavior of the controller of the drones

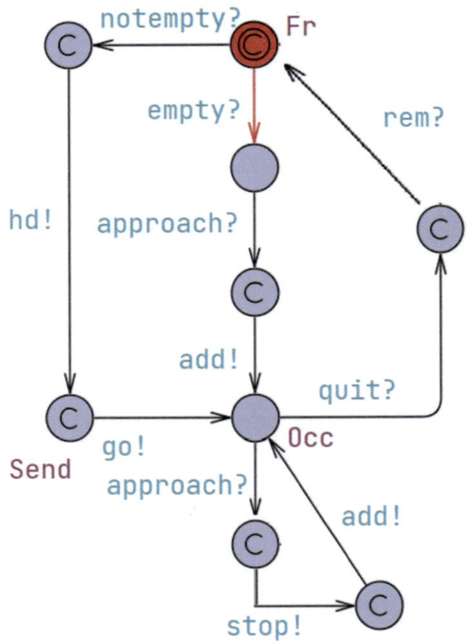

screenshot of the verification and strategy generation for the traveller trip. Also, limitations were observed always related to the computation time and the necessary resources. More information about this example can be found in [22].

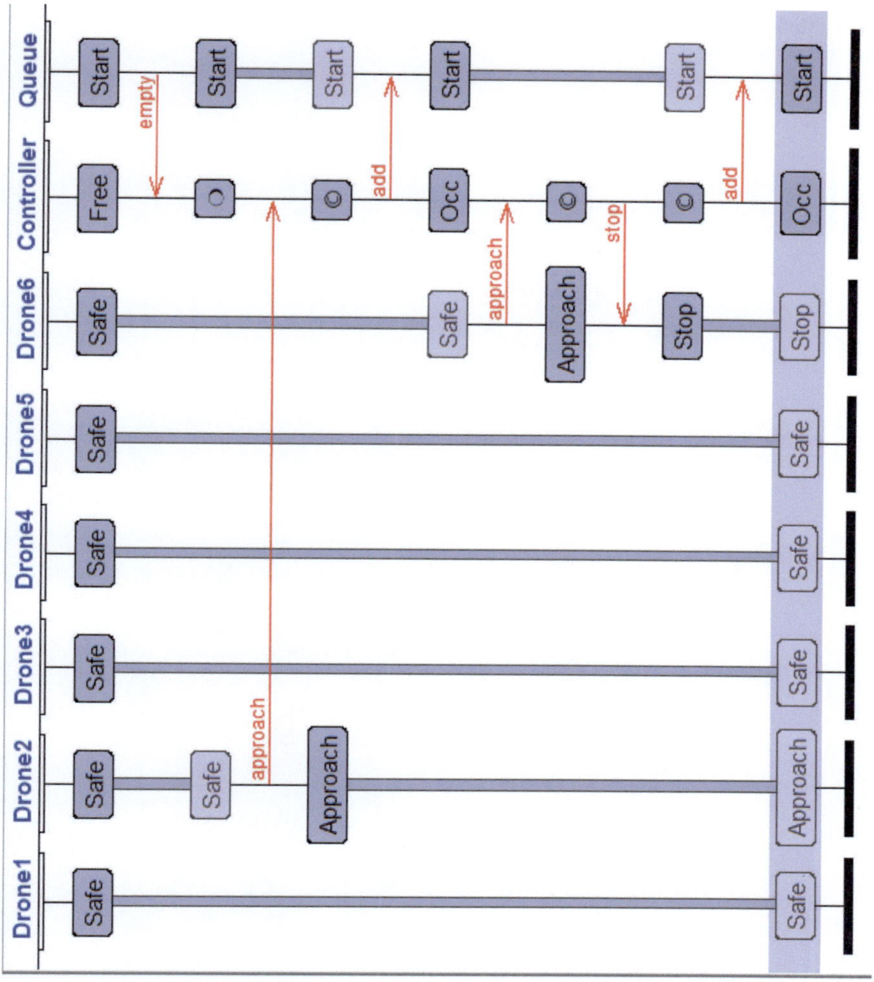

Fig. 6 Symbolic simulation of the model of the CPS

5 Challenges and Limitations

Despite their great utility, formal methods still present certain limitations and difficulties, including the following:

- The problem of state explosion and high computational cost: These methods are too demanding in terms of computing resources, given that they are generally based on the calculation of the system's global state space, which is generally very large, hence the need for significant computing and storage resources. This is all the more critical for systems consisting of several interacting blocks and for models that take into account the very fine details of the system under consideration.

Fig. 7 Model of the possible Ways between the two Cities

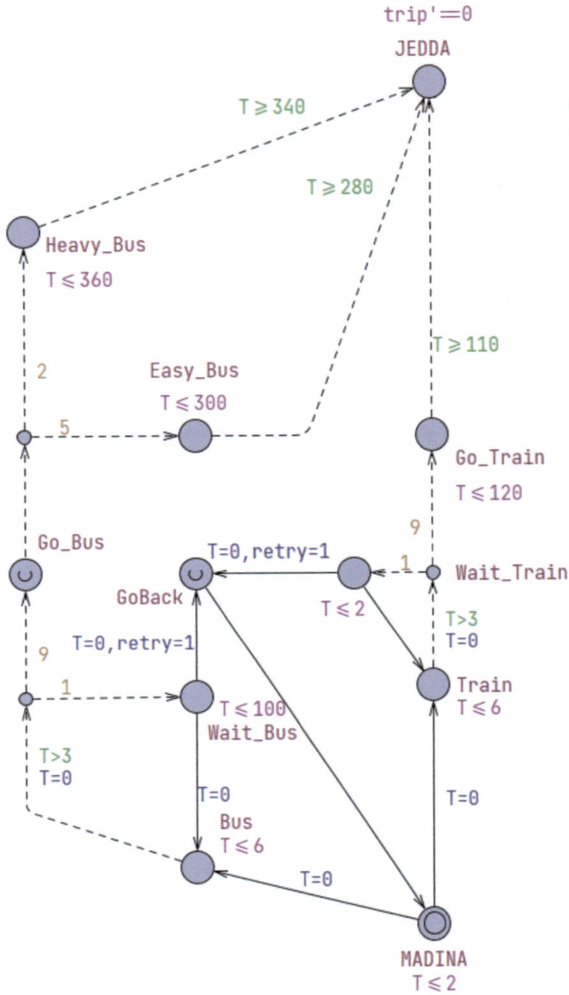

- The complexity of modeling and writing the properties to be verified: Often the use of formal methods requires a great deal of expertise from the users of these techniques since they are based on very specific and advanced mathematical notions that are not accessible in most cases to the general public. Even the available tools are often difficult to use and require advanced theoretical knowledge. In addition, the users of the techniques need a respectable knowledge of the application domain to which the system under consideration belongs. Hence the need to have in certain cases teams formed by experts in varied and complementary fields.

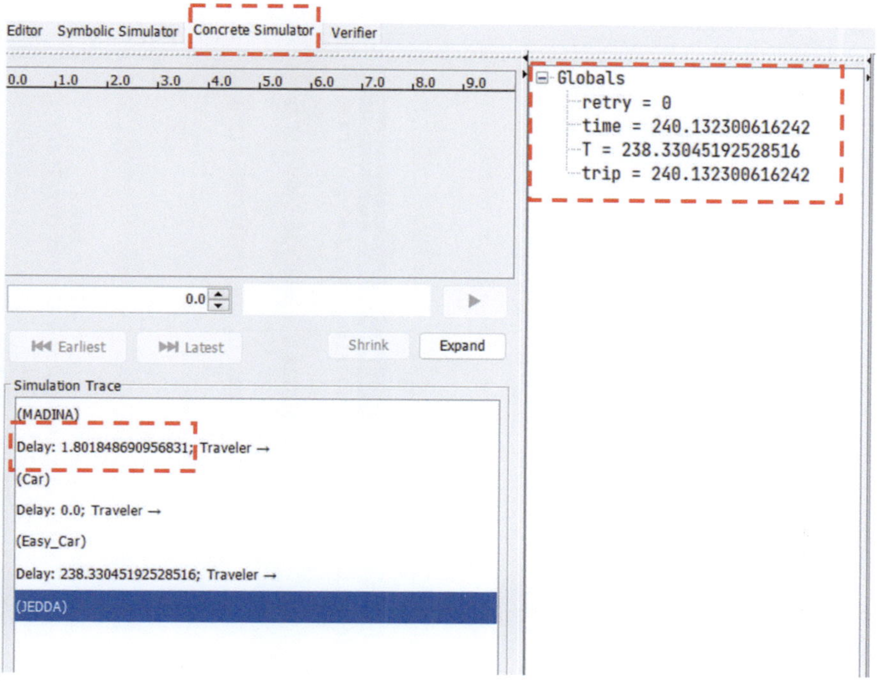

Fig. 8 An example of concrete simulation

6 Future Directions

Regarding future directions for the use of formal methods for CPSs, several interesting avenues can be explored. Among these, we cite:

- The use of new optimization techniques to reduce the high cost of computation and storage. At the same time, formal methods can also benefit from the use of quantum computing techniques, which offers enormous promise for improving computing capabilities.
- The use of new modeling and proof tools that can facilitate the adoption of formal methods by non-experts in the field. Thus, these methods will be used on a larger scale and in a shorter timeframe, further reducing the deployment time of the solutions deployed.
- The integration of artificial intelligence techniques into all stages of the use of formal methods for CPSs. This will facilitate, among other things, the use of these methods and reduce the number of members of the work teams, thus ensuring greater profitability and greater speed for the entire verification process.

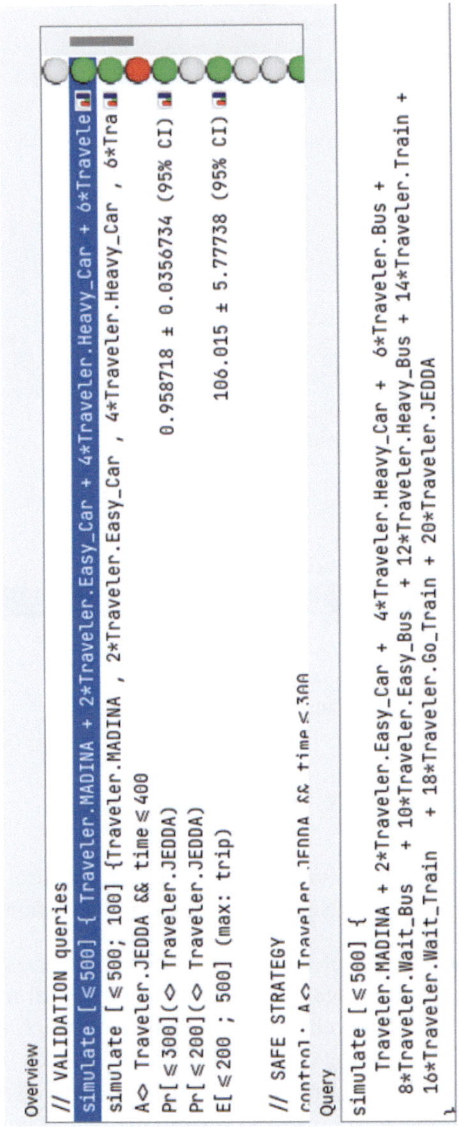

Fig. 9 Verification and strategy generation

7 Conclusion

In conclusion, in this chapter, we have given an overview of the use of formal methods for CPS. These methods can be very useful in this context by facilitating error detection in the preliminary phases of development of these systems. However, these methods have certain limitations mainly related to their high computational and storage costs and also related to the difficulty of using these methods by non-experts in the field. Among the possible solutions to overcome these difficulties, we have cited the use of adequate optimization techniques, quatum computing techniques, and the integration of artificial intelligence techniques. Thus, the future of the use of formal methods for CPSs remains promising and will require a lot of additional efforts in the industrial and research fields.

References

1. Amin S, Salahuddin T, Bouras A (2020) Cyber physical systems and smart homes in healthcare: current state and challenges. In: 2020 IEEE International conference on informatics, IoT, and enabling technologies (ICIoT). IEEE, USA, pp 302–309
2. Askarpour M, Ghezzi C, Mandrioli D, Rossi M, Tsigkanos C (2019) Formal methods in designing critical cyber-physical systems. In: From software engineering to formal methods and tools, and back: essays dedicated to Stefania Gnesi on the occasion of her 65th birthday, pp 110–130
3. Bagade P, Banerjee A, Gupta SK (2017) Validation, verification, and formal methods for cyber-physical systems. In: Cyber-physical systems. Elsevier, pp 175–191
4. Burch JR, Passerone R, Sangiovanni-Vincentelli AL (2003) Modeling techniques in design-by-refinement methodologies. In: System specification & design languages. Springer, Berlin, pp 283–292
5. Cousot P (2001) Abstract interpretation based formal methods and future challenges. Informatics 138–156
6. Dafflon B, Moalla N, Ouzrout Y (2021) The challenges, approaches, and used techniques of cps for manufacturing in industry 4.0: a literature review. Int J Adv Manuf Technol 113:2395–2412
7. Deshmukh JV, Sankaranarayanan S (2019) Formal techniques for verification and testing of cyber-physical systems. Des Autom Cyber-Phys Syst 69–105
8. Duo W, Zhou M, Abusorrah A (2022) A survey of cyber attacks on cyber physical systems: Recent advances and challenges. IEEE/CAA J Automatica Sinica 9(5):784–800
9. Dusadeerungsikul PO, Nof SY, Bechar A, Tao Y (2019) Collaborative control protocol for agricultural cyber-physical system. Procedia Manuf 39:235–242
10. Geuvers H (2009) Proof assistants: History, ideas and future. Sadhana 34(1):3–25
11. Gleirscher M, Marmsoler D (2020) Formal methods in dependable systems engineering: a survey of professionals from europe and north america. Empir Softw Eng 25(6):4473–4546
12. Gosain A, Sharma G (2015) Static analysis: a survey of techniques and tools. In: intelligent computing and applications. Springer, Belrin
13. Guo P, Dusadeerungsikul PO, Nof SY (2018) Agricultural cyber physical system collaboration for greenhouse stress management. Comput Electron Agric 150:439–454
14. Hasuo I (2017) Metamathematics for systems design: comprehensive transfer of formal methods techniques to cyber-physical systems. N Gener Comput 35(3):271–305
15. Hoffmann MW, Malakuti S, Grüner S, Finster S, Gebhardt J, Tan R, Schindler T, Gamer T (2021) Developing industrial cps: a multi-disciplinary challenge. Sensors 21(6):1991
16. Jones A, Kong Z, Belta C (2014) Anomaly detection in cyber-physical systems: a formal methods approach. In: 53rd IEEE conference on decision and control. IEEE, USA, pp 848–853
17. Kayan H, Nunes M, Rana O, Burnap P, Perera C (2022) Cybersecurity of industrial cyber-physical systems: a review. ACM Comput Surv (CSUR) 54(11s):1–35
18. Kim S, Park KJ, Lu C (2022) A survey on network security for cyber-physical systems: from threats to resilient design. IEEE Commun Surv Tutor 24(3):1534–1573
19. Krichen M (2010) A formal framework for conformance testing of distributed real-time systems. In: International conference on principles of distributed systems. Springer, Berlin, pp 139–142

20. Krichen M (2018) Contributions to model-based testing of dynamic and distributed real-time systems. Ph.D. thesis, École Nationale d'Ingénieurs de Sfax (Tunisie)
21. Krichen M (2024) Timed automata-based strategy for controlling drone access to critical zones: a uppaal modeling approach. Electronics 13(13):2609
22. Krichen M, Harbaoui A (2024) Optimizing traveler behavior between madina and jedda using uppaal stratego: a stochastic priced timed games approach. Mathematics 12(21):3421
23. Michael JB, Drusinsky D, Wijesekera D (2021) Formal methods in cyberphysical systems. Computer 54(9):25–29
24. Müller-Olm M, Schmidt D, Steffen B (1999) Model-checking. In: International static analysis symposium. Springer, Berlin, pp 330–354
25. Munir A, Koushanfar F (2018) Design and analysis of secure and dependable automotive cps: a steer-by-wire case study. IEEE Trans Dependable Secure Comput 17(4):813–827
26. Oks SJ, Jalowski M, Lechner M, Mirschberger S, Merklein M, Vogel-Heuser B, Möslein KM (2024) Cyber-physical systems in the context of industry 4.0: a review, categorization and Outlook. Inf Syst Front 26(5):1731–1772
27. Pnueli A, Ruah S, Zuck L (2001) Automatic deductive verification with invisible invariants. In: International conference on tools and algorithms for the construction and analysis of systems. Springer, Berlin, pp 82–97
28. Ramesh MV, Devidas AR, Rangan V et al (2017) Using cps enabled microgrid system for optimal power utilization and supply strategy. Energy Build 145:32–43
29. Sanwal MU, Hasan O (2013) Formal verification of cyber-physical systems: coping with continuous elements. In: International conference on computational science and its applications. Springer, Berlin, pp 358–371
30. Song J, Guo Y, Yu W (2025) Edge intelligence in smart home cps. In: Edge intelligence in cyber-physical systems. Elsevier, pp 247–264
31. Tiwari S, Agarwal S (2022) Data stream management for cps-based healthcare: a contemporary review. IETE Tech Rev 39(5):987–1010
32. Tyagi AK, Sreenath N (2021) Cyber physical systems: analyses, challenges and possible solutions. Internet Things Cyber-Phys Syst 1:22–33
33. Wang F, Cao Z, Tan L, Zong H (2020) Survey on learning-based formal methods: taxonomy, applications and possible future directions. IEEE Access 8:108561–108578
34. Weyer S, Meyer T, Ohmer M, Gorecky D, Zühlke D (2016) Future modeling and simulation of cps-based factories: an example from the automotive industry. Ifac-Papersonline 49(31):97–102
35. Yin X, Li S (2020) Recent advances on formal methods for safety and security of cyber-physical systems. Control Theor Technol 18:459–461
36. Yu Z, Gao H, Cong X, Wu N, Song HH (2023) A survey on cyber-physical systems security. IEEE Internet Things J 10(24):21670–21686
37. Zahid F, Tanveer A, Kuo MM, Sinha R (2022) A systematic mapping of semi-formal and formal methods in requirements engineering of industrial cyber-physical systems. J Intell Manuf 33(6):1603–1638
38. Zhang K, Shi Y, Karnouskos S, Sauter T, Fang H, Colombo AW (2022) Advancements in industrial cyber-physical systems: an overview and perspectives. IEEE Trans Industr Inf 19(1):716–729
39. Zhang Y, Qiu M, Tsai CW, Hassan MM, Alamri A (2015) Health-cps: healthcare cyber-physical system assisted by cloud and big data. IEEE Syst J 11(1):88–95
40. Zhong S, Deng S, Zhao J, Wang Y, Su P (2019) A case study of operation optimization on a renewable energy building by e-cps method: from both sides of supply and demand. Energy Procedia 158:6145–6151

MIX
Papier aus verantwortungsvollen Quellen
Paper from responsible sources
FSC® C105338

If you have any concerns about our products,
you can contact us on
ProductSafety@springernature.com

In case Publisher is established outside the EU,
the EU authorized representative is:
Springer Nature Customer Service Center GmbH
Europaplatz 3, 69115 Heidelberg, Germany

Printed by Libri Plureos GmbH
in Hamburg, Germany